T0319247

Introduction to Protein Mass Spectrometry

Introduction to Protein Mass Spectrometry

Pradip K. Ghosh

Former Professor of Chemistry
Indian Institute of Technology Kanpur, India

Academic Press is an imprint of Elsevier
125 London Wall, London, EC2Y 5AS, UK
525 B Street, Suite 1800, San Diego, CA 92101-4495, USA
225 Wyman Street, Waltham, MA 02451, USA
The Boulevard, Langford Lane, Kidlington, Oxford OX5 1GB, UK

Copyright © 2016 Elsevier Inc. All rights reserved.

No part of this publication may be reproduced or transmitted in any form or by any
means, electronic or mechanical, including photocopying, recording, or any information
storage and retrieval system, without permission in writing from the publisher. Details
on how to seek permission, further information about the Publisher's permissions
policies and our arrangements with organizations such as the Copyright Clearance
Center and the Copyright Licensing Agency, can be found at our website:
www.elsevier.com/permissions.

This book and the individual contributions contained in it are protected under
copyright by the Publisher (other than as may be noted herein).

Notices
Knowledge and best practice in this field are constantly changing. As new research and
experience broaden our understanding, changes in research methods, professional
practices, or medical treatment may become necessary.

Practitioners and researchers must always rely on their own experience and knowledge
in evaluating and using any information, methods, compounds, or experiments
described herein. In using such information or methods they should be mindful of their
own safety and the safety of others, including parties for whom they have a professional
responsibility.

To the fullest extent of the law, neither the Publisher nor the authors, contributors, or
editors, assume any liability for any injury and/or damage to persons or property as a
matter of products liability, negligence or otherwise, or from any use or operation of
any methods, products, instructions, or ideas contained in the material herein.

Library of Congress Cataloging-in-Publication Data
A catalog record for this book is available from the Library of Congress

British Library Cataloguing in Publication Data
A catalogue record for this book is available from the British Library

ISBN: 978-0-12-805123-8

For information on all Academic Press publications
visit our website at http://store.elsevier.com/

Publisher: Sara Tenney
Senior Acquisitions Editor: Jill Leonard
Editorial Project Manager: Fenton Coulthurst
Production Project Manager: Chris Wortley
Designer: Mark Rogers

Typeset by SPi

Printed and bound in the United States of America

Working together
to grow libraries in
developing countries

www.elsevier.com • www.bookaid.org

To my children:

Anirvan Ghosh
Tannishtha Reya

Preface

The field of biological mass spectrometry is in the midst of a period of explosive growth. It is the major driver for proteomic data, which is central to modern efforts in cell and molecular biology, as well as systems biology and the identification of clinical biomarkers. It is very likely that the impact of mass spectrometry on biological understanding over the coming decades will be no less than the contribution that DNA and RNA sequencing technologies have had in modern biology.

The capability of mass spectrometry to determine large molecular masses was developing gradually, when, powered by certain technical developments it rapidly became possible to determine molecular masses of very large biological molecules like proteins and lipids, and from that derive their chemical identities. This transformed the investigation of biochemical processes in the laboratory, and enabled other applications such as identifying protein biomarkers in diseases. Initially the technology was employed for the assessment of macromolecular complexes but as it became possible to identify hundreds or even thousands of proteins in a sample, it rapidly emerged as a core driver of systems biology efforts. It is now being widely used to understand intra and intercellular signaling networks that regulate the function of cells and tissues.

Protein mass spectrometry is a powerful analytical technique albeit a complex one. Unlike many other methods, it does not yield a unique protein name as its output, because it generally involves a deduction of protein identity from determination of peptide fragmentation products. This requires sophisticated analytical approaches and chemical/biological knowledge to drive decision-making between alternate conclusions. Until the field evolves further and it becomes more definitive, it is important to understand and appreciate how determinations about protein identity from mass spectrometric data are made, because this could have a bearing on the application for which it is being used. It is for this reason it is incumbent on every user to have a good understanding of the method to be able to understand the limitations that may be present in the final conclusion.

This book is a general-purpose introduction to protein mass spectrometry for scientists and medical researchers and should satisfy at least three requirements. First, for someone just entering the field, it should provide a concise introduction to all branches of the biological mass spectrometric method while citing the most recent and relevant work in the field. Second, it should provide biologists and medical researchers not only with information on the capabilities of the technique but also give an appraisal of the present limitations of the computer-analyzed mass spectrometric results. Third, in it specialists working in one branch of protein mass spectrometry should find brief, state-of-the-art information on other branches of the technique. This is what is attempted in the present book.

Over the years I have had the privilege of interactions with many of the scientists working in the field. I would like to thank them for clarifications of their work and to their publishers for the kind permission to use copyrighted material from their publications. To Profs Brian Chait, Alan Marshall, Akos Vertes, James Jorgenson, and David Fenyö I would like to express my special thanks. Kevin Blackburn and Dr John Tran were very helpful in answering my numerous queries.

It is a pleasure to acknowledge the support I received in various ways from Prof. N. Sathyamurthy, Director, Indian Institute of Science Education and Research, Mohali, and Prof. M. S. Hegde, former Dean of Faculty of Science, Indian Institute of Science, Bangalore, over the past few years, which enabled completion of this work. Prof. G. N. Rao of Adelphi University got me started on the project with his initial support.

February 2015   P. K. G.
IIT Kanpur

Contents

1 Introduction   **1**

2 Sample Preparation and Ionization for
   Mass Spectrometry   **9**
   2.1 Separation of protein mixtures 9
       2.1.1 Ultracentrifugation 9
       2.1.2 Gel electrophoresis 11
       2.1.3 Capillary electrophoresis 15
       2.1.4 Liquid chromatography 19
   2.2 Multidimensional separations 27
       2.2.1 2D SCX-RPLC-based separation in *Shotgun*
            proteomics 28
       2.2.2 Difference gel electrophoresis (DIGE) 30
       2.2.3 GELFrEE fractionation 32
   2.3 Proteolysis 35
   2.4 Ionization of proteins and peptides 37
       2.4.1 Electrospray ionization — ESI 37
       2.4.2 Matrix-assisted laser desorption ionization —
            MALDI 58
       2.4.3 Laser-induced liquid beam/bead ionization
            desorption — LILBID 62
       2.4.4 Laser ablation with electrospray
            ionization — LAESI 66

3 Instruments   **68**
   3.1 Mass analysis 68
       3.1.1 Mass analyzers: time-of-flight (TOF),
            2D-quadrupole (Q), Linear Trap (q,LT), 3D-ion
            trap (IT), Fourier transform ion cyclotron
            resonance (FT-ICR), Orbitrap, ion mobility (IM) . . 68
       3.1.2 Theoretical ion mobility cross-sections using
            PA, PSA 109

        3.1.3   MS/MS instrumental supplements: CAD/CID, ECD,
               ETD, EDD, HCD, IRMPD 114

3.2 Some examples of instruments 118
        3.2.1   Agilent 6560 IM QTOF 118
        3.2.2   Bruker Daltonik — Impact II, maXis II^{TM} 120
        3.2.3   JEOL — JMS-S3000 SpiralTOF 122
        3.2.4   SCIEX TripleTOF® 6600 123
        3.2.5   Thermo Fisher Scientific — Orbitrap Fusion 124
        3.2.6   Thermo Fisher Scientific – Hybrid FT-ICR at
               14.5 tesla 127
        3.2.7   IMS-IMS-MS 130
        3.2.8   Waters — Synapt G2-S 132

4 Mass Spectrometry of Peptides and Proteins   **135**
4.1 Proteins 135
        4.1.1   Primary, secondary, tertiary, and quaternary
               structures 135
        4.1.2   Amino acid data 138
        4.1.3   Nomenclature of ions generated in peptide
               fragmentation 141
        4.1.4   Immonium ions 148
        4.1.5   Molecular weight determination of an intact
               protein from its ESI spectrum 148

4.2 Alternative analysis methods 150
        4.2.1   SRM, MRM 150

4.3 Quantitation 151
        4.3.1   Label-free, ICAT, SILAC, iTRAQ, TMT, AQUA . . 151

4.4 Peptide fragmentation: Experiments 165
        4.4.1   MS/MS methods — CAD (CID), ECD,
               ETD, EDD 165
        4.4.2   Use of H-D exchange 167

4.5 Data mining scheme for identifying
    peptide structural motifs 173

4.6 Peptide fragmentation: mechanism 175
        4.6.1   Fragmentation pathways: mobile proton model . . . 175

4.7 Possible pathways in the competition model 177
        4.7.1 b_x-y_z pathway 178
        4.7.2   Diketopiperazine pathway 179
        4.7.3   Histidine effect 186
        4.7.4   Proline effect 187
        4.7.5   Sequence scrambling in fragmentation 189
        4.7.6   Gas-phase fragment ion isomer analysis 190

4.7.7   Role of gas phase conformations in ETD
         fragmentation   192
4.7.8   Fragmentation patterns as a measure of
         antioxidant capacity   194
4.8   Studying post translational modifications   195
4.9   Chemical cross-linking/mass spectrometry   200
4.10 Manual *de novo* sequencing of peptides   206
      4.10.1 Procedure in manual sequencing   207
      4.10.2 Examples of manual sequencing   213

5   Examples from Biological Applications   220
5.1   Mapping intact protein isoforms using
      top-down proteomics   220
5.2   Quantitative analysis of intact
      apolipoproteins in human HDL   228
5.3   Rapid sequence analysis of some
      conotoxins — combination of *de novo*,
      bottom-up methods   232
5.4   Mass spectrometry of ribosomes   237
5.5   Proteins in Purkinje cell post-synaptic
      densities   243
5.6   Neurexin-LRRTM2 interaction effect
      in synapse formation   245
5.7   Rapid analysis of human plasma
      proteome — an IMS-IMS-MS application   248
5.8   Topology of two transient virus capsid
      assembly intermediates   252
5.9   Mass spectrometry of intact V-type
      ATPases reveals bound lipids and the
      effects of nucleotide binding   260
5.10 Mass spectrometric imaging of
      biological material   268

6   Mass Spectrometry-Based Bioinformatics   272
6.1   Peptides to proteins   273
      6.1.1   Peptide mass fingerprinting   273
6.2   MS/MS fragments to peptides to
      proteins   274
      6.2.1   Peptide sequence tag   274
6.3   Data-dependent acquisition   277
      6.3.1   SEQUEST   277
      6.3.2   MASCOT   280
      6.3.3   PEAKS DB   281
      6.3.4   X! Tandem   283

        6.3.5   OMSSA 286
        6.3.6   Scaffold 287
    6.4 Targeted acquisition covering SRM and PRM 289
    6.5 Data-independent acquisition 290
        6.5.1   MS$^\mathrm{E}$ 290
        6.5.2   ARM — All-Reaction Monitoring 291
        6.5.3   AIF — All-Ion Fragmentation 292
        6.5.4   SWATH-MS 293
        6.5.5   Multiplexed MS/MS MSX 295

Index   **297**

CHAPTER 1
INTRODUCTION

Mass spectrometry is the method of choice at present for fast, accurate characterization of proteins. This is both in terms of obtaining their molecular masses and for deriving their primary structures — amino acid sequence information. And the method is generally applicable to all proteins. At the present state of the art, mass spectrometry, in quantitative terms, can routinely analyze low femtomoles, and increasingly, attomoles of samples; and this is with a dynamic range of four orders of magnitude.

In a broader context, this analytical ability in the post-genomic era of biology is playing an especially important role. In particular, a major focus of attention of biologists is now on the cell proteome. The cell proteome — the aggregate of all proteins expressed in a cell — is complex in its composition and it changes with time, reflecting the cell condition. In the human genome there are 20,300 protein-coding genes; at first consideration, it would seem, quantification of the whole human proteome is achievable. But the complexity of this undertaking is enormous; each such protein may be expressed at a given time in a given cell in one or more of its multiple isoforms, splice variants, and post-translational modifications. Whereas the totality of the problem is immense, the compositions however fully incorporate those normal as well as any abnormal states of the cells, overall holding clues to details of cell biological processes. For determination of these compositions, however, comprehensive analysis is required. Only a few analytical methods at present can take on this challenge and deliver satisfactory results. Among these, mass spectrometry, even with its limitations, is definitely the frontrunner. The present availability of a wide array of mass spectrometry platforms besides the antibody-based Human Protein Atlas project and ProteomeXchange — a single point of submission of proteomics data to integrate proteomics-based knowledge bases — was probably a compelling reason for the Human Proteome Organization's (HUPO's) recent launching of the global Human Proteome Project (HPP). The goal of the HPP is to identify and characterize at least one protein product from each of the protein-coding genes.

The impact of mass spectrometry on biological understanding may be best illustrated by a few examples. Toward higher-order structure determination of membrane proteins and their complexes, hydrogen bonding and stability have been probed via site-directed mutagenesis experiments using the membrane protein bacteriorhodopsin. Myoglobin conformation has been examined in the presence of lipid bilayer and how it is preserved within reverse micelles in the gas phase of the mass spectrometer. It has been possible to compare the conditions that result in the release of transporters ($BtuC_2D_2$, MacB, MexB) from gas phase micelles with those typically used

Introduction to Protein Mass Spectrometry
http://dx.doi.org/10.1016/B978-0-12-805123-8.00001-1

Copyright © 2016 Elsevier Inc.
All rights reserved.

for soluble complexes of comparable molecular mass. By controlled release of complexes from intact micelles, subunit stoichiometry and lipid-binding properties have been unequivocally determined. There have been experiments on quaternary structures, and the gating mechanism of the ion channel KirBac3.1 has been determined. It has been shown that rotary adenosine triphosphatases (ATPases)/synthases can remain intact in vacuum along with membrane and soluble subunit interactions. Incorporation of protein subunits into the EmrE membrane complex and the effects of their PTM status upon dimer formation have been examined. Using targeted proteomics approaches it has been shown that low-abundance proteins, such as the tau protein in spinal fluid, or cytokines, can be detected in a proteome. A recent mass spectrometric work identified 1,043 gene products from human cells that are dispersed into more than 3,000 protein species created by PTMs, RNA splicing, and proteolysis. The present status of the field has been summarized in several papers.[1,2,3,4,5,6,7,8,9,10]

Mass spectrometric analysis requires the sample molecules — rather their ionic forms — be present in gaseous state, something that was difficult to achieve for large molecules by traditional means without causing concurrent uncontrolled fragmentation. Two technical innovations about a decade before the turn of the twenty-first century made soft ionization process

[1]Nilsson, T., Mann, M., Aebersold, R., Yates, J. R. III., Bairoch, A., and Bergeron, J. M. (2010) Mass spectrometry in high-throughput proteomics, ready for the big time. *Nature Methods*, **7**, 681–685.

[2]Chait, B. T. (2011) Mass spectrometry in the postgenomic era. *Annu. Rev. Biochem.*, **80**, 239–246.

[3]Barrera, N. P. and Robinson, C. V. (2011) Advances in the mass spectrometry of membrane proteins: from individual protein to intact complexes. *Annu. Rev. Biochem.*, **80**, 247–271.

[4]Cox, J. and Mann, M. (2011) Quantitative high-resolution proteomics for data-driven systems biology. *Annu. Rev. Biochem.*, **80**, 273–299.

[5]Harkewicz, R. and Dennis, E. A., (2011) Applications of mass spectrometry to lipids and membranes. *Annu. Rev. Biochem.*, **80**, 301–325.

[6]Vestal, M. L. (2011) The future of biological mass spectrometry. *J. Am. Soc. Mass Spectrom.*, **22**, 953–959.

[7]Lamond, A., Uhlen, M., Horning, S., Makaraov, A., Robinson, C. V., Serrano, L., Hartl, F. U., Baumeister, W., Werenskiold, A. K., Andersen, J. S., Vorm, O., Linial, M., Aebersold, R., and Mann, M. (2012) Advancing cell biology through Proteomics in Space and Time (PROSPECTS), *Molecular and Cellular Proteomics.* Published on February 6, 2012 as Manuscript O112.017731.

[8]Gillette, M. A. and Carr, S. A. (2013) Quantitative analysis of peptides and proteins in biomedicine by targeted mass spectrometry. *Nat. Methods*, **10**, 28–34.

[9]Marcoux, J. and Robinson, C. V. (2013) Twenty years of gas phase structural biology. *Structure*, **21**, 1541–1550.

[10]Wilhelm, M., Schlegl, J., Hahne, H., Gholami, A. M., Lieberenz, M., Savitski, M. M., Ziegler, E., Butzmann, L., Gessulat, S., Marx, H., Mathieson, T., Lemeer, S., Schnatbaum, K., Reimer, U., Wenschuh, H., Mollenhauer, M., Slotta-Huspenina, J., Boese, J-H., Bantscheff, M., Gerstmair, A., Faerber, F., and Kuster, B. (2014) Mass-spectrometry-based draft of the human proteome. *Nature*, **509**, 582–587.

conveniently realizable and ushered in large-molecule thus biological mass spectrometry: one was the electrospray ionization (ESI), and the other matrix assisted laser desorption ionization (MALDI). In electrospray ionization, an aqueous solution of the sample is passed through a capillary and vaporized in the presence of a strong electric field which results in protons adhering to the molecules and forming — usually multiply charged — molecular ions. In MALDI, the sample mixed with a solid matrix that has high ability of absorbing laser radiation is exposed to laser; the energy deposited by the radiation causes both desorption of the sample molecules and ionization, forming, mostly singly charged ions. In most protein mass spectrometry these are the two methods of ionization that are used. Laser-induced liquid bead ionization desorption (LILBID) and laser ablation with electrospray ionization (LAESI) are recently introduced methods of ionization.

In determining protein structure, after a first stage of mass spectral analysis or *protein mass fingerprinting* (PMF), fragmenting the intact molecular ion in a controlled manner bit by bit — usually by some collisional activation — and then from the identity of the fragments thus generated (through a second stage of mass spectral analysis, that is, MS/MS) reconstruct the structure of the parent ion is the basis of the *top-down* method. The advantage of this intact protein analysis is a simpler tracking of the fragmentation process hence more dependable reconstruction of the parent structure; its disadvantage is that it is more time consuming and often not possible for the analyzer to handle very large molecular mass ions. Its antipode is by far the more popular *bottom-up* method, in which the initial step is an enzymatic proteolysis of the sample. Then from fragmentation of the peptides thus generated, the peptides are first identified, from which the structure of the original protein is inferred. Here, the advantage is high speed; the disadvantage is the substantially increased complexity in reconstructing the original structure. A new intermediate course is now making appearance, being referred to as the *middle-down* method. The *de novo* method is deriving the peptide sequence without any prior information on the sequence or use of DNA database.

In most routine applications, including analysis of mixtures of proteins, the bottom-up method is commonly used. The most commonly used protease is trypsin. The bottom-up method coupled with high-performance liquid chromatography (HPLC) before the sample enters the mass spectrometer is referred to as the *shotgun* method. In *MuDPIT* — multidimensional protein identification technology — SCX and RP stationary phases are packed together in the same microcapillary column; the peptides get separated there by 2D chromatography, then they are directly eluted into the mass spectrometer. Inside the mass spectrometer, the tryptic peptides are first sorted out through a mass analysis step, which in tandem

is followed by fragmentation studies of individual peptides. Fragmentation
is most commonly effected by collisional activation (collision-induced, collisionally activated dissociation, CID, CAD) — helium is mostly used as
the collisional agent — but photodissociation (such as infrared multiphoton
dissociation IRMPD) also is used. Additional fragmentation techniques include electron capture dissociation (ECD) in which low-energy electrons
injected from an external source are made to interact with peptide ions
which cause fragmentation; in electron transfer dissociation (ETD), radical
cations, instead of free electrons, are employed. Also in use is the electron-
detachment dissociation (EDD) method which mimics positron capture; in
this, bombardment of anionic species by moderate energy electrons causes
electron detachment followed by backbone dissociation.

Once the fragmentation data have been obtained, the next and the final
daunting task is to retrace steps and infer from the fragment ion data the
amino acid sequence in the peptides and then, in turn, the protein structure. This is commonly done using one or the other available proprietary
software packages, mostly commercial, which are created, in the first place,
based on available data on fragmentation patterns of known peptides. While
subjecting the mass spectral output to software application, the additional
inputs usually sought are the taxonomy of the sample, the protein mass
range, the protease employed, and the database to be searched; the output
is a list of names of probable proteins, along with their expectation values,
that are compatible with spectral data. Because of limitations of the software, there is no guarantee that the protein actually present in the sample
would turn up at the top of the list. Further, since all packages may not
point to the same set of proteins, additional considerations are required to
arrive at the final conclusion.

Often, in experiments, after the qualitative identification of the proteins in a sample has been accomplished, the next objective is to follow
their variations with respect to changes, which could be over a wide range
of biological conditions of the sources of the samples. It is then expedi-
ent to concentrate attention on specific peptides related to the proteins
of interest. To this objective, operational modes are then set — through
software control — to track specific peptides, or specific ions that are fragmentation products of those peptides. Narrowing scans to relatively much
smaller region of mass spectrum increases the speed of analysis drastically;
it also increases the dynamic range. In the single reaction monitoring (SRM)
procedure, no mass spectra are recorded; two mass analyzers are used in
tandem, both tantamount to filters, which track a specific fragment ion
arising from its precursor ion. In the multiple reaction monitoring (MRM)
procedure too, no mass spectra are recorded; the scanning here is between
multiple sets of precursor and fragment ion pairs. Both are targeted approaches; the paramount gains here are greater speed of analysis and higher
sensitivity.

There has been significant progress also in quantitative proteomics. Here just the main methods are named. In the label-free method no isotope labeling is used. ICAT (isotope-coded affinity tags) and SILAC (stable isotope labeling by amino acids in cell culture) are mass-difference methods which use isotope labeling. In the ICAT method two different cell states are treated with isotopically light and heavy (often ^{13}C or D) ICAT reagents; then they are combined, digested, and analyzed by mass spectrometry. In the SILAC method two populations of cells are cultivated with their growth medium in their cell culture having two different isotopes in their amino acids prior to mass spectrometric analysis. iTRAQ (isobaric tag for relative and absolute quantification) and TMT (tandem mass tag) employ isobaric labeling. Here peptides are labeled with isobaric chemical groups, but on MS/MS fragmentation reporter ions of different masses result. iTRAQ tags are available in 4-plex and 8-plex forms; TMT tags are available in duplex and 6-plex forms. In AQUA (absolute quantification), a synthetic tryptic peptide — a few daltons heavier than the peptide of interest and which incorporates one stable isotope labeled amino acid — is added as an internal standard to the biological sample prior to digestion and HPLC-MS analysis. From the peak ratios of the ion chromatograms, the quantity of the native peptide is obtained. In this method high-attomole detection limits have been claimed. The protein mass spectrometry field has witnessed extensions in nomenclature. The area of analysis where protein composition in the sample was the primary objective used to be known as analysis in *discovery* mode; analysis of change in concentration of a few specific proteins used to be known as analysis in *targeted* mode. Currently data acquisition modes are also referred to as *data-dependent acquisition* (DDA) or *data-independent acquisition* (DIA). Both DDA and DIA can be used for discovery — that is, without testing a hypothesis. In targeted analysis, a set of hypotheses are needed. Usually the experiment is performed by only collecting data for a set of peptides, but DIA which attempts at collecting data about *all* peptides can be analyzed in a targeted way once the data has been acquired.

The instrumentation in the field is configurationally traditional in that it has the same set-up: ion source (in this case mainly ESI and MALDI), analyzer, and detector. The options in analyzers however are many: the time-of-flight, the 2D quadrupole (with its trapping version, the linear trap), the 3D RF trap, the ion cyclotron resonance (ICR) cell which requires a high magnetic field, along with a significant new addition to the array, the RF Orbitrap. Along with it are the implements needed for MS/MS: that is, for techniques like CAD/CID, ECD, ETD, EDD, IRMPD. A characteristic of protein mass spectrometry instrumentation is that the designs of the instruments are modular — that is, the overall assembly structure of a given instrument is one of many possible combinations of the basic implements mentioned above to achieve a certain desired result. And almost overwhelmingly, every research laboratory in the field uses commercial instruments including those which have demonstrated capability of designing their own original instruments. An advantage of this is that in many cases results across a number of laboratories can be compared conveniently because of their use of the same or similar instruments; a severe limitation is that the way an experiment can be conducted is entirely determined by what the manufacturer of an instrument makes feasible operationally; the user is limited to a relatively narrow range of options. A major recent introduction to this instrumentation field is the coupling of ion mobility spectrometry and mass spectrometry; ions, in the first segment, undergo ion mobility analysis, then, in tandem, the mobility-separated ions are subjected to mass analysis. This adds a powerful new dimension to the analyses; in this mobility mode it is possible to obtain, superposing theoretical collision cross-section estimates, information on the shapes of the molecular ions in transit. And, in the MS/MS stage, sometimes, additionally, information on the stoichiometry of components constituting a molecular ion emerges. The fact that electrospray ionization yields multiply charged ions — sometimes the ions carrying charges of 100+ — provides fortuitously an important dividend. A mass analyzer having a maximum mass-to-charge ratio m/z range 5000, when the ionic charge z is +100, its molecular mass determination potential extends to 500,000 Da. To give an example, for a ribosomal protein carrying a 90+ charge, its molecular mass of higher than 2 MDa has been determined. It is perhaps relevant here to mention that both the ion physics of the mass analysis process, and the software handling of the mass spectral data to evolve a list of likely proteins, remain rather opaque to an ordinary user. Both are formidable, but details of the latter are practically beyond reach because of the proprietary nature of the software.

Despite substantial capabilities and achievements of the technique, deficiencies exist, which surface when it is subjected to stringent tests. Reproducibility of results on different platforms is one such domain. In a recent study it was observed that 20 out of 27 laboratories participating in the study involved in identifying proteins in a simple mixture had difficulty in the identification task. The quality of biological replicates, technical replicates, sampling handling, all matter. Not long ago in a study involving seven different plasma proteins, using three replicates, there was a 25% quantitative variation in the results of the eight laboratories that participated in the study.

Certainly, with time the analysis speed will increase to make faster routine analysis feasible and MS tools will grow more robust for relatively more convenient use of the technique by biologists. Improved handling of the sample prior to its introduction into the mass spectrometer and its better utilization inside the instrument will increase sensitivity; the dynamic range is likely to increase to $\sim 10^6$ from the present $\sim 10^4$ — needed for efficient analysis of low-abundance proteins, and for 1 protein per cell analysis, the number of cells required will come down from the present $\sim 10^9$ to $\sim 10^6$. For obtaining more effective simultaneous positional and temporal information in cell biological work, more efficient collection of samples — capture of information in very high resolution both in the space and the time domains — will be required.

This is perhaps also the place to mention *chemoproteomics*.[11] Whereas proteomic approaches have now been widely used to understand biological processes, an area of emerging interest is the application of proteomics for drug discovery. Drug discovery has traditionally been driven by pharmaceutical approaches, including screen for compounds that have a desired biological activity. The biological activity can be a molecular event, such as activation of a receptor or enzyme, or a cellular process, such as inhibition of proliferation of a cancer cell. One of the challenges of the pharmaceutical approach is that the target profile of the compound under development is often not fully understood, which can make it difficult to optimize the compound being studied. This uncertainty in the biological basis of the effect of a compound is largely absent in molecular genetic approaches where one can precisely target a gene of interest, and in recent years, such genetic evidence has been the driver of new targets in drug discovery programs.

The advent and application of chemoproteomic approaches could play a transformative role in drug development by illuminating the mechanism of drug action. For example, drugs that are in use for neuropsychiatric indications, such as anti-psychotics, were historically discovered based on their action in animal models. Whereas the primary target of these drugs is often known, for instance, the action of atypical antipsychotics on Dopamine D2 receptors, the other receptors on which the drug acts, and to what extent those activities contribute to the biological effects, are not well understood. In a chemoproteomic approach, the pharmacological agent is used as a bait to capture interacting proteins, which can then be analyzed by mass spectrometry. Chemoproteomic approaches to identify molecular targets of known antipsychotics could provide valuable insight into the polypharmacology of these compounds and could help identify targets signatures that correspond to clinical efficacy. Such knowledge could influence the development and characterization of next generation medicines that are optimized based not on activity in animal models, but on a better understanding of the desired molecular pharmacological profile.

---

[11]Moellering, R. E. and Cravatt, B. F. (2012) How chemoproteomics can enable drug discovery and development. *Chemistry & Biology* **19**, January 27.

An example of quantitative chemical proteomics making use of mass spectrometry is the work on mechanisms of action of clinical ABL kinase inhibitors.[12] A subproteome of hundreds of endogenously expressed protein kinases and purine binding proteins was captured and the bound proteins were quantified in parallel by mass spectrometry using iTRAQ. The binding of drugs to their targets in cell lysates and cells then were assessed by measuring the competition with the affinity matrix. Signaling pathways downstream of target kinases were analyzed by mapping drug-induced changes in the phosphorylation state of the captured proteome.

This book introduces the reader to all essential aspects of protein mass spectrometry. It is to be borne in mind that the book is on an *analytical technique* for protein structure analysis the technique ingeniously combining special strengths of a wide range of scientific disciplines. The preparation of the protein sample is in the domain of analytical chemistry, proteolysis in the domain of chemistry/biochemistry, sample ionization and peptide fragmentation in physical chemistry and chemical physics, ion mass-charge ratio analysis in gaseous ion physics, provision of electrical and magnetic fields and signal processing via electronics, and data analysis based on computer science. It takes the reader from liquid chromatography of protein mixtures all the way to data analysis going through tandem mass spectrometry of peptide fragments. Examples provided from the literature illustrate the recent state of the art and also indicate potential future trends. As it is intended to be a compact volume, many of the findings are briefly mentioned with literature references provided so that readers may follow them up on their own. Also, it is to be mentioned that although the chapters are arranged in the logical order of workflow, they can be used independently.

In Chapter 2, the next chapter, we consider the two initial stages of protein mass spectrometry, the preparation of the sample and its ionization. Mass spectrometric instruments that are in use for mass-to-charge ratio analysis of the ions are presented in Chapter 3. Chapter 4 starts with peptide fragmentation preliminaries, then various methods for deriving peptide and protein structure from their fragmentation patterns are presented. Also included in this chapter are a number of methods of quantitation. In Chapter 5, a few application examples are presented in some detail. Finally, bioinformatics as applied to mass spectrometry is introduced in Chapter 6. Together, the contents should provide the reader with a comprehensive review of protein mass spectrometry as currently applied.

---

[12]Bantscheff, M., Eberhard, D., Abraham, Y., Bastuck, S., Boesche, M., Hobson, S., Mathieson, T., Perrin, J., Raida, M., Rau, C., Reader, V., Sweetman, G., Bauer, A., Bouwmeester, T., Hopf, C., Kruse, U., Neubauer, G., Ramsden, N., Rick1, J., Kuster, B., and Drewes, G. (2007) Quantitative chemical proteomics reveals mechanisms of action of clinical ABL kinase inhibitors. *Nat. Biotechnol.*, **25**, 1035–1044.

CHAPTER 2
SAMPLE PREPARATION AND IONIZATION FOR MASS SPECTROMETRY

Several processes are involved before a protein sample is introduced into a mass analyzer, either as intact protein ions or as peptide ions after proteolysis of sample proteins. For a protein mixture,[13] it first needs to be separated into individual proteins; this is usually accomplished by means of some electrophoresis. If peptide ions are to be introduced, then the proteins are first subjected to a proteolysis step. In either case they are then taken through a high-performance liquid chromatography procedure. Following that the sample — now separated into individual proteins or peptides — is ready to be ionized. There are several alternative methods for ionization, the chief ones being electrospray and matrix-assisted laser desorption and ionization.

2.1 Separation of protein mixtures

2.1.1 Ultracentrifugation

Before chromatographic separation can be undertaken, a clean sample of protein mixture (free of unbroken cells, lipids, and particulate matter) is desired. Ultracentrifugation is one method to achieve that. Proteins — which are macromolecules — because of their constant Brownian motion in a solution, ordinarily do *not* sediment; only by subjecting them to enormous acceleration — many times the acceleration due to gravity g ($g = 9.81$ $\mathrm{m\,s^{-2}}$) — they can be made to sediment. By spinning the sample solution in a tube (with the axis of spinning perpendicular to the sample tube axis) with the use of a rotor, high centrifugal force can be generated. This is the principle of ordinary *centrifuge*. In *ultra*centrifuges, centrifugal forces even of the order of 1,000,000 g can be generated. Under such conditions, individual components of a sample in a solution can be made to *differentially* sediment — the actual rate of sedimentation depends

---

[13]There are numerous protocols for protein extraction from biological sources, which are widely available, each requiring some special treatment depending on the nature of the biological source. A sample protocol for pulldown with LRRTM2-ecto-Fc from synaptosome, presynaptic fraction: Phillips, G. R., Huang, J. K., Wang, Y., Tanaka, H., Shapiro, L., Zhang, W., Shan, W. S., Arndt, K., Frank, M., Gordon, R. E., Gawinowicz, M. A., Zhao, Y., and Colman, D. R. (2001) The presynaptic particle web: ultrastructure, composition, dissolution, and reconstitution. *Neuron*, **32**, 63–77.

Introduction to Protein Mass Spectrometry
http://dx.doi.org/10.1016/B978-0-12-805123-8.00002-3

Copyright © 2016 Elsevier Inc.
All rights reserved.

on sample poperties and the density and viscosity of the sample solution.

The *sedimentation coefficient s* is defined as

$$s = \frac{\nu}{\omega^2 r} = \frac{1}{\omega^2}\left(\frac{d\ln r}{dt}\right) = \frac{M(1 - \overline{V}\rho)}{Nf} \qquad (2.1)$$

where $\nu = dr/dt$ is the rate of migration of the sedimenting particle of mass m that is located a distance r from a point about which it is revolving with angular velocity ω (in radians s^{-1}). The sedimentation force is the centrifugal force ($m\omega^2 r$) on the particle *less* the buoyant force ($V_P\rho\omega^2 r$) exerted by the solution, where V_P is the particle volume and ρ is the density of the solution. The sedimentation force is opposed by the frictional force νf, where f is the frictional coefficient. M is the mass of 1 mol of particles, $M = mN$ (N is Avogadro's number), and $\overline{V} = V_P/m$. The sedimentation coefficient is usually expressed in units of 10^{-13} s known as svedbergs S.

To give a few examples of magnitudes of sedimentation coefficient ($20°$ C w): bovine heart cytochrome c (13.4 kDa) s 1.71 S; horse heart myoglobin (16.9 kDa) s 2.04 S; human fibrinogen (340 kDa) s 7.63 S; Turnip yellow mosaic virus protein (3013 kDa) s 48.80 S. A particle's mass can be determined from its sedimentation coefficient and the solution density if the frictional coefficient and its partial specific volume are known.

There can be *analytical* ultracentrifugation[14] (by means of analytical centrifuges, the physico-chemical properties of a sedimenting particle or molecular interactions of macromolecules and their possible subunits, respectively, can be unraveled) and *preparative* ultracentrifugation. In *preparative* ultracentrifugation (preparative ultracentrifuges are usually used for the fractionation of fine particulate samples such as tissue homogenates aiming to isolate subcellular organelles, macromolecules, bacteria, or viruses), in which we are interested here,[15] the *density gradient* of the solution is used. There are two major types of applications: (i) *zonal* ultracentrifugation, and (ii) *equilibrium density gradient* ultracentrifugation. In zonal ultracentrifugation, a macromolecular solution is carefully layered on top of a density gradient (usually sucrose solution); this separates similarly shaped macromolecules largely on the basis of their molecular masses. After centrifugation, the bottom of the celluloid centrifuge tube is punctured with a needle and the individual zones are collected in order. In the equilibrium density gradient ultracentrifugation, the sample is dissolved in a relatively fast diffusing substance (such as $CsCl$ or Cs_2SO_4) and is spun at high speed

---

[14]Cole, J. L., Lary, J. W., Moody, T., and Laue, T. M. (2008) Analytical ultracentrifugation: sedimentation velocity and sedimentation equilibrium. *Methods Cell. Biol.*, **84**, 143–179.

[15]Voet, D. and Voet, J. G. (2010) Chapter 6: Techniques of Protein and Nucleic Acid Purification. In *Biochemistry*, Wiley Publications.

until the solution achieves equilibrium; the sample components form *band* at positions where their densities are equal to that of the solution.

EMBL recommends, for some applications it is sufficient to use a low centrifugation step: 15 min at 10,000 *g* ($\sim$ 15,000 rpm using a Sorvall SS-34 rotor). When a very clear sample is needed, such as for refolding or when the sample is directly applied to an expensive chromatography column, an ultracentrifugation step becomes necessary: 30 min at $> 100,000$ *g* (45,000 rpm using a Beckman 45 Ti rotor).

2.1.2   Gel electrophoresis

In *gel electrophoresis*, the analyte sample is placed on a gel and is subjected to electrophoresis. Under electrophoresis, constituent analyte molecules migrate at different rates depending on their mass and charge; from this differential migration, separation of analyte constituents takes place. Bands are formed; they can be visualized in a number of ways, and can be excised, digested, and separated. Here we first present a procedure for forming a sample gel, along with some general information on gel electrophoresis, then present SDS-PAGE, which is extensively used for separation of proteins. The core equipment in gel electrophoresis is a platform on which the gel is placed submerged in a buffer, along with ways to connect the gel to the anode ($-$) and cathode ($+$) of a power supply (50–150 V).

Gel preparation/casting: Usually for setting the gel, some simple casting support is used. Depending on the gel to be used, preparation method varies; here we present the preparation of agarose gel; later a procedure for preparation of the gel used in SDS-PAGE will be described. To prepare agarose gel, agarose (purified form of agar) powder (a linear polymer — D-galactose, 3,6 anhydrous L-galactose) is dissolved in a hot buffer solution, and the hot solution ($\sim$ 60°C) is poured into a casting tray into which gel combs have already been placed; the combs cause formation of wells into which the sample would be placed eventually. The gel solution is allowed to sit and cool for about 30–45 min during which it sets; after the gel sets, the combs are removed.

Electrophoresis: The gel is then placed in the electrophoresis electrode chamber, sufficient buffer is added so that the gel remains just submerged (about 1–2 mm under the buffer surface). The analyte sample, is mixed with about six times the buffer containing an indicator (such as Bromophenol Blue — 3′,3″,5′,5″ tetrabromophenolsulfonphthalein; this dye carries negative charge; so, in the case that the sample is protein (or DNA), in the electrophoretic field they would both move in the same direction — that is, the positive anode) and density increasing materials such as glycerol or sucrose, are then pipetted into the wells of the gel. It is common to add a standard; for protein analysis it would be a protein *ladder* standard (which would contain a mixture of proteins with a wide range of masses); for easy tracking of the extent of electrophoretic separation, the movement of the indicator is helpful.

Staining: After the electrophoretic process, staining of the gel is done. The staining can be of more than one kind: one that uses a staining compound which binds with the sample and fluoresces under ultraviolet radiation; the other which uses a compound that makes the sample visible in ordinary light. Though not much used in electrophoresis, green fluorescence protein (GFP) is a stainer for uv visualization of proteins; Coomassie Brilliant Blue R-250 ($C_{45}H_{44}N_3NaO_7S_2$) helps visualization of proteins in ordinary light. Staining exposure could take a time of about 30 min; after staining, excess stainer is removed with the solvent, which could be water. Coomassie Blue, which non-specifically binds to proteins, is an anionic dye, is commonly used in methanolic solution containing acetic acid. Silver staining raises sensitivity from microgram range (as it would be when Coomassie blue is used) to nanogram range. In silver staining, the gel is first cleared of interfering substances such as amino acids, ampholytes, Tris, SDS, etc. It is then sensitized (STS, dithionite, glutaraldehyde) which is followed by impregnation by the silvering agent (silver nitrate 0.2%, 37% HCHO 0.7 mL/L 30–120 min). In the acidic methods of silver staining, silver nitrate solution is used; in the basic methods a basic silver-ammonia (or silver-amine) complex is used. The image is usually developed by using a dilute formaldehyde solution (37% HCHO 0.25 mL/L, potassium carbonate 3%, STS 10 mg/L 5–15 min) and then stopped (Tris 50g/L, AcOH 25 mL/L). Silver staining is sensitive to the purity of chemicals used; therefore use of pure reagents and highest purity water is called for.

SDS PAGE

Polyacrylamide gel (PAG) is chemically inert, thermally stable, can with stand high-voltage gradients, alternative staining-destaining procedures, also withstands digestions associated with extracting fractions that have been separated. It provides *uniform* pore size; polyacrylamide gel electrophoresis (PAGE), that is, electrophoresis using polyacrylamide gel, can be used to separate proteins in the range 5–2,000 kDa. By controlling the relative concentrations of acrylamide (acrylamide is a potent neurotoxin — care needs to be exercised in its handling) and bis-acrylamide powder when the gel is prepared, the pore size can be controlled. Five percent bis-acrylamide concentration yields the smallest pore size. An increase, or decrease, of bis-acrylamide concentration from 5% increases pore size — the pore size is a parabolic function of bis-acrylamide concentration. For initiating polymerization, free radicals and stabilizer are required; this need can be satisfied by using approximate equimolar concentration (1–10 mM) of ammonium persulfate and N,N,N′,N′-tetramethylenediamine (TEMED).

Sodium dodecyl sulfate (SDS, $C_{12}H_{25}NaO_4S$) is an anionic detergent; it can denature secondary and non-disulfide-linked tertiary structures of proteins, linearize them, and in proportion to the mass of the protein can add negative charge to each protein. Denaturation of protein and binding of SDS to protein are augmented by heating the sample to 60° or higher. For further denaturation and reduction of disulfide linkages, the analyte is heated to near boiling with a reducing agent such as dithiothreitol or 2-mercaptoethanol, to break up quaternary structure and tertiary folding. Because of its convenience, SDS is the most common choice in separation of proteins — it has been in use since the early 1960s. It is even possible to separate different protein isoforms. The binding of SDS to polypeptides is at a constant weight ratio: 1.4 g of SDS for every g of polypeptide. *Native* PAGE is PAGE *without* using SDS. In Native PAGE, proteins with similar mass but different structure and different m/z would migrate at different rates because of their different isoelectric points. This problem is overcome by SDS through binding with and denaturing the protein, linearizing, and along the polypeptide backbone providing nearly uniform negative charge.

Polyacrylamide gel is prepared by mixing acrylamide (which can be varied in the concentration range 5%–25%; lower concentrations are better for resolving high molecular weight proteins and higher concentrations for resolving low molecular weight proteins), bis-acrylamide (~ 1 part in 35 parts of acrylamide, assists in formation of cross-linking between polyacrylamide molecules), and SDS, in the presence of a buffer. In DISC (discontinuous) electrophoresis (for better resolution, two gels of two different concentrations of polyacrylamide — the one of lower concentration placed on top of the other having higher concentration), Tris — tris(hydroxymethyl)amino-methane — $(HOCH_2)_3CNH_2$, pK_a 8.07 at 25°C, Bis-Tris — Bis(2-hydroxyethyl)amino-tris(hydroxymethyl)methane—$C_{18}H_{19}NO_5$, pK_a 6.5 at 25°C, or imidazole are common buffers that are used. Different buffers can be used at anode and cathode.

For staining the protein sample after electrophoresis, Coomassie Brilliant Blue R-250 is the most common choice. (Since, like the dye, SDS too is anionic, large volumes of the staining solution — order of ten times more than the gel — need be used.) Silver staining is also used. There are a number of protein standards (protein ladders) from various vendors such as SeeBlue, BenchMark, HiMark, Novex, etc. We mention a couple of them: one (ideal for large proteins), HiMark prestained protein standard of high MW range 20-260 kDa, nine bands two colors, low band sharpness, gel compatibility: Tris-Acetate, Tris-Glycine; and the other (easiest to interpret MW; best size estimation) Novex sharp prestained protein standard, 12 bands 12 colors, MW range 2.5–260 kDa, gel compatibility: NuPAGE® Bis-Tris Tricine. Tris-Glycine.

Coupling gel electrophoresis directly to mass spectrometry by ESTASI technique. Girault *et al.*[16] have extended the scope of possible applications of the ESTASI technique[17] by coupling gel electrophoresis to mass spectrometry. They have used *electrostatic spray ionization* for *in situ* ionization of proteins and peptides inside a surfactant-free polyacrylamide gel. The samples are first separated by isoelectric focussing in a gel, then detected by scanning the gel by ESTASI; charged samples in the gel are directly extracted by an electric field and then electrosprayed for mass spectrometric detection. Detection of sample by ESTASI-MS in the presence of salts, SDS, and surfactants is difficult. In the first of their strategies, after gel electrophoresis, the gel strip is placed above an insulating (PMMA poly(methyl methyl acrylate) gel) plate, a drop of 50% methanol, 49% water, and 1% acetic acid mix (buffer droplet pH = 2) placed on the gel point to be analyzed, and 6.5 kV is applied from below the insulating plate using a high-voltage electrode. The sample gets protonated and the protonated sample directly enters the mass spectrometer. Negative ions could be sampled by reversing the polarity of the high voltage. In the second, for sampling at various pH points, a plastic cover patterned with holes ($d = 1\,\mathrm{mm}$, space between two adjacent holes $2\,\mathrm{mm}$, thus resolution $\approx 3\,\mathrm{mm}$) is placed on the gel strip and $1\,\mu\mathrm{L}$ of acetic buffer is used for droplet size uniformity. Microfabrication of the holes on the plastic cover placed on top of the immobilized pH gradient IPG gel strip was performed in order to increase the spatial resolution. Limits of detection in the range of low nanograms were observed with IEF in IPG gel. Small amounts of target proteins or peptides as low as 20 ng or 1 ng in a band, respectively, were detectable (the limit of detection was found as 1.3 ng of angiotensin I peptide and 20 ng of cytochrome c protein). Peptide identification ability was checked with enzymatic digestion of bovine serum albumin BSA (digest loaded on an IPG strip, samples were then directly taken from the wells, around selected areas of the strip — in their study, around pH 5.8 and pH 6.2 areas). ESTASI-MS was also employed to separate *Escherichia coli* protein extract labeled with myoglobin and cytochrome c. The efficiency of peptide identification in IEF-ESTASI-MS was checked against two parallel control experiments with IEF-MALDI-MS and OFFGEL-ESI MS. In the IEF-ESTASI-MS method, 74% of the BSA digest could be identified, in comparison to 47% in IEF-MALDI-MS and 48% in OFFGEL-ESI-MS.

---

[16]Qiao, L., Tobolkina, E., Baohong, L., and Girault, H. H. (2013) Coupling isoelectric focusing gel electrophoresis to mass spectrometry by electrostatic spray ionization. *Anal. Chem.*, **85**, 4745–4752.

[17]Qiao, L., Sartor, R., Gasilova, N., Lu, Y., Tobolkina, E., Liu, B., and Girault, H. H. (2012) Electrostatic-spray ionization mass spectrometry. *Anal. Chem.*, **84**, 7422–7430.

2.1.3 Capillary electrophoresis

In capillary electrophoresis[18,19,20] (also known as capillary zone electrophoresis), the analyte solution is filled in a capillary tube and is subjected to an electric field. The field causes two kinds of flow: *electrophoretic* flow and *electroosmotic* flow. Migration velocities of the analyte ions depend on both the electrophoretic and electroosmotic mobilities of the ions and the ions get separated — by differential migration — because of differences in their charge and hydrodynamic radii.

Electroosmotic flow originates from the field exerted by fixed charges on interior wall of the capillary acquired from the buffer solution — charge accumulates on the inner surface of the capillary leading to a double layer. (A potential applied across any fluid conduit — such as a capillary, membrane, microchannel — causes electroosmotic flow, a few mL/s, movement of the net charge in electric double layer; the force is Coulomb force owing to the applied electric field.) Migration velocity v of an ion in an electric field can be expressed in general as

$$
\begin{aligned}
v\,(\mathrm{ms^{-1}}) &= u_p + u_\mathrm{o} = (\mu_p + \mu_\mathrm{o})E \\
&= \mu_p(\mathrm{m^2V^{-1}s^{-1}})E(\mathrm{Vm^{-1}}) + \mu_\mathrm{o}(\mathrm{m^2V^{-1}s^{-1}})E(\mathrm{Vm^{-1}})
\end{aligned}
\tag{2.2}
$$

where μ_p is the electrophoretic mobility, μ_o is the electroosmotic mobility, and E is the electric field strength (dimensions in parentheses). The first term on the right side of eqn 2.2 represents the electrophoretic velocity and the second term the electroosmotic velocity. Electrophoretic mobility μ — which is dependent on pH besides frictional forces in the medium and size and shape of the ion — is given by

$$
\mu = \frac{z(\text{ion valence})\, q(\text{charge C})}{6\pi\eta(\text{solution viscosity Nsm}^{-2})\, r(\text{Stokes radius m})}
\tag{2.3}
$$

where z is ionic valence, q is ionic charge in coulombs, η is solution viscosity, and r is Stokes (hydrodynamic) radius. The Stokes radius is given by

$$
r = \frac{k_B T}{6\pi\eta D}
\tag{2.4}
$$

[18]Dolnik, V. (2008) Capillary electrophoresis of proteins 2005-2007. *Electrophoresis*, **29**, 143–156.

[19]Kašička, V. (2008) Recent developments in CE and CEC of peptides. *Electrophoresis*, **29**, 179–206.

[20]Mischak, H., Coon, J. J., Novak. J., Weissinger, E. M., Schanstra, J. P., and Dominiczak, A. F. (2009) Capillary electrophoresis — mass spectrometry as a powerful tool in biomarker discovery and clinical diagnosis: an update of recent developments. *Mass Spectrom. Rev.*, **28**, 703–724.

where k_B is the Boltzmann constant, T the temperature, and D the diffusion coefficient. Electroosmotic mobility is given by

$$\mu_\circ = \frac{\epsilon \zeta}{\eta} \qquad (2.5)$$

where ϵ is the relative permittivity of the buffer solution and ζ is the zeta potential of the capillary wall.

The number of theoretical plates N in capillary electrophoresis—which can be of the order of several hundred thousand and indicates resolution—is given by

$$N = \frac{\mu(\mathrm{m^2V^{-1}s^{-1}})V}{2D_a(\mathrm{m^2s^{-1}})} \qquad (2.6)$$

where μ is the 'apparent' mobility and D_a is the diffusion coefficient of the analyte. The *resolution* is given by

$$R_s = \frac{1}{4}\left(\frac{\Delta\mu_p\sqrt{N}}{\mu_p + \mu_\circ}\right) \qquad (2.7)$$

where $\Delta\mu_p$ is the differential electrophoretic mobility of one molecule with respect to another of a different characteristic charge or length. The components of the denominator indicate that maximum resolution is reached when the electrophoretic and electroosmotic mobilities are of comparable magnitude and are opposite in sign.

In a capillary electrophoresis set-up, the essentials are: a source reservoir, a destination reservoir, and a connecting capillary, all filled with buffer. Sample is introduced into the capillary by dipping the source end of the capillary into a separate sample reservoir and then returning that capillary end to the source reservoir. A potential difference is then applied across the source and the destination reservoirs; analyte ions separate based on their different electrophoretic mobilities along with common electroosmotic mobility, and are detected at the capillary's exit end. Since $\mu_p = v/E = (L/t_r)(L_t/V)$, from a knowledge of the migration time t_r to reach a distance L, along with $E = L_t/V$, where L_t is the total length of the capillary, the electrophoretic mobility can be determined experimentally. Neutral analytes, in capillary electrophoresis, are not separated well.

Electroosmotic mobility (which drives ions toward the negatively charged cathode—its magnitude can be obtained by measuring migration of *neutral* analytes where the only force active is electroosmotic force) is considerably greater in magnitude than electrophoretic mobilities (the latter can be positive or negative depending on ion polarity and increases with ion charge). Since electroosmotic mobility is overwhelmingly large, it causes—deluging the effect of electrophoretic mobility—all ions to move toward the cathode, their relative velocities determined by sign and charge of the ions: multiply charged cations more speedily and multiply charged anions slowly.

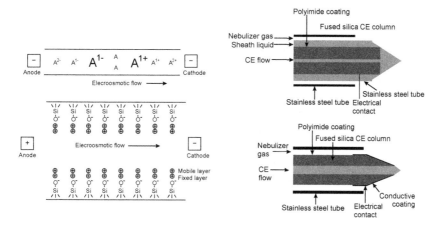

Figure 2.1: Left: (top) Separation of charged and neutral analytes (A) according to their respective electrophoretic and electroosmotic flow mobilities; (bottom) Interior of a fused-silica-gel capillary in the presence of a buffer solution. The flow profile of EOF-driven systems is flat, rather than the rounded laminar flow profile characteristic of the pressure-driven flow in chromatography columns. (Source: http://en.wikipedia.org/wiki/Capillary_electrophoresis.) Right: Capillary Electrophoresis – Mass spectrometry. Schematics of commonly used CE-ESI-MS interfaces: sheath-flow and sheathless coupling. Upper panel: In sheath-flow coupling, a sheath liquid is applied on the outside of the capillary, that circumflows the end of the capillary and closes the electrical circuit. Lower panel: In sheathless coupling the electrical field is established using an outer metal or graphite coating of the capillary as electrical pole. Reproduced by permission of Wiley Periodicals, Inc. from *Mass Spectrometry Reviews*, Vol. 28, 703–724, 2009.

An electric double layer is formed on the inner surface of the capillary in the presence of a buffer solution. If the capillary is made of fused silica, at a somewhat high pH value (>3), negatively charged silanoates (Si-O-) are present on the wall originating from silanol groups (Si-OH). These negatively charged centers attract two layers of positive ions, one (fixed layer) relatively firmly sitting on the negative charge and the other (mobile layer) loosely sitting. On application of a potential across the cathode and anode, the mobile layer moves. The origin of electroosmotic flow of the buffer solution is movement of the mobile layer of the double layer formed on the capillary wall preceded by adsorption of the ions on the wall. Running a basic solution (NaOH, KOH) first before loading the buffer solution enhances the ionization. Charge density on the wall — proportional to the pH of the buffer — contributes directly to the electroosmotic force, and, until all silanol on the wall are fully ionized, the electroosmotic force increases with the pH.

The main advantage[20] of CE in comparison to LC is the former's basic robustness, simple separating principle with high reproducibility, ability to recondition fast with NaOH, and in MS interfacing no problems, unlike in LC, with any buffer gradient — buffer composition does not change in CE analysis — in case of changes, in order to maintain optimal electrospray ionization, ionization parameters required changes. Limited loading capacity in CE analysis is a disadvantage; a maximum of *ca.* 1 μL can be loaded in a CE capillary column, whereas in an LC column, mL quantities can be loaded. The CE-MS approach, however, seems to have suffered because no complete CE-MS system is available commercially.

There have been several alternative types of internal capillary coatings tried to reduce interaction of proteins and peptides with the electroosmotic flow as well physical interaction with the capillary wall, but at pH conditions typically used in peptide separation — they were of little use; in fact, some decreased the resolution of the CE and the sensitivity of MS detection.

In CE coupling to MALDI targets, sheath-flow coupling (in this, a sheath liquid is applied on the outside of the capillary that circumflows the end of the capillary and closes the electrical circuit; ionization in this is comparable to the nanoflow ionspray; the detection limit is inferior to sheathless interface, but stability is better) is straightforward as is the data interpretation; however, there are loss of resolution, signal suppression, and considerable matrix effects; because of this, CE-ESI coupling is preferred. Sheath-flow (Fig. 2.1)[20] interfacing is relatively more stable and is a sensitive detection device in the amol range. Sheathless coupling provides an improved detection limit, but shows reduced stability (this is of disadvantage when using a large number of samples) — therefore sheath-flow interfacing is preferred.

Some techniques related to CE may be mentioned. Micellar electrokinetic chromatography (MEKC)—capillary electrophoresis (CE) is where the analytes are separated by partitioning between a pseudo-stationary micellar phase and a surrounding aqueous buffer solution which is the mobile phase. This can be used to separate neutral analytes or analytes where analyte electrophoretic mobilities do not differ sufficiently. In MEKC, a surfactant (commonly SDS) is added to the solution at a concentration (> Critical Micellar Concentration) where there is equilibrium between surfactant monomers and micelles.

For separation of charged DNA species, capillary gel electrophoresis (high-resolution alternative to slab gel electrophoresis) — capillary filled with a gel matrix, retards longer strands — can be employed.

Some capillary electrophoresis systems can also be used for microscale liquid chromatography or capillary electrochromatography.

A number of modifications of capillary electrophoresis exist. *Isotachophoresis* is electrophoresis which by using a discontinuous electrical field creates sharp boundaries between the sample constituents. *Isoelectric focusing* — usually performed on proteins in a gel — is like zone electrophoresis; separation of molecules occurs by differences in their isoelectric point (pI) and it takes advantage of the fact that overall charge on the molecule of interest depends on the pH of its environment. In *affinity electrophoresis* the value of $\mu_{P,L}$ is measured for a protein as a function of ligand concentration in a buffer where both the protein and the ligand are present. Ion charge ranges from -1 to -3 electrons in amino acid separations. However, since the dye label dominates the size, changes in charge relative to change in size have a significant effect on mobility.

The separation efficiency in capillary electrophoresis separations is much higher than that in HPLC. And since the flow profile in EOF-driven systems is flat, in contrast to pressure-driven chromatography, it does not significantly contribute to band broadening.

2.1.4 Liquid chromatography

Here we first present the basics of HPLC — high-performance liquid chromatography, sometimes also referred to as high-pressure liquid chromatography — then present some recent advances in liquid chromatography in general. Some LC basics are given in the footnote.[21] A reference on matrix effects in LC also is given below.[22]

Basic principle of separation: In liquid chromatography, a solution of the analyte sample in some solvent, which constitutes the *mobile phase*,

---

[21]Pryde, A. and Gilbert, M. T. (1979) *Applications of high performance liquid chromatography.* Chapman and Hall, London. *Column capacity ratio k'* — indicates solute retention: $k' = \frac{t_r - t_0}{t_0}$ where t_0 is the retention time of the solvent peak, t_r is that of the peak of interest. *Number of theoretical plates* (the efficiency of a chromatographic column): $N = 16(t_r/W)^2$ where W is the base width of the peak; in terms of the width of the peak at half-height: $N = 5.54(t_r/W_{HH})^2$. Height equivalent of a theoretical plate/ or the plate-height: $H = L/N$ where H is the column length and N the plate number. Smaller values of H mean more efficient columns. Plate height varies with linear velocity of the eluent u. Plate-height curve is a plot of H against u (or h non-dimensional reduced plate height $h = H/d_p$ versus ν the reduced fluid velocity $\nu = ud_p/D_m$ where d_p is the particle size and D_m is the diffusion coefficient of the solute in mobile phase). Efficiencies of different columns and packing materials can be compared from plate-height curves. *Plate height equation*: $h = (B/\nu) + A\nu^{0.3} + C\nu$. Estimates of D_m (diffusion coefficient of the solution in the mobile phase) can be obtained from Wilke-Chang equation. Wilke-Chang equation: $D_m = 7.4 \times 10^{-12} T \sqrt{(\psi M_{\text{eluent}})}/(\eta V_{\text{solute}}^{0.6})$. D_m is in m^2s^{-1}, η eluent viscosity is in mPas, 1 mPas = 1cP = 1mNsm^{-2}, V_{solute} is solute molar volume, mL, mol. wt./density, M_{eluent} is eluent molecular weight (g), ψ is an eluent association factor, 1 for non-polar solvents, 1.5 for ethanol. 1.9 for methanol, 2.6 for water. *Resolution R_s* is expressed as $R_s = 2((t_2 - t_1)/(W_1 + W_2))$.
[22]Trufelli, H., Palme, P., Famiglini, G., and Cappiello, A. (2011) An overview of matrix effects in liquid chromatography. *Mass Spectrom. Rev.*, **30**, 491–509.

is passed through a chromatographic column which contains a packing of
some solid material, the *stationary phase*, which separates the analytes
based on their, either adorption properties (*adsorption chromatography*),
or size (*size exclusion chromatography*), or ionic charge (*ion-exchange chromatography*). For substances that are soluble in water and ionic in nature,
the course is to employ ion-exchange chromatogaphy. For other analytes
if the molecular size is small ($< 30\,$nm) and they are soluble in water or
in organic solvents, the customary course is use of bonded reversed-phase
chromatography (using stationary phase C4, C8, C18); for sizes in the range
of 30–400 nm molecular exclusion chromatography is used.

Basic instrumentation: HPLC Instrument: Commonly used HPLC instruments consist of pumps, sampling contrivances, and an appropriate detector. Samplers and pumps control the composition of the mobile phase,
carry the analyte sample to the column, and force the mobile phase through
the latter. Column bodies are commonly made of stainless steel. They,
usually are of 4.6 mm i.d., and come in various lengths (commonly in the
range 10–25 cm); the specific length to use depends on the degree of resolution wanted. In protein separation, the adsorption mode is used most.
In adsorption chromatography, the stationary phase is some material on
which the solutes can adsorb; in size-exclusion chromatography the pore
size of the stationary phase is the controlling parameter in separation —
the chromatography releases first the large molecules, and in ion-exchange
chromatography the stationary phase used is some ion-exchange resin with
charge opposite to that of analyte ions; there, those with weaker charge are
released first.

Adsorption chromatography: In adsorption chromatography, the column
is packed with some adsorbent material. When the sample solution flows
through the column, analyte molecules get adsorbed on the stationary phase
with strengths depending on their mutual chemical affinity. The adsorption
step is followed by *elution* which consists of further passage of the mobile phase — now only the solvent — through the column during which the
adsorbed analyte molecules get dislodged from the stationary phase and
ultimately emerge from the column. It is the *differential* release of the adsorbed analyte molecules from the column that results in chromatographic
separation.

Types of elution: Elution can be of more than one kind. In *isocratic*
elution, the composition of the mobile phase — usually a single solvent —
remains constant throughout the elution process. In *gradient* elution, the
composition of the mobile phase is changed continuously on purpose. (In
gradient elution, the polarity and surface tension of the mobile phase are
automatically continually reduced by adding a less polar solvent to the
mobile phase. This results in reduction of retention time; less peak tailing
occurs; peak shapes and overall separation efficiency increase.) *Retention
time* in liquid chromatography is defined as the time taken by an analyte

to travel through the column, which depends on the nature of the analyte besides operating conditions such as the composition of the column and eluent properties. Various analytes must have different, but not too long, retention times so that they come out from the column in separate groups — which would be the chromatographic peaks — yet the overall flow should be fairly rapid so that the column is fully cleared quickly. In HPLC, use of fine packing material requires high head pressure to force the liquid through the capillary.

Particle size, head pressure: Pressure required to maintain a given flow velocity through a column packed with particles varies inversely with the square of the particle diameter. Head pressures are in the range 50–350 bar for typical column diameters of 2.1–4.6 mm and column lengths of 30–250 mm with average particle size 2–5 μm. Waters ACQUITY UPLC particles are of 1.7 μm diameter. The common HPLC detector, when only analyte separation is the objective, is a simple uv-visible photodiode (detection at 254 nm is most common, the range is 180–350 nm) or a photodiode array. We will indicate later how the output of an LC is coupled to a mass spectrometer input.

Properties of stationary phase, mobile phase: In normal phase chromatography, the stationary phase employed is highly polar, such as silica/alumina, or silica-based compositions. Differential affinity of the analytes in binding with such polar stationary surface — through dipole-dipole or hydrogen bonding interaction — is the basis of separation, binding strength increase takes place with analyte polarity increasing. Non-polar, non-aqueous mobile phases (such as n-hexane, methylene chloride, tetrahydrofuran, or their mixtures) are used, thus they are effective for analytes which are soluble in nonpolar solvents. As steric factors play role in the binding, for separation of structural isomers this is a good method. Elution efficiency increases with use of more polar solvents — they decrease retention times, the least polar constituents are eluted first; retention times, however, can have poor reproducibility when water or protic organic solvent layers are present on the stationary phase surface.

In reversed phase chromatography, in contrast to normal phase chromatography, the stationary phase is non-polar and the mobile phase is moderately polar (methanol, water, acetonitrile mixture — the less polar the solvent, the higher the eluent strength). In elution, the most polar constituent comes out first. The stationary phase contains usually a silica base whose surface contains silane-like silicon compounds such as $(C_{18}H_{37})$-$(CH_3)_2$-Si-Cl or (C_8H_{17})-$(CH_3)_2$-Si-Cl with linear side chains. With this, polar molecules are not retained and therefore elute early. Less polar molecules are retained longer with such stationary phase; this can be further augmented by making the mobile phase more polar such as by adding water. Likewise, retention times of the less polar molecules can be decreased by adding less polar solvents—such as CH_3OH, CH_3CN—to the mobile phase, which reduce the effect of surface tension of water (water 7.3×10^{-6} J/cm^2, methanol 2.2×10^{-6} J/cm^2). Similarly, addition of inorganic salts causes increase in surface tension (ca. 1.5×10^{-7} J/cm^2 per Mol for NaCl, 2.5×10^{-7} J/cm^2 per Mol for $(NH_4)_2SO_4$) and causes an increase in retention time. In general, analytes with non-polar bonds in their structure—with large hydrophobic interaction surface—have longer retention times; however, this is qualified by any steric hindrances, just as those with polar groups in their structure; since they interact positively with aqueous mobile phase, they have lower retention times. Because the hydrophobic character is affected by $[H^+]$, the mobile phase uses a buffer to ensure a certain pH. Buffers control the pH, neutralize any charge on the analyte and the stationary phase.

Ammonium formate, which forms adducts, is often added for detection of certain analytes. Addition of trifluoroacetic acid enhances retention of carboxylic acids.

Reversed phase columns are generally robust and they stand aqueous acids reasonably well. Although they may corrode metallic parts of the flow system, aqueous bases act adversely on the silica particles of the stationary phase.

In the following paragraphs, through projection of trends, we present some data and observations by Jorgenson[23] from liquid chromatography research, even though not all the improvement possibilities, projections have reached routine HPLC. At present in column packing technology, the state-of-the-art HPLC packing material diameter is of the order of $2\,\mu$m porus–$1.5\,\mu$m nonporus, column length 5–3 cm, steady pressure 200–230 bar, migration time 0.5–0.3 min, and the efficiency in the range 10,000–8,000 plates, respectively.

Head pressure: For operation superior than ordinary HPLC, it is necessary to go to higher pressures. As a function of particle diameter, required head pressure varies inversely with the cube of the particle diameter—therefore if particle-size is halved the required pressure would need to be increased eightfold. This means that pressure would need to be taken certainly to 20,000 psi level, preferably to a much higher level.

Working at high pressure such as 7,000 bar (100,000 psi), with particle diameter $1\,\mu$m, column length 50 cm, column i.d. about 4.6 mm, and flow rate 2.0 mL/min—the power generated that would need to be dissipated is 24 watts, which is a substantial amount of heat. Working at the same pressure, reducing the column i.d. to 1 mm—near-capillary size, flow rate 95 μL/min—the power to dissipate comes down to 1.1 watts. This amount of heat can be taken out rather quickly.

---

[23]Jorgenson, J. W. (2011) LC and MS: A match made in heaven. *Proc. 59th Annual ASMS Conference on Mass Spectrometry and Allied Topics*, Denver, June 5–9.

Theoretical plates: How does the number of theoretical plates vary with particle diameter? There have been calculations. Taking dead time (t_m) = 2 min, viscosity of water as 1 cP, and diffusion coefficient of *small organic analytes* as 8×10^{-6} cm^2/s, one obtains, presented here in the order: operating pressure : optimum particle diameter : column lengths : theoretical plates, as: 100 bar: 2.4 μm: 13 cm : 30,000; 300 bar, which is very high HPLC pressure: 1.8 μm : 17 cm : 50,000; 1,000 bar, which is UPLC[24] condition : 1.3 μm : 23 cm : 100,000 ; 3,000 bar, which would be a good target pressure of work: 1.0 μm : 30 cm : 170,000. Calculations with *peptide-like structures* $(D_m = 3 \times 10^{-6}$ cm^2/s$)$ give correspondingly : 100 bar: 1.4 μm: 7.7 cm : 30,000; 300 bar, which is very high HPLC pressure: 1.1 μm : 10.5 cm : 50,000; 1,000 bar UPLC condition : 0.8 μm : 14 cm : 100,000 ; 3,000 bar, which would be a good target pressure of work: 0.6 μm : 19 cm : 170,000.

Nature of ordering of column material: Near the inner surface of the column wall, the particles sit well-ordered, but in the interior, order disappears. Confocal laser scanning microscopy data[25,26] of 30 μm i.d. packed capillary columns show that the first four layers from the wall are quite ordered, but at interior particles sit randomly with interparticle volume about 40%. Packing particles do not have to be spherical, although they may be the most efficient. Many alternative shapes have been tried: cubes, parallelopipeds, disks, hex, arbitrary shapes in μm dimensions — all with two parallel flat surfaces.

Superficially porous particles: The advantage with porous particles is that there is more surface area for adsorption available in a given adsorbent volume. One disadvantage, however, of an *entirely* porous particle is that solute molecules may take too much time to get out of the porous particle back into the mobile phase. One solution to this problem is to employ *superficially* porous particles. Initial developers of this kind of particles were Kirkland *et al.*;[27] several vendors are now marketing a range of such particles. Commercial particles available are (from technical literature of Advanced Materials, Phenomenex, Agilent, and Supelco): Advanced Materials – HALO - Peptide – 2.7 μm; Phenomenex – 2.7 μm and 1.7 μm;

[24]Swartz, M. E. (2005) Ultra performance liquid chromatography: an introduction. *Separation Science Redefined*, 8–14.

[25]Bruns, S., Grinias, J., Blue, L., Jorgenson, J., and Tallarek, U. (2012) Morphology and separation efficiency of low-aspect-ratio capillary UHPLC columns, *Anal. Chem.*, **84**, 4496–4503 DOI: 10.1021/ac300326k.

[26]Bruns, S., Franklin, E. G., Grinias, J. P., Godinho, J. M., Jorgenson, J. W., and Tallarek, U. (2013) Slurry concentration effects on the bed morphology and separation efficiency of capillaries packed with sub-2 micron particles, *J. Chromatogr. A.*, **1318**, 189–197. DOI: 10.1016/j.chroma.2013.10.017.

[27]Kirkland, J. J., Truszkowski, F. A., Dilks, Jr., C. H., and Engel, G. S. (2000) Superficially porous silica microspheres for fast high-performance liquid chromatography of macromolecules. *J. Chromatography A*, **890**, 3–13.

Supelco – Ascentis Express – 2.7 μm; Agilent – Poroshell 120/300 – 2.7 μm, 5.0 μm.

Preparation of superficially porous particles: The inner core — 1 μm diameter — is solid silica. Around this solid core, silica nanospheres — tens of nm in diameter — are laid. The superficial layer may be about 0.1 μm thick. It has been estimated that such a superficially porous particle of 1–2 μm diameter does have about 42% of the surface area of a totally porous particle of the same diameter. Blue and Jorgenson have prepared superficially porous particles with 1.5 μm non-porous silica core and five coating steps with 20 nm colloidal silica; they have also reported on 1.1 μm superficially porous particles.[28]

Monolithic LC columns: Tanaka *et al.*[29,30,31] have invented a method of *preparing* monolithic LC columns *right inside a column*. Utilizing information from phase diagram, various physical properties such as pore size and pore volume can be obtained by deciding on the composition of the components: solvent, silica, and PEO polyethylene oxide. More PEO, smaller pore size; more solvent, higher pore volume. The most important property is that particle size and pore size are *not* linked — thus for the mobile phase, low flow resistance can be assured. For example, 2.5 μm particles can be synthesized in a system that has flow resistance of systems usually associated with 5 μm particles. Although the question of reproducibility of the columns and column properties is not answered entirely definitively, the advantage of monolithic columns is that it gets around the problems of slurry packing, and flow resistance is low, hence it provides columns with improved separation efficiency.

Operating temperatures: How does the operating temperature affect LC operation? Since viscosity decreases and diffusivity of solutes increases with increase of temperature, the dominant result is faster flow of the mobile phase and hence faster analysis — provided the solute remains stable, undecomposed at higher temperature. (There has been one tryout at least in which a mixture of three isomers of dihydroxybenzene along with methyl catechol was tested from 22°C to 85°C. As the sample went through the col-

[28]Blue, L. E. and Jorgenson, J. W. (2011) 1.1 μm superficially porous particles for liquid chromatography. Part I: Synthesis and particle structure characterization, *J. Chromatography A*, **1218**, 7989–7995.

[29]Tanaka, N., Kobayashi, H., Nakanishi, K., Minakuchi, H., and Ishizuka, N. (2001) Monolithic LC columns. *Anal. Chem.*, **73**, 420A–429A.

[30]Miyamoto, K., Hara, T., Kobayashi, H., Morisaka, H., Tokuda, D., Horie, K., Koduki, K., Makino, S., Nunez, O., Yang, C., Kawabe, T., Ikegami, T., Takubo, H., Ishihama, Y., and Tanaka, N. (2008) High-efficiency liquid chromatographic separation utilizing long monolithic silica capillary columns. *Anal. Chem*, **80**, 8741–8750.

[31]Miyazaki, S., Takahashi, M., Ohira, M., Terashima, H., Morisato, K., Nakanishi, K., Ikegami, T., Miyabe, K., and Tanaka, N. (2011) Monolithic silica rod columns for high efficiency reversed-phase liquid chromatography. *J. Chromatography A*, **1218**, 1988–1994.

umn at 45°C and 65°C, the chromatograms observed were intact, except that the analysis took less time with increase of temperature. However, in the 85°C chromatogram, methyl catechol was missing — it had decomposed.) Up to 100°C probably it is all right; 200°C may be too high. An increase of temperature from 20°C to 100°C brings down the viscosity of water from 1 cP to 0.28 cP, and this can produce a very significant effect.

Theoretical plates and temperature: It is instructive to look at the results of calculations on theoretical plates as a function of temperature. We give here data on two temperatures, 20°C and 100°C. The viscosities are 1 cP and 0.28 cP, and diffusion coefficients are 8×10^{-6} cm^2/s and 2.8×10^{-6} cm^2/s, respectively. Results show (particle size, column length, theoretical plates): at 100 bar 20°C — 2.4 μm and 13 cm and 30,000, at 100 bar 100°C — 3.2 μm and 33 cm and 60,000; at 1,000 bar 20° C — 1.3 μm and 23 cm and 100,000, at 1,000 bar and 100° C — 1.8 μm and 59 cm and 180,000. If both pressure and temperature are increased, theoretical plates can increase from 30,000 to 180,000.

Observations, trends: Adsorbent particles' size has already gone below 2 μm (ACQUITY 1.7 μm) — superficially porous particles of 1.5 μm diameter core have reached; this is likely to go near 1 μm diameter size. Packing becomes the limiting criterion as the size becomes smaller. Pressures in the domain of 20,000 psi are already in vogue. One foresees this going to the range of 50,000 psi. Beyond that, the present technology of *valves and pump seals* is not adequate. Increase of operational temperature to 100°C seems feasible, above that the main concern is about the stability of the analyte. Use of higher temperature would necessarily require use of capillaries for efficient heat transfer to the column. Superficially porous particles going down to 1 μm is a possibility; the superficially porous particles provide the strength of the core for uniformity in size, yet the porosity provides larger effective surface area for interaction. Monolithic columns are a possibility. Experience with open-tubular columns tells that smaller diameters in the domain of 5 μm would be an improvement. Microfluidic devices where a large number of components are integrated on the chip would be efficient. As separations become more efficient yielding sharper peaks at lower flow rates and smaller samples, MS will have to improve its both speed and sensitivity. With the requirement of substantially higher sensitivity, detection by UV absorption cannot be depended upon.

Before closing this section we would like to mention some recent work on slip flow and very high efficiency chromatography. This has been reviewed.[32] Wirth *et al.* studied[33] pressure-driven protein chromatography

---

[32]Rogers, B. A., Wu, Z., Wei, B., Zhang, X., Cao, X., Alabi, O., and Wirth, M. J. (2015) Submicrometer particles and slip flow in liquid chromatography. *Anal. Chem.*, **87**, 2520–2526.

[33]Wei, B., Rogers, B. J., and Wirth, M. J. (2012) Slip flow in colloidal crystals for ultraefficient chromatography. *J. Am. Chem. Soc.*, **134**, 10780–10782.

in capillaries packed with silica colloidal crystals. Their observations were surprising: the back-pressure was much *lower* than was predicted, and the plate height was very much *lower* than was predicted. Why was this so?

The velocity at the wall would be expected to be zero *only if* the attractive interactions between the mobile and stationary phase molecules were exactly as strong as those between the mobile phase molecules themselves. However, for RPLC, since the functional groups on the surface are chosen to be hydrophobic, there is *little interaction* with the mobile phase. This results in *non*zero velocity of the mobile phase *at the wall*. Thus, in this *slip flow*, the average velocity at the wall is higher with respect to the conventionally expected nonzero velocity, by an additive amount. This means that in slip flow the needed pressure is less.

It was observed that smaller particles give extraordinary improvements in plate height. The improvement in plate height observed was about an order of magnitude better than predicted from theory. For a capillary packed with particles of $0.47\,\mu$m, they measured plate heights as small as $0.015\,\mu$m for BSA, which is nearly 30-fold smaller than the particle diameter. The peaks observed were extremely narrow, narrower than the $75\,\mu$m width of the capillary. One normally thinks about peak width in the time domain, and a $5\,$s base width seems quite narrow for a protein, but it is the high flow rate that makes it intuitively seem sharp. On this basis, the much narrower physical width submicrometer particles will not only yield better resolution, but will also provide better sensitivity because the peak is less diluted by mobile phase.

The instrument also contributes significantly to broadening, which includes the parabolic flow profile through the tubing and the detector volume. They presented data on base widths — for peaks inside the medium — based on Gaussian plots: $64\,\mu$m for no instrument contribution, $400\,\mu$m in typically images observed using a nanoLC for injection with MS for detection, and $2.5\,$mm width calculated for the current best-performing UHPLC instrument with a $5.5\,\mu$L dispersion volume considering no contribution from the column.

The homogeneity of packed submicrometer particles could still have an impact, since, due to packing heterogeneity, columns do contribute broadening beyond the instrument contribution. NanoLC instruments might begin competing in this arena because less peak broadening means greater sensitivity. For comparable sensitivity, narrower peak width could fully offset the short path length across a capillary.

In small molecule separations, likelihood of effectiveness of submicrometer particles remains an open question. The selectivity would suffer as the surface area of $0.5\,\mu$m particles is much lower than that of small-pore silica gel. While use of smaller particles would offset this, but there are other factors, such as the injected width and the diffusion rate on the stationary phase; they become important. Submicrometer particles would be expected

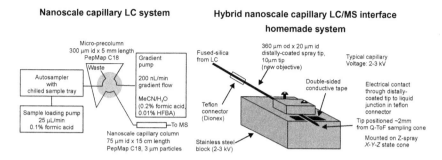

Figure 2.2: Left: Nanoscale Capillary LC System. Right: Hybrid Nanoscale Capillary LC/MS Interface. Courtesy: M. Arthur Moseley, Proteomics and Metabolomics Core Facility, Duke University School of Medicine. Reproduced by permission.

to receive a preferential boost in instruments with higher pressure. Since peptide separations require wider pore silica, on the intermediate scale of molecular weights, it is possible that the submicrometer particles would be advantageous in bottom-up proteomics.

Nano-LC / LC-MS coupling

There have been many efforts toward LC-MS coupling.[34,35,36,37] A nanoscale capillary LC/MS interface and another hybrid nanoscale capillary LC/MS interface are shown in Fig. 2.2.

2.2 Multidimensional separations

A number of *multi*dimensional separation platforms can be constructed by combining one or more independent one-dimensional separation procedures. In this section we consider a few multidimensional separation platforms for protein/peptide samples' separation prior to their ionization and subsequent injection to mass spectrometer analyzer. First we present

---

[34]Henion, J. D. (1980) Direct injection LC/MS/COM of total HPLC eluents applied to drugs and metabolism studies. *Adv. Mass Spectrom.*, **9**, 1241–1249.

[35]Blakley C. R. and Vestal, M. L. (1983), Thermospray interface for liquid chromatography/mass spectrometry. *Anal. Chem.*, **55**, 750–754.

[36]Willoughby, R. C. and Browner, R. F. (1984) Monodisperse aerosol generation interface for coupling liquid chromatography with mass spectrometry. *Anal. Chem.*, **56**, 2625–2631.

[37]Whitehouse, C. M., Dreyer, R. N., Yamashita, M., and Fenn, J. B. (1985) Electrospray interface for liquid chromatographs and mass spectrometers. *Anal. Chem.*, **57**, 675–679.

SCX-RPLC-based separation, then differential gel electrophoresis DIGE, and finally GELFrEE fractionation.

2.2.1   2D SCX-RPLC-based separation in *Shotgun* proteomics

The first report of multidimensional chromatography coupled to a mass spectrometer was by Opitcek *et al.*[38] The first report of using the SCX/RP approach for mass spectrometry of a complex peptide mixture was by Link *et al.*[39] Use of 2D separation — by SCX (strong cation exchange resin) and RPLC (C18 material) — of already-proteolyzed cell lysate followed by MS, or MS/MS, is the basis of 'shotgun' proteomics. We consider the online method; the 2D separation is carried out in tandem. After the SCX-RPLC separation (Fig. 2.3) the separated peptides are ionized and then subjected to m/z analysis. The 2D SCX-RPLC separation followed by MS/MS and database search has been called MudPIT — multidimensional protein identification technology.[40] Fournier *et al.*[41] have reviewed shotgun proteomics with a wide range of applications of MudPIT.

SCX separates on the basis of charge; RPLC separates on the basis of hydrophobicity. In online SCX/RP use employing a single column (typically, fused silica column of $100\,\mu$m i.d. $\times$ $365\,\mu$m o.d.) packed in *tandem* with *both* SCX and RP (that is, *bi*phasic column), the peptide mixture is loaded onto the column, interfaced at the *input* with a quaternary HPLC (the *output* acts as a nanospray ion source to the mass spectrometer). The peptides are then eluted off the SCX resin onto the RP resin by using salts (usually ammonium acetate; also ammonium formate — one limitation of MudPIT has been the inability to use phosphate buffers containing KCl/NaCl) in steps (salt bumps) from low to high (2–3 M) concentration of salt causing peptides to get released from the SCX and bind to C18. After each salt bump an acetonitrile gradient is performed for output (usually to mass spectrometer); separation capability increases with the number of salt bumps used. An improvement in which a small amount of RP is packed prior to SCX in the column (which makes the column RP/SCX/RP — *tri*phasic) desalts the sample in its first dimension. A split-*three*-phase column (250 μm i.d. $\times$ 365 μm o.d. fused silica column on the inlet side, intermediate

[38]Opiteck, G. J., Lewis, K. C., Jorgenson, J. W., and Anderegg, R. J. (2007) Comprehensive on-line LC/LC/MS of proteins. *Anal. Chem.*, **69**, 1518–1524.

[39]Link, A. J., Eng, J., Schieltz, D. M., Carmack, E., Mize, G. J., Morris, D. R., Garvik, B. M., and Yates, J. R., 3rd. (1999) Direct analysis of protein complexes using mass spectrometry. *Nat. Biotechnol.*, **17**, 676–682.

[40]Washburn, M. P., Wolters, D., Yates, J. R. III. (2001) Large-scale analysis of the yeast proteome by multidimensional protein identification technology. *Nat. Biotechnol.*, **19**, 242–247.

[41]Fournier, M. L., Gilmore, J. M., Martin-Brown, S. A., and Washburn, M. P. (2007) Multidimensional separation-based shotgun proteomics. *Chem. Rev.*, **107**, 3654–3686.

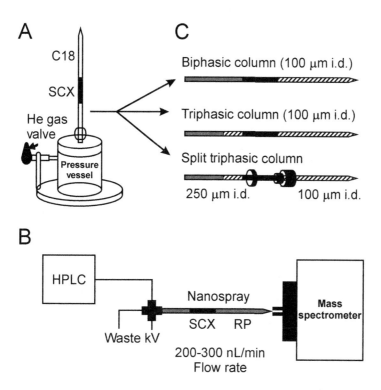

Figure 2.3: MudPIT system initially consists of packing a bi- or triphasic column with RP and SCX resin. (A) The resin is packed into a 100 μm i.d. column offline by using a pressure vessel. The resin is pushed through the MudPIT column using helium at high pressure vessel. Once the column is packed, the pressure vessel is used to load the sample. (B) During a MudPIT experiment, a series of salt bumps, followed by acetonitrile gradients, is used to elute the peptides directly into the mass spectrometer. The MudPIT approach requires the use of volatile salts such as ammonium acetate and ammonium formate. (C) Different columns can be used for MudPIT analysis. A biphasic column consists of only RP resin followed by SCX. Initially, samples that were loaded onto a biphasic column had to be desalted offline before MudPIT analysis. The triphasic MudPIT is similar to the biphasic column with the addition of RP resin added after the SCX resin. This allowed the samples to be desalted on-line before entering the mass spectrometer and results in decreased sample handling. The split triphasic column was designed for holding more RP and SCX resin for larger samples. Two parts of the split column are connected by a microfilter assembly. One side of the column consists of 250 μm i.d. fused silica, and the other side consists of a 100 μm tip. Reproduced by permission of the American Chemical Society from *Chemical Reviews*, Vol. 107, 3654–3686 (2007).

M-520 microfilter assembly, and a 100 μm i.d. $\times$ 365 μm o.d. tip at the exit) has been used for loading larger amount of RP/SCX resin and sample (employed in an analysis of microbial biofilm). When an autosampler is used in MudPIT, maintaining zero dead volumes is a challenge. KetaSpire® PEEK (polyetheretherketone) plastic cross filled with packing material — a vented fully triphasic column, which permits use of higher flow rates during loading — has been used. Applications of MudPIT using SCX-RPLC separation have been extensive.

High-pH reversed-phase liquid chromatography (RPLC) with fraction concatenation does provide better peptide analysis. It has been shown,[42] in shotgun proteomics analyses of a digested human protein sample, that high-pH RPLC increased identification of peptides by 1.8-fold and proteins by 1.6-fold.

There are certain advantages of *off-line* separations, such as the option to use a wider range of buffers and elution conditions particularly suitable for a certain separation step. For instance, utilizing phosphate buffers with salts NaCl/KCl in SCX elution and then desalting before the sample is taken to the next stage (such as mass spectrometer) is one such use.

2.2.2   Difference gel electrophoresis (DIGE)

Difference Gel Electrophoresis (DIGE) — a variation of 2D PAGE – was first reported by Minden *et al.*,[43] later developed and marketed by Amersham Biosciences (now by GE Ettan™).[44,45] It is two-dimensional gel electrophoresis in which the protein mixtures (up to three in number) are labeled with spectrally resolvable fluorescent cyanine dyes Cy2, Cy3, and Cy5,[46,48] *prior to* electrophoretic separation, mixed, and are analyzed on the same 2D PAGE gel by laser-scanning the gel images. The first dimension is isoelectric focusing (IEF) — the separation is based on pI; the second dimension is this special SDS-PAGE. After electrophoresis, the gel is scanned sequentially with the excitation wavelengths of the three dyes, and protein abundances are obtained. A given protein from different samples, regardless of the labeling dye used reaches the same position on the gel;

---

[42]Yang. F., Shen, F., Camp, D. G., and Smith, R. D. (2012) High-pH reversed-phase chromatography with fraction concatenation for 2D proteomic analysis. *Expert Review of Proteomics*, **9**, 129–134.

[43]Ünlü, M., Morgan, M. E., and Minden, J. S. (1997) Difference gel electrophoresis. A single gel method for detecting changes in protein extracts. *Electrophoresis*, **18**, 2071–2077.

[44]Lilley, K. S. and Friedman, D. B. (2004) All about DIGE: quantification technology for differential-display 2D-gel proteomics. *Expert Rev. Proteomics*, **1**, 401–409.

[45]Ettan DIGE Manual 18-1173-17 Edition AB.

[46]Cy2 cyanine, typical excitation 488 nm, emission filter 520 nm, BP (Bandpass) 40; Cy3 Indocarbocyanine, excitation 532 nm, emission filter 580 nm, BP 30, Cy5 Indodicarbocyanine, excitation 650 nm, emission filter 670 nm, BP 30 (ref.[48]).

thus problems of inter-gel variations do not interfere. Proteins of the order of 25 pg can be detected. The method is not affected by any pH change, and during labeling, separation, and scanning, there is negligible loss of signal/sensitivity. Two labeling methods are used: minimum labeling and saturation labeling.

In *minimum* labeling CyDye DIGE fluors Cy2, Cy3, Cy5 dyes — matched for mass and charge — are used which contain N-hydroxy succinimidyl ester; they react with primary amino groups, such as ε-amine groups of *lysine* side chains (and add 450 Da to the protein mass). In minimum labeling, only 2–5% of the total number of lysine residues get labeled. This type of labeling restrains multiple dye additions and thus also later multiple spots in the second dimension. The labeling is very sensitive; detection can be 25 pg (compared to 1 ng in silver staining) and the dynamic range (linear response) is over five orders of magnitude (compared to two orders of magnitude in silver staining). This labeling — labeled only once — is compatible with downstream quantification of the protein in MS analysis. However, since because of the dyes influencing migration rate, movement in the second dimension may get affected, gels that are minimally labeled are often *post*-stained with a total protein stain, such as with SyproRuby, so that most of the separated protein is recovered in excision for in-gel digestion preparatory to MS.

In *saturation* labeling, every *cysteine sulfhydryl*, from cysteine residue, following its reduction by TCEP (tris (2-carboxyethyl) phosphine) in the protein, is labeled by maleimide reagents (maleimide reactive group present in the dye forms a covalent bond with the thiol group of cysteine in proteins via a thioether linkage) Cy3 and Cy5, which add approx. 677 Da to the mass of the protein. Saturation labeling is a much more sensitive technique (detection lower than 25 pg); increased sensitivity results because, owing to the nature of the labeling, *more* fluorophors are now integrated into each protein. In the Ettan^{TM} system, multiplexing up to two samples is possible. It is important that reduction of cysteine residues is complete, also the protein:dye ratio must be appropriate; overlabeling can lead to reaction with lysine residues, adding complications. Saturation dye-labeled proteins have been reported to be compatible with MS.[47] One intrinsic limitation of this type of labeling is that the protein must contain cysteine; therefore some proteins, though this does not constitute a large percent, cannot be assayed using this method.

---

[47]Tonge, R., Shaw, J., Middleton, B., Rowlinson, R., Rayner, S., Young, J., Pognan, F., Hawkins, E., Currie, I., and Davison, M. (2001) Validation and development of fluorescence two-dimensional differential gel electrophoresis proteomics technology. *Proteomics*, **1**, 377–396.

A key element in DIGE is the use of *pooled internal standard*.[48] Equal amounts of *each* of the analyte samples being compared constitute the pool. The pooled standard ensures all proteins present in the samples are represented (every protein from every sample), it makes up the average of all the samples under analysis, and normalizes abundance measurements across multiple gels used in an experiment. This results in a highly similar spot pattern in each gel image, which improves confidence of inter-gel spot matching and quantification.

The Ettan^{TM} system comprises CyDye^{TM} DIGE Fluor minimal dyes, Typhoon^{TM} Variable Mode Imager, and DeCyder^{TM} 2D software. Alternatively, Ettan DIGE Imager^{TM} can be used instead of Typhoon Variable Mode Imager, and ImageMaster^{TM} 2D Platinum software can be used instead of DeCyder 2D software.

2.2.3 GELFrEE fractionation

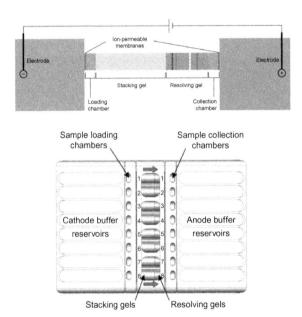

Figure 2.4: Top: Schematic of the GELFrEE fractionating system; Bottom: eight-sample capacity cartridge. Courtesy: Expedeon, Inc. Reproduced by permission.

---

[48]Alban, A., David, S. O., Bjorkesten, L., Andersson, C., Sloge, E., Lewis, S., and Currie, I. (2003) A novel experimental design for comparative two-dimensional gel analysis: two-dimensional difference gel electrophoresis incorporating a pooled internal standard. *Proteomics*, **3**, 36–44.

For size-based separation of proteins, SDS-PAGE is a powerful method, but recovery of the protein from the gel is a problem. To surmount this, electrophoretic elution of proteins from SDS-PAGE is an attractive possibility, but the available eluter devices offer only restricted number of fractions, and it has been felt they are difficult to multiplex. A viable strategy to overcome this challenge is elution of the proteins from the end of a gel column by continuous application of electric field, wherein proteins are trapped by a molecular weight cut-off (MWCO) membrane and are subsequently collected. A new protein fractionation device,[49,50] based on continuous elution tube gel electrophoresis, employs a gel column, in the first stage to separate proteins; that is followed by elution of the proteins from the column and collection — then free of the gel — in the solution phase. Originators of this separation technique named it GELFrEE (*G*el-*E*luted *L*iquid *Fr*action *E*ntrapment *E*lectrophoresis). High recovery at submicrogram to milligram sample loadings, and proteome separation over mass range from below 10 to 250 kDa in 1 h have been reported. Under optimized conditions, it yields fast, reproducible, broad mass range proteome separation, under a high-yield format.

The device has four main components: a cathode chamber, the gel column, a collection chamber, and an anode chamber. Eluted samples are trapped and recovered in the collection chamber. A 3.5 kDa MWCO membrane is sandwiched — pressure sealed — between the collection chamber and the anode chamber. An access port on the top of the collection chamber allows multiple fractions to be removed without requiring disassembly of the device. There are three stages of operation: sample loading, separation, and collection. The electrode chambers and also the void volume above the gel column are filled with electrode running buffer (0.192 M glycine, 0.025 M Tris, 0.1% SDS). Separation takes place at constant 240 V; after the dye front has visibly entered the collection chamber, power supply is paused, and the first fraction is collected. Restart, stop, restart processes are repeated over the course of separation.

A commercial realization of the GELFrEE principle by Expedeon is shown in Fig. 2.4. GELFREE 8100 Fractionating System, which is a benchtop instrument, enables preparative scale fractionation of analytes based on electrophoresis. Its two major components are: (i) the fractionating station, which is a programmable power supply that can accurately apply potentials, and can be made to stop at intervals chosen by the user for collection of fractions, and, (ii) GELFREE Cartridge kit (with eight-channels, which allows simultaneous fractionation of eight samples), having a gel size of 2.5 cm length, 8 mm o.d., sample volume 250 μL, reservoir volume 9.5 μL. The cartridge buffers are: running buffer 500 mL, sample buffer 1.9 mL.

[49]Tran, J. C. and Doucette, A. A. (2008) Gel-eluted liquid fraction entrapment electrophoresis: an electrophoretic method for broad molecular weight range proteome separation. *Anal. Chem.*, **80**, 1568–1573.

[50]Tran, J. C. and Doucette, A. A. (2009) Multiplexed size separation of intact proteins in solution phase for mass spectrometry. *Anal. Chem.*, **81**, 6201–6209.

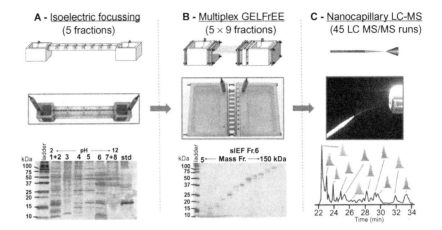

Figure 2.5: The four-dimensional platform for high-resolution fractionation of protein molecules. Schematics (top) and photographs (middle) are shown for (a) a custom device for sIEF, (b) a custom device for multiplexed GELFrEE and (c) RPLC coupled to mass spectrometry. Representative one-dimensional gels of fractions collected from two electrophoretic devices are shown below their pictures; note the resolution attainable at the level of intact proteins. The combined resolution of RPLC with Fourier-transform mass spectrometry is depicted by the chromatogram along with selected isotopic distributions for protein ions measured during the run. Reproduced by permission of Macmillan Publishers Limited from *Nature*, Vol. 480, 254–258, 2011.

A custom device used for multiplexed GELFrEE, part of a 4-dimensional separation,[51] is shown in Fig. 2.5.

For top-down proteomics with limited sample availability, there is a need for a decreased column diameter and smaller collection chamber volume. In a later work involving miniaturization,[52] those were decreased — with respect to that of the commercial system — the column diameter three fold — to 2 mm, the collection chamber volume reduced to 15 μL, which enabled use of higher voltage, leading to three-fold increase in separation speed, also resulted in requiring a tenfold decrease in loading amount. Column length change from 1 cm to 3 cm, 700 Da mass difference was resolvable for proteins

[51]Tran, J. C., Zamdborg, L., Ahlf, D. R., Lee, J. E., Catherman, A. D., Durbin, K. R., Tipton, J. D., Vellaichamy, A., Kellie, J. F., Li, M., Wu, C., Sweet, S. M., Early, B. P., Siuti, N., LeDuc, R. D., Compton, P. D., Thomas, P. M., and Kelleher, N. L. (2011) Mapping intact protein isoforms in discovery mode using top-down proteomics. *Nature*, **480**, 254–258.

[52]Tran, J. C., Chu, P. Y., Catherman, A. D., Wu, C., Durbin, K. R., Compton, P. D., Ahlf, D., Fellers, R., Early, B., Thomas, P. M., and Kelleher, N. L. (2012) Miniaturization of a size-based separation device for top-down proteomics analysis of low microgram samples. *Proc. 60th Annual ASMS Conference on Mass Spectrometry and Allied Topics* Vancouver, May 20–24.

$<25\,$kDa along with 5–10-fold resolution improvement across 10–100 kDa. With only $50\,\mu$g sample loading amounts, 400 unique HeLa S3 could be identified.

2.3 Proteolysis

A few proteases for polypeptide chain cleavage and cleaving properties:[53]

Trypsin	Arg or Lys on P_1 when P_1' is not Pro. Tris-HCl or ammonium bicarbonate buffers; pH 7.0–9.0, optimal at 8.2. Often used in modified form: methylated (Promega), mannosylated (Roche). Best if used in the presence of TPCK.
Chymotrypsin	Tyr, Phe, Trp on P_1 when P_1' is not Pro. Leu, Met, Ala, Asp, and Glu on P_1 are cleaved with correspondingly slower reaction rates. Tris-HCl or ammonium bicarbonate buffers; pH 7.0–9.0. Best if used in the presence of TLCK.
Factor Xa	Arg on P_1 in the sequence Ile-(Glu, Asp)-Gly-Arg when P_1' is not Arg or Pro. pH 7.5–8.5, optimal 8.0–8.2.
Endoproteinase Arg-C	Arg on P_1 when P_1' is not Pro; may also cleave Lys-Lys at P_1-P_2 or any other combination of two basic residues in a row. pH 7.0–9.0, optimal 8.0.
Endoproteinase Lys-C	Arg or Lys on P_1 when P_1' is not Pro. pH 7.0–9.0, optimal 8.2.
Endoproteinase Asp-N	Asp on P_1' when P_1 is not Pro, may also cleave Asp-Asp at P_1'-P_2'. pH 6.0–8.5.
Endoproteinase Glu-C	Glu or Asp on P_1 when P_1' is not Pro at pH 4; or Glu on P_1 when P_1' is not Pro at pH 7. pH 4.0–7.8.

---

[53]The general nomenclature of cleavage site positions of the substrate: $...P_3...P_2...P_1...|...P_1'...P_2'...P_3'...$ (| is the cleavage site); P_1, P_2, ... numbers increase toward the N-terminal direction of the cleavage site; P_1', P_2', ... numbers increase toward the C-terminal direction of the cleavage site. For a listing and some properties of the 20 essential amino acids including abbreviations used here, see Chapter 4. Two major references for all proteolytic sequencing or sequencing-related enzymes: Enzyme structure, in *Methods in Enzymology*, Hirs, C. H. W., ed., **11**, 3–988, 1967; Hill, R. L. (1965) Hydrolysis of proteins, *Adv. Prot. Chem.*, **20**, 37–107. For enzymatic hydrolysis and mass spectrometry applied to the field of proteomics: Medzihradszky, K., In-solution digestion of proteins for mass spectrometry, in *Methods in Enzymology*, Burlingame, A. L., ed., **405**, 50–65 (2005), Elsevier.

Endoproteinase K	Cleavage after aliphatic and aromatic amino acids. Ala on P_2 enhances cleavage. Stable in the pH range 4–12; optimal pH 8.0.
Thermolysin	Leu, Met, Phe, Trp on P_1'. pH 5.0–8.5.
Carboxypeptidase A	Decreasing reaction rates for acidic residues on P_1, followed by neutral, hydrophobic, then basic residues when P_1' is not Pro. Acetate and phosphate buffers, pH 4.0–6.0, optimal at 5.6.
Carboxypeptidase B	Arg or Lys on P_1 when P_2 is not Pro. Tris-HCl or ammonium bicarbonate buffers. pH 7.0–9.0, optimal at 8.2.
Carboxypeptidase Y	Most residues on P_1, reaction rates considerably slower when Pro is on P_1. Pro on P_2 also shows removal of amino acids. Acetate buffers, pH 4.0–6.0.
Aminopeptidase M	Most amino acids on P_1'; Pro on P_2' slows reaction rates. Tris-HCl buffer. pH 7–9, optimal 8.5.
Ficin	Nonspecifically cleaves at exposed residues favoring the aromatic residues.
Papain	Nonspecifically cleaves at exposed residues.
Pepsin	Nonspecifically cleaves at exposed residues favoring the aromatic residues. pH 1.0–3.5.

A middle-down protease has been used by Wu *et al.*[54] The middle-down method is between the top-down, which provides better sequence coverage, and the bottom-up, which provides more number of identifications. It detects both the high molecular weight proteins and also gives a better chance of discovering multiple post-translational modifications on one spectrum. Wu *et al.* developed a method for restricted enzymatic proteolysis using the outer membrane protease T (OmpT) to produce large peptides (> 6.3 kDa on average). Applying this approach to analyze prefractionated high-mass HeLa proteins, they identified 3,697 unique peptides from 1,038 proteins. They demonstrated the ability of large OmpT peptides to differentiate closely related protein isoforms and to enable detection of many post-translational modifications.

For the top-down workflow, since the samples are always not digested, always a front end separation (RPLC or ion exchange-RPLC or size-separation-RPLC) is done for a complex mixture, but for simple pure

---

[54]Wu, C., Tran, J. C., Zamdborg, L., Durbin, K. R., Li, M., Ahlf, D. R., Early, B. P., Thomas, P. M., Sweedler, J. V., and Kelleher, N. L. (2012) A protease for "middle-down" proteomics. *Nat. Methods*, **9**, 822–826.

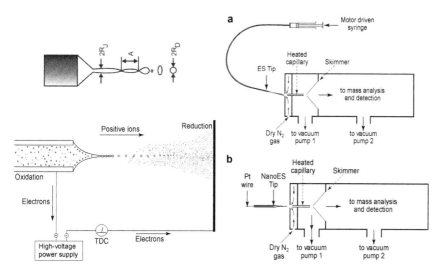

Figure 2.6: Left: Schematic of electrospray ionization and Taylor cone formation. (Left Top: Cone jet mode of spray; $2R_J$ jet diameter, R_D droplet radius. Left Bottom: Electrospray ionization process; TDC: Total droplet current.) Right: Input configurations for normal electrospray (top) and nanospray (bottom). Redrawn as composite. Reproduced by permission of Wiley Periodicals, Inc. from *Mass Spectromtery Reviews*, Vol. 28, 898–917, 2009.

compounds, simply a direct infusion into the mass spectrometer is often done. A size-based separation is especially useful for top-down because it is the larger intact proteins which mask the signals of the smaller proteins by forming adducts and water clusters more readily resulting in higher baseline noise.

2.4 Ionization of proteins and peptides

The two major methods for ionization of proteins and peptides are the electrospray ionization (ESI) and the matrix-assisted laser desorption ionization (MALDI). Two relatively new and much less used methods of ionization are: laser-induced liquid beam ion desorption (LILBID) and laser ablation with electrospray ionization (LAESI). We consider the electrospray ionization first.

2.4.1 Electrospray ionization — ESI

In *electrospray* ionization, gas phase ions are generated from molecules in solution by passing an aqueous solution of the sample through a capillary and spraying it into a strong electric field, toward a counterelectrode

(Fig. 2.6). The tip of the capillary is held at a high voltage, typically at
2–3 kV — positive for *positive ion mode* and negative for *negative ion mode*
— with respect to the counterelectrode, which is located about 1–3 cm from
the capillary tip. The capillary diameter is of the order of ~ 1 mm. On the
counterelectrode there can be an orifice to let the ions directed toward it enter the mass spectrometer. As the charged droplets move across both a field
gradient and a pressure gradient toward the counterelectrode, their sizes decrease as the solvent evaporates. With evaporation of the solvent, charge
concentration in the droplets increases, ultimately resulting in fission of the
droplets causing further size decrease and desolvation. This is the normal
electrospray. The initial experiments on electrospray were by Dole *et al.*;[55]
the technique was fully developed by Fenn *et al.*[56] *Nanospray*, developed by
Wilm and Mann,[57,58] uses a much smaller diameter capillary tip (of the order of 1–10 μm) and is placed much closer (to about 0.5–2 mm) to the mass
spectrometer orifice. It requires much less analyte solution, and unlike the
syringe-forced flow in normal electrospray, the pull exercised by the applied
electric field causes, for non-viscous solutions, a self-flow controlled by the
capillary tip diameter. ESIMS is a very sensitive method; very low concentration of analyte — like 10^{-7}–10^{-3} mol/L (M) — is adequate. Therefore,
even such solvents in which solubility of the analyte is low, can be used.
Typically a polar solvent is used; commonly used solvents are: methanol,
acetonitrile, methanol-water, acetonitrile-water mixtures. Toluene too can
be used even though the solubility of electrolytes is low in toluene, because
even low concentrations such as $\sim 10^{-7}$ M suffice. The technique has been
reviewed.[59,60,61,62] For some additional details on what follows, see the reference[62] in the footnote.

The electric field E_c at the tip of the capillary is very high. If V_c is
the applied potential, r_c the outer radius of the capillary, d the distance
between the capillary tip and the counterelectrode, then the field E_c at the
capillary tip is given by:

---

[55]Dole, M., Mack, L. L., Hines, R. L., Mobley, R. C., Ferguson, L. D., and Alice, M.
B. (1968) Molecular beams of macroions. *J. Chem. Phys.*, **49**, 2240–2249.

[56]Fenn, J. B., Mann, M., Meng, C. K., Wong, S. F., and Whitehouse, C. M. (1989)
Electrospray ionization for mass spectrometry of large biomolecules. *Science*, **246**, 64–71.

[57]Wilm, M. S. and Mann, M. (1994) Electrospray and Taylor-cone theory, Dole's
beam of macromolecules at last? *Int. J. Mass Spectrom. Ion Proc.*, **136**, 167–180.

[58]Wilm, M. and Mann, M. (1996) Analytical properties of the nanoelectrospray ion
source. *Anal. Chem.*, **68**, 1–8.

[59]Hamdan, M. and Curcuruto, O. (1991) Development of the electrospray ionisation
technique. *Int. J. Mass Spectrom. Ion. Proc.*, **108**, 93–113.

[60]Gaskell, S. J. (1997) Electrospray: Principles and Practice. *J. Mass Spectrom.*, **32**,
677–688.

[61]Karas, M., Bahr, U., and Dülcks, T. (2000) Nano-electrospray ionization mass spectrometry: addressing analytical problems beyond routine. *Fresenius J. Anal. Chem.*, **366**,
669–676.

[62]Kebarle, P. and Verkerk, U. H. (2009) Electrospray: From ions in solution to ions
in the gas phase, what we know now. *Mass Spectrom. Rev.*, **28**, 898–917.

$$E_c = \frac{2V_c}{r_c \ln(4d/r_c)} \tag{2.8}$$

For $V_c = 2{,}000$ V, $r_c = 5 \times 10^{-4}$ m, and $d = 0.02$ m, $E_c \approx 1.6 \times 10^6$ V/cm.

Taylor cone formation. Since the capillary is positively charged and the counterelectrode located downfield is negative, polarization effect of the solvent near the meniscus of the liquid draws positive ions near the surface of the meniscus and negative ions away from the meniscus. The meniscus undergoes distortion because of the downfield force and it assumes the form of a cone (Taylor cone). The surface tension forces on the liquid act to contain the cone, but if the applied voltage is above a threshold then the electric field dominates over the force of surface tension, the tip becomes unstable resulting in release of a fine jet from the cone tip. As a result, a mist of highly charged droplets emerges. Vertes *et al.*[63] have obtained, by using fast time-lapse imaging, details of the evolution of the Taylor cone into a cone jet.

Mechanism of droplet formation. The droplets, which are approximately of the same size since the size depends on the jet diameter, drift through the ambient air toward the counterelectrode with constant evaporation of the solvent. With reduction in size, charge concentration increases, and when the electric field dominates over the force due to surface tension, fission — Coulomb fission or Coulomb explosion — takes place. Evaporation of the solvent continues from the smaller droplets thus generated and repeated fissions that take place result in very small droplets and finally in gas phase ions. The Coulomb fission takes place when the *Rayleigh limit*[64] on the droplet charge is crossed, the limit Q_{Ry} given by

$$Q_{\mathrm{Ry}} = 8\pi(\varepsilon_0 \gamma R^3)^{1/2} \tag{2.9}$$

where ε is the electrical permittivity, γ the surface tension of the solvent, and R is the radius of the droplet. Solvent dependence is not significant. On fission the loss of mass is $\sim 1\%$, but the charge loss is much more, ~ 15–25% of the droplet. Time-dependent observations of Taylor cone functioning have shown pulsations of the cone-jet, which manifest as spray current fluctuations. This signal can be used to stabilize the jet, thus improve the signal-to-noise ratio in mass spectra obtained. The cone-jet method is most commonly used, and it is the most efficient. Typical meniscus shapes and disintegration in three axial spraying modes are shown in Fig. 2.7.[65]

---

[63]Marginean, I., Parvin, L., Hefferman, L., and Vertes, A. (2004) Flexing the electrified meniscus: the birth of a jet in electrospray. *Anal. Chem.*, **76**, 4202–4207.

[64]Rayleigh, L. (1882) On the equilibrium of liquid conducting masses charged with electricity. *Phil. Mag.*, **14**, 184–186.

[65]Nemes, P., Marginean, I., and Vertes, A. (2007) Spraying mode effect on droplet formation and ion chemistry in electrosprays. *Anal. Chem.* **79**, 3105–3116.

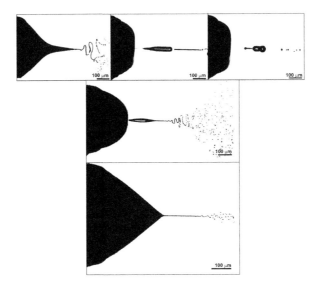

Figure 2.7: Taylor cone. Comparison of typical meniscus shapes and disintegration in three axial spraying modes. The top three panels show the different phases
of liquid ejection in a single current spike of the burst mode (axial I). Initially the
liquid filament exhibits kink instabilities and produces relatively small droplets.
The majority of the ejected liquid is carried away by a few very large droplets that
vary in initial shape between a spindle and a bottle. These large droplets require
additional generations of fission to become sufficiently small for ion production;
thus, they do not contribute significantly to the ion signal. The middle panel is an
example of liquid ejection in pulsating Taylor cone mode (axial II). A spindlelike
large droplet is accompanied by a significant amount of smaller ones produced
by lateral kink instabilities at the end of the jet. The steady cone-jet (axial III)
mode is presented in the bottom panel. This mode is stationary with no major
changes in the shape of the meniscus. It produces much smaller droplets that
downstream become undetectable by the Phase Doppler Anemometer. Ion production is most efficient in this mode. Reproduced by permission of the American
Chemical Society from *Analytical Chemistry*, Vol. 79, 3105–3116, 2007.

Normal electrospray, nanospray comparison. The electrospray configurations are different in normal electrospray and nanospray. In the normal
electrospray, a syringe pushes the analyte solution through a flexible glass
capillary and the emerging droplets drift toward the counterelectrode on
which the orifice — the mass spectrometer inlet — is mounted. Inside the
mass spectrometer, in the first chamber, the droplets experience a weak
counterflow of dry nitrogen gas that further desolvates them. The ions then
enter the next chamber through a heated capillary under the influence of
an electric field and a pressure gradient in which, before the ions enter next

the analyzer chamber through a skimmer, final desolvation takes place. In nanospray, where only μL of analyte solution is involved and which we will discuss in some detail later, instead of a flexible capillary and syringe, a short nano-ES tip capillary is used. A metal film coating on the outside of the capillary or a thin platinum wire introduced from the large end of the capillary supplies the electric potential.

Electrospray, electric discharge comparison. Electrospray is different from electric discharge. In electric discharge, currents are higher than 10^{-6} A. Electric discharge can take place in neat water between the spray-capillary tip and the counterelectrode. Appearance of ions like $H_3O^+(H_2O)_n$ from water or $CH_3OH_2^+(CH_3OH)_n$ in the positive ion mode are indicative of discharge. In contrast, with acidified solvent in the capillary — under that condition H_3O^+ and $CH_3OH_2^+$ ions are present in the solution — protonated solvent ions at high abundance are observed in electrospray. When the discharge currents are high, ions generated in the discharge appear at high intensities whereas the electrospray ions appear at low intensities. Electric discharge is more likely to occur in the negative ion mode and also when the ambient gas has free electrons, the latter being caused by cosmic rays or other background radiation. Air is a suitable ambient gas for electrospray as oxygen molecules scavenge electrons by electron attachment. SF_6 can be a more efficient free electron remover and therefore better for electrospray, but that can introduce spuriously F^- ions into the mass spectra.

Onset voltage of electrospray. The onset voltage V_{on} for electrospray depends on r_c, surface tension γ, half-angle of the Taylor cone θ, the distance d between the capillary tip and the counterelectrode, and is given by

$$V_{on} \approx \left(\frac{r_c \gamma \cos \theta}{2\varepsilon_0} \right)^{1/2} \ln \left(\frac{4d}{r_c} \right) \tag{2.10}$$

where ε_0 is the permittivity of vacuum. Substitution of $\varepsilon_0 = 8.8 \times 10^{-12}$ $J^{-1}C^2$ and $\theta = 49.3°$ gives:

$$V_{on} = 2 \times 10^5 (\gamma r_c)^{1/2} \ln \left(\frac{4d}{r_c} \right) \tag{2.11}$$

In order to obtain V_{on} in volts using the above equation, γ must be in N/m and r_c in m. For stable operation of the electrospray a few hundred volts higher than V_{on} is needed. Eqn 2.10 has been experimentally verified. Approximate onset voltages V_{on} for ESI with solvents having a range of surface tensions γ are given in the following Table[62] for a spray tip of 0.1 mm radius and a distance of 4 cm between the spray tip and the counterelectrode:

Solvent	CH_3OH	CH_3CN	$(CH_3)_2SO$	H_2O
$\gamma(N/m)$	0.0226	0.030	0.043	0.073
$V_{on}(Volt)$	2200	2500	3000	4000

For water, because it has the highest surface tension among the liquids mentioned in the above table, it would be most difficult to stretch the surface into a cone, and thus it has the highest V_{on}.

Three analyte standards, spanning a wide mass range have been suggested for ESI characterization: dodecyltrimethylammonium bromide (m/z 308.34) having a high surface activity therefore high sensitivity/efficiency, leucine encephalin (m/z 556 (555.62268)) which generates singly charged ion, and apo-myoglobin (17 kDa (16952.27)) for protein analytes.

Charge Residue Model and Ion Evaporation Model for ion formation. There are two models on the mechanism of gas phase ion formation from very small and highly charged droplets in electrospray. In the *charge residue model* (CRM) it is assumed that among those tiny droplets some, on the surface of the droplet, would contain only one analyte molecule along with some charges. When such a droplet loses its solvent molecules, a gaseous ion is formed where the charges are those which were on the surface of the droplet. According to the *ion evaporation model* (IEM) on the other hand, when droplet radius shrinks beyond 10 nm, ion emission from the droplet would occur. It was observed that in an assembly of $[(NaCl)_n(Na)_m]^{m+}$ ions, the abundance decreased rapidly as n decreased, although $Na^+(n = 0$, $m = +1)$ and hydrated $Na(H_2O)_k^+$ ($k = 1$–3) were of highest abundances. It was thought that the abundant Na^+ and Na^+ hydrates were formed via a different mechanism where, from the surface of the droplets, the Na^+ ions directly escape. For small organic and inorganic ions, IEM is experimentally well supported. For large macromolecular ions such as proteins, CRM is considered more reasonable and valid than IEM.

In the course of search for reasons why proteins/macromolecules are multiply charged, an empirical correlation was found[66] — from a study of some multiply branched alkyl amine polymers with rigid structures like those of globular proteins — between average charge Z^{obs} observed on the macromolecule/protein and its molecular mass M

$$Z^{obs} = aM^b \qquad (2.12)$$

where a and b are constants. In a separate work, over a large number of non-denatured proteins[67] a bestfit was observed for $b = 0.52$–0.55. With additional data from the literature on non-denatured proteins the relationship was found[68] to hold; it was also shown that the relationship could be

[66]Tolic, R. P., Anderson, G. A., Smith, R. D., Brothers, H. M., Spindler, R., and Tomalia, D. A. (1997) Electrospray ionization Fourier transform ion cyclotron resonance mass spectrometric characterization of high molecular mass starburst (TM) dendrimers. *Int. J. Mass Spectrom. Ion Proc.* **165**, 405–418.

[67]Chernuschevich, I. V., Ens, W., and Standing, K. G. (eds) (1998) *New methods for the study of biomolecular complexes.* Dordrecht, Boston, London: Kluwer Academic Publishers. p. 101.

[68]Fernandez de la Mora, J. (2000) Electrospray ionization of large multiply charged species proceeds via Dole's charged residue mechanism. *Anal. Chim. Acta,* **406**, 93–104.

derived from the CRM. Studies on evaporating charged droplets of 5–35 μm diameter show that between before and after Coulomb fission, the droplet charge stays within 95–75% of the Rayleigh limit, which is the maximum charge a liquid droplet can hold. It was assumed that when all the water disappears, the protein would be neutral — ion pairing between positive and negative ions could cause this as the solvent disappears.

The radius R of the protein can be obtained from the following equation, by applying the equation to the radius at Rayleigh limit, relating molecular mass M of the protein to the droplet volume

$$\frac{4}{3}\pi R^3 \varphi N_A = M \qquad (2.13)$$

φ being the density of the protein and $N_A = 6 \times 10^{23}$ (molecules/mole). The non-denatured protein and water are assumed to be of the same density — mobility measurements support this assumption. The number of charges on the protein can be taken as the number of charges Z on a droplet of water, of radius R, that contains the protein at the Rayleigh limit. Z can be obtained by using the above equation, the Rayleigh limit equation (eqn 2.9) and taking $Q = Z \times e$. That gives:

$$Z = 4 \left(\frac{\pi \gamma \varepsilon_0}{e^2 N_A \varphi} \right)^{1/2} \times M^{1/2} = 0.078 \times M^{1/2} \qquad (2.14)$$

using which Z can be obtained for M, the latter in megadaltons. Z increases with square root of M. Experimental results are in good agreement with the equation; the experimentally observed exponent is 0.53. This could be taken as a strong support for the CRM explaining the formation of globular proteins and protein complexes. It is to be mentioned that in the calculations surface tension of water was used although the sample solution actually contained methanol-water (methanol γ is 22.6×10^{-3} N/m, water 73×10^{-3} N/m). The justification was, that methanol, having much higher vapor pressure, would be lost rapidly in the progeny droplet formation and the final droplets would mostly contain water. In a plot of average charge of proteins observed by Kaltashov and Mohimen[69] versus molecular mass shows a very good fit with respect to eqn 2.14 — the plot up to protein mass about 1 MDa is presented in Ref.[62] Barran *et al.* have shown[70] that for structured proteins M correlates with median z; disordered proteins have a higher median z (Fig. 2.10).

[69]Kaltashov, I. A. and Mohimen, A. (2005) Estimates of protein areas in solution by electrospray ionization mass spectrometry. *Anal. Chem.*, **77**, 5370–5379.

[70]Beveridge, R., Covill, S., Pacholarz, K. J., Kalapothakis, J. M. D., MacPhee, C. E., and Barran P. E. (2014) A mass-spectrometry-based framework to define the extent of disorder in proteins. *Anal. Chem.*, October 3, 2014 (Perspective) DOI: 10.1021/ac5027435.

Role of additives in CRM. Chemical reactions by which charging of proteins occurs depend on the nature of additives present in the analyte solution. For example, with 1% acetic acid in the analyte solution, it would be H_3O^+ ions at the surface of the droplets which would get transferred to protein sites (functional groups) having relative basicities greater than that of H_2O. Here gas-phase basicities would be quite useful indicators, since in this situation most of the solvent would be gone. Likely sites would be amide groups or other basic residues of the peptide backbone (on the surface of the protein. The following Table[62] gives some gas-phase basicities.

Table 2.1: Some gas-phase basicities of bases for reaction: $BH^+ = B+H^+$. [a]GB(B) = gas phase basicity; GB(B)[a]=. All values from NIST database.

Base	GB(B)[a] (kcal/mol)	Base	GB(B)[a] (kcal/mol)
H_2O	157.7	NH_3	195.7
$(H_2O)_2$	181.2	CH_3NH_2	206.6
CH_3OH	173.2	$C_2H_5NH_2$	210.0
$(CH_3OH)_2$	196.3	$(CH_3)_2NH$	214.3
C_2H_5OH	178.0	$(CH_3)_3N$	219.5
$(CH_3)_2O$	179.0	N-methyl acetamide	205.0
$(C_2H_5)_2O$	182.7	Pyridine	214.8

This CRM model (Z–M correlation) is widely accepted. Proteins that diverge much from the spherical shape, however, have a charge substantially higher than that predicted by eqn 2.14. This is understandable; such non-spherical proteins would require larger precursor droplets which would imply higher charge. (See also Konermann *et al.*[74])

Effects of additives on the sprays and mass spectra. When salt additives (buffers) are present in the analyte solution the charge states of native proteins are lower. They are lower in the presence of ammonium acetate, and much lower in triethylammonium acetate. To explain this it has been suggested that here IEM precedes CRM; surface active ions like triethyammonium would require less energy to escape, thus leaving behind the protein with fewer charges.

The CRM mechanism finds support from mass spectra observed in the presence of salt additives. The case of the ubiquitin spectrum in the presence of NaI can be taken. Evaporation increases concentration of both the analyte and the additives in the solution. Increased concentration of salt causes ion pairing — pairing of positive and negative ions — including pairing with ionized protein residues at or near the surface of the protein. In the presence of NaI there would be two kinds of addition: one in which Na^+ would be expected to bind, replacing all H^+; for ion pairing reactions, Na^+ would go to the ionized acidic residues, and I^- ions would go to the ionized basic residues at the surface of the protein.

The spectrum of native ubiquitin, obtained using with $25\,\mu$M ubiquitin, $1\,$mM NaI, and nanospray, under condition of little or no collisional activation (CAD) effect (ensured by small potential difference between the spray nozzle and the skimmer), shows in the 1400 m/z region a series of peaks from ubiquitin with $Z = +6$ charge state (+6 charge state of ubiquitin, mass 8564.8448 Da, would appear at 1427.4741). The peaks appear in groups, the peaks in a given group originate from how many Na^+ ions and how many Na^+I^- ion pairs are attached. Denoting each such peak by m,n (where m represents how many Na^+ ions and n how many Na^+I^- ion pairs are attached), the first group shows (Fig. 2.8, top left) peaks 3,0;4,0;5,0;6,0;7,0, the second group 4,1;5,1;6,1;7,1, etc. This is in the absence of any CAD effect. In the presence of CAD, HI formation would be expected to take place from the ion pairs. Thus, under strong CAD conditions all Na adducts would be expected to remain and all NaI would disappear. That is what is observed in the spectrum under strong CAD; in the 1427 m/z region only one series of peaks appear (Fig. 2.8, bottom left) as a result of multiple HI losses, which could be assigned to 0,0–15,0 with the 6,0 peak having the maximum intensity.

Additional experiments with NaCl and NaAc showed that for dissociation the energy required decreased in the order HI, HCl, HAc. Experiments were also conducted with high concentration of NaAc to ensure that there were enough Na^+ ions for all the ionized acidic sites and Ac^- ions for all the basic sites. A series of peaks was observed at high CAD in the m/z range 1420–1510 (Fig. 2.8, bottom right), which could be assigned to (1,0)–(23,0). A sharp fall in peak intensity beyond (18,0) indicated that $m = 18$ has 6 Na^+ charges and 12 ionized acidic sites paired to Na^+ contributing to 18; the acetate ions associated as ion pairs with the basic sites have fallen off owing to high CAD. This experiment and the CRM mechanism predict 12 acidic sites on the protein; the known structure of ubiquitin gives the same number.[71]

When a protein is the analyte, ammonium acetate in mM concentration is often used as an additive. Ammonium acetate acts as a buffer, and also, for multiply protonated proteins, it helps produce very clean mass spectra. When ammonium acetate is present in excess, its ion pairing that involves only acetate and ammonium ions prevents Na^+ forming adducts (Na-salt is a common contaminant in protein samples, it causes mass shifts in mass spectra); the ammonium acetate ion pairs can be dissociated by CAD (the dissociation is facile); overall, neutralization of positive and negative groups takes place.

---

[71]Verkerk, U. H. and Kebarle, P. (2005) Ion-ion and ion-molecule reactions at the surface of proteins produced by nanospray. Information on the number of acidic residues and control of the number of ionized acidic and basic residues. *J. Am. Soc. Mass Spectrom.*, **16**, 1325–1341.

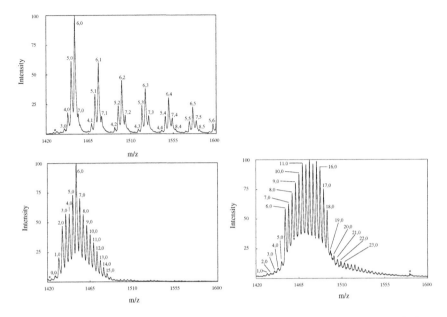

Figure 2.8: Top, left: Mass spectrum of the major $Z = +6$ charge state of ubiquitin with an aqueous solution of $25\,\mu$M ubiquitin containing 1 mM NaI. Bottom, left: Mass spectrum of ubiquitin under the same condition but with high collisional activation (CAD). Right: Mass spectrum showing $Z = +6$ charge state obtained with ubiquitin $25\,\mu$M, and a very high, 5 mM NaI. Reproduced by permission of Elsevier, Inc. from *J. Am. Soc. Mass Spectrom.*, Vol. 16, 1325–1341, 2005.

$$^{+}H_3N(CH_2)_2-Prot-(CH_2)2COO^{-} + NH_4^{+} + CH_3COO^{-} =$$
$$CH_3COO^{-}\ H_3N^{+}(CH_2)_2-Prot-(CH_2)2COO^{-}NH_4^{+} \rightarrow$$
$$CH_3COOH + H_2N(CH_2)_2-Prot-(CH_2)_2COOH + NH_3$$

Anions of strong acids like phosphate or trifluoroacetate — present in the mobile phase in LCMS — cause ion suppression and thus loss of sensitivity; they can form strongly bonded adducts with ionized basic residues. Excess ammonium acetate prevents that formation.

There are now some additional findings from gas-phase, solution-phase protein structure studies (IM-MS). Konermann *et al.*,[72,73] using IM-MS,

---

[72]Vahidi, S., Stocks, B., and Konermann, L. (2013) Do electrosprayed protein ions retain memory of their solution phase structure: insights from ion mobility mass spectrometry. *Proc. 61st Annual ASMS Conference on Mass Spectrometry and Allied Topics*, Minneapolis, June 9–13.

[73]Vahidi, S., Stocks, B. B., and Konermann, L. (2013) Partially disordered proteins studied by ion mobility-mass spectrometry: implications for the preservation of solution phase structure in gas phase. *Anal. Chem.*, **85**, 10471–10478.

addressed the question whether the solution structure of a protein is retained in the gas phase. They took myoglobin in solution at different pH and subjected the protein (holo-Mb at pH 7, and apo-Mb at pH 7, 4, and 2) to IM-MS and measured gas phase collision cross-sections of the species. Holo-Mb was observed to lose the heme group early — at pH 7, the 9+ charge ion drops to a lower mass 9+ apo ion, indicating loss of the heme group. With increasing acidification the mygolobin structure undergoes a transition from tightly folded form to random coil. The peak maxima of gas phase cross-sections versus the ionic charge state measured by them are presented in Fig. 2.9 (left). The cross-section behavior seems to indicate that the structure is mainly determined by the ionic charge.

They explain the observed trend of the cross-sections as a function of ionic charge by invoking a new mechanism for formation of ESI ions. In their explanation,[74] most proteins exist in some sort of globular form in neutral aqueous solutions, in which most charged and polar residues of the protein face the outside and have lyophilic interactions with the solvent, whereas the nonpolar hydrophobic components of the protein face inwards (buried, are solvent-inaccessible and form a hydrophobic core). For most globular proteins, ESI ion formation is by IEM and by CRM. In the third mechanism of ESI ion formation, they propose (which they call the Chain Ejection Model CEM), they suggest that (nonpolar polymer) chain conformation governs the ESI behavior; that unfolded (or partially unfolded) proteins can form ESI ions in an alternative manner.

Their molecular dynamics simulation[75,76] showed that, in solution, protein unfolding can be triggered by an acidic phase (which could even be by the mobile LC phase). This condition causes extensive disordering of conformers, and the previously sequestered nonpolar residues in the core become *accessible* to the solvent. This unfolding causes switch/transform of property from compact/hydrophilic to extended/hydrophobic state; this hydrophobicity makes their continued residence inside the Rayleigh-charged nanodroplet interior *untenable*; chains unfold, they migrate immediately to the droplet surface. Once one chain terminus gets expelled into the vapor phase, then stepwise sequential ejection of the remaining protein takes

[74]Konermann, L., Ahadi, E., Rodriguez, A. D., and Vahidi, S. (2013) Unraveling the mechanism of electrospray ionization. *Anal. Chem.*, **85**, 2–9.

[75]Ahadi, E. and Konermann, L. (2012) Modeling the behavior of coarse-grained polymer chains in charged water droplets: implications for the mechanism of electrospray ionization. *J. Phys. Chem. B*, **116**, 104–112.

[76]Konermann, L., Rodriguez, A. D., and Liu, J. (2012) On the formation of highly charged gaseous ions from unfolded proteins by electrospray ionization. *Anal. Chem.*, **84**, 6798–6804.

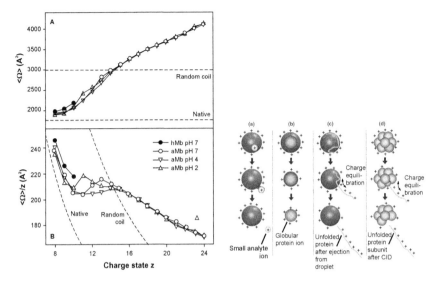

Figure 2.9: Left: Average collision cross-section $< \Omega >$ of ions generated by electrospraying hMb and aMb at different pH as a function of the ionic charge state. In B, the same data as in panel A but after division of $< \Omega >$ values by the corresponding charge state z. Reproduced by permission of the American Chemical Society from *Analytical Chemistry*, Vol. 85, 10471–10478, 2013; Right: Schematic representation of IEM, CRM, and CEM. (a) IEM: small ion ejection from a charged nanodroplet, (b) CRM: release of a globular protein into the gas phase, (c) CEM: ejection of an unfolded protein, and (d) Collision-induced dissociation of a gaseous multiprotein complex. Reproduced by permission of the American Chemical Society from *Analytical Chemistry*, Vol. 85, 2–9, 2013.

place, ultimately resulting in separation from the droplet. The CEM, which is completely distinct from the CRM (Fig. 2.9 (right)), has elements in common with the IEM (Fig. 2.9 (right)). The CEM applies to polymer chains that are (i) disordered, (ii) partially hydrophobic, and (iii) capable of binding excess charge carriers. On the question whether electrosprayed protein ions retain memory of their solution phase structure, their conclusion from their IM-MS experiments: compact solution phase structures tend to form compact ions, and expanded solution phase structures tend to form expanded ions. Charge state of ion governs ion structure; and this is significantly affected by ionization mechanism (CEM vs CRM). And, whereas ESI ions undergo relaxation prior to IMS separation, intermediate charge states retain *some* memory of the solution phase conformation.

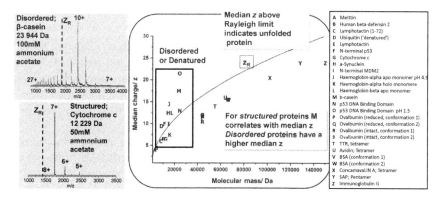

Figure 2.10: Median charge state vs. Molecular mass. Left (top): Mass spectrum of β-casein; left (bottom) mass spectrum of Cytochrome C. Courtesy: Perdita Barran, Manchester Institute of Biotechnology, University of Manchester. Data published in *Anal. Chem.*, Vol. 86, 10979–10991, 2014. Permission from the American Chemical Society taken for reproduction.

Beveridge *et al.*[77] have examined the structured-disordered property of proteins by ESI-IM-MS. They selected 20 different proteins, both monomeric and multimeric, ranging in mass from 2,846 Da (melittin) to 150 kDa (immunoglobulin G) and using IM-MS distinguished — using range of charge states occupied by the protein (Δz) and the range of collision cross-sections in which the protein is observed (ΔCCS) — which of the proteins are structured (melittin, human beta defensin, truncated human lymphotactin, cytochrome c, holo hemoglobin-α, ovalbumin, human transthyretin, avidin, bovine serum albumin, concavalin, human serum amyloid protein, and immunoglobulin G) and which (human lymphotactin, N-terminal p53, α-synuclein, N-terminal MDM2, and p53 DNA binding domain) have at least some regions of disorder, or which (ubiquitin, apo-hemoglobin-α, apo hemoglobin-β) are denatured due to solvent conditions. They also presented a methodology to determine if a given protein is structured or disordered. Results on median charge vs. molecular mass are shown in Fig. 2.10. Results on range of CCS vs. molecular mass are shown in Fig. 3.15.

They observed that while structured proteins have a median charge lower than the Rayleigh[64]/de la Mora[68] limit, disordered proteins have a higher median charge and, importantly, display a *wider* charge state distribution (CSD) than structured proteins. For proteins with a mass of M < 150 kDa, structured proteins present with $z \leq 7$ states or less, whereas

---

[77]Beveridge, R., Covill, S., Pacholarz, K. J., Kalapothakis, J. M. D., MacPhee, C. E., and Barran P. E. (2014) A mass-spectrometry-based framework to define the extent of disorder in proteins. *Anal. Chem.* **86**, 10979–10991.

for disordered proteins $z \geq 7$. The low Δz for structured proteins is very weakly dependent on mass.

Comparison of collision cross-sections (CCSs) show that, although there is little distinction in terms of *median* CCS between structured and disordered systems, the *range* of CCSs that a given protein presents reveals its conformational flexibility; structured proteins will have a CCS range of $< 750\,\text{Å}^2$, whereas for disordered systems the corresponding range will be $> 750\,\text{Å}^2$.

For initial characterization to obtain information about the extent of structure or disorder in unknown proteins using MS and IM-MS Beveridge *et al.* propose a workflow along the following lines. By carrying out MS at low pH if wide CSD (charge state distribution) (≥ 5) is observed, it means a large number of protonation sites. After adding a reducing agent along with a thiol capping group if the CSD widens (indicating denaturing) it means intramolecular disulfides, the number of capping groups indicates that number of disulfides. With the oxidized form if MS under high ionic strength conditions CSD narrows — with respect to that at low pH — this means the protein has some structured regions; if CSD remains the same, it means the protein is intrinsically with no structure, that is, is disordered. From the shape of the CSD for intrinsically disordered proteins: if there are significantly higher intensity peaks at lower charge states, it means both structured and disordered regions. If intensity similar across most of the charge states, this means disordered. For IM-MS under physiological conditions, a disordered protein will present a large ΔCCS, a structured protein will present narrow ΔCCS, and a protein with both structured and disordered regions will cause large ΔCCS, with higher intensity of more compact conformations.

Role of the timeframe of measurements. It has been pointed out[78] that in deriving information about the molecular ion entering the mass spectrometer it is important when and how quickly its structure is examined, because a number of factors contribute. An interesting example is the finding on calmodulin binding with 18-crown-6-ether (18C6) in ESI, studied by the selective noncovalent adduct protein probing[79] (SNAPP) method. Calmodulin has four calcium binding sites with association constants $\geq 10^6\,\text{M}^{-1}$; in solution calcium is strongly bound to calmodulin. When a solution of calmodulin is electrosprayed in the positive ion mode, the number of calcium ions that continue to remain bound, varies. 18C6 binds little in bulk water; adduct formation takes place sometime during the ESI process. On

[78]Hamdy, O. M. and Julian, R. R. (2012) Reflections on charge state distributions, protein structure, and the mystical mechanism of electrospray ionization. *J. Am. Soc. Mass Spectrom.*, **23**, 1–6.

[79]Ly, T. and Julian, R. R. (2008) Protein-metal interactions of calmodulin and α-synuclein monitored by selective noncovalent adduct protein probing mass spectrometry. *J. Am. Soc. Mass Spectrom.*, **19**, 1663–1672.

the basis of solution-phase binding, in SNAPP data for calmodulin in the +11 charge state calcium adducts (zero, one, and two calcium attached), the mass spectral peak with two calcium (Calm + 2Ca^{2+} adduct (H$_2$O, Ca^{2+})) would be expected to be ~50 times more intense than that due to the naked protein (Calm (H$_2$O, Ca^{2+})). But experimental data show that that is not the case, which implies significant calcium loss during ESI. The loss of calcium should cause structural rearrangement, but the intensity distribution of naked protein observed over five 18C6 attached ligands show similarity with those with calcium ions still attached to the protein. This indicates that the structural arrangement has not taken place. However, with fewer calcium a shift toward the true apo distribution is discernible. The reason why the loss of calcium does not show up in change in structure arrangement is probably because of the short timeframe. The holo structure is retained for some time.

Nanospray. The flow rate in nanospray is < 100 nL/min. In the flow rate range 1000–4 nL/min the efficiency increase is linear; the efficiency is ~1% at its maximum. Smaller droplet size leads to increased ionization efficiency, increased charge per analyte molecule, therefore increased sensitivity. At a flow of 25 nL/min, femtomoles of analyte can be analyzed with less than 1 μL of sample. In normal electrospray, ion suppression and matrix effects result in inefficiency; in nanospray, at lower flow rates (< 50 nL/min) these adverse effects are nearly absent, conversion efficiency of analyte ions into gas-phase ions is very high. The efficiency however varies strongly from emitter to emitter; emitter tip shape/architecture materially affects the physical properties of the Taylor cone and therefore the stability and the efficiency of the ionization process. Only in a narrow range of applied potential does the cone-jet mode operate satisfactorily; the exact potentials depend on the flow rate, the solvent composition, and the nature of the analyte and emitter properties.[80]

The nature of electrospray depends in a complex way on the applied voltage, surface tension, emitter architecture, and physical placement of the emitters. For cone-jet mode operation the total spray current is proportional to the square root of the number of emitters $I_{total} = \sqrt{n} I_s$ where n is the number of emitters, I_s is the spray current of an individual emitter — this ion current does not proportionally increase with respect to the spray current. Performance testing should be carried out over a range of solvents — from highly aqueous to highly organic (MeOH or acetonitrile). For testing emitter clogging something like Hanks' salt solution (rich in bicarbonate ions) should be used.

Emitter geometry. In single electrospray emitter, the basic design by Wilm and Mann[57,58] is still used, which consists of a needle-like electrode —

[80]Gibson, G. T. T., Mugo, S. M., and Oleschuk, R. D. (2009) Nanoelectrospray emitters, trends, and perspective. *Mass Spectrom. Rev.*, **28**, 918–936.

drawn out of a glass tube, both its inner diameter and outer diameter tapering to a very fine point. This is able to produce a stable Taylor cone down to flow rates of 20 nL/min. In practice, however, much thinner fused silica capillary tubing is used; the tip opened up with HF etching. Optimal flow rate depends on the outer diameter, channel taper, and the solvent used. The inner diameter is the most effective and the easier one to vary; because of fewer fissions, smallest inner diameter tips produce ions that closely resemble those present in the solution. However, small inner diameters result in clogging; less than 10 μm inner diameter tips clog too rapidly. With etching — when inner diameters are not tapered, the outer diameter is tapered — the clogging problem reduces; the thin wall helps stabilization of the Taylor cone.

Other tip modifications. Tapered design maintains a stable Taylor cone; one reason for that is that the solvent does not wet the tip face. Having a hydrophobic end with coatings such as polymers or something like a trimethylsilyl monolayer have showed improved sensitivity and Taylor cone stability, resulting in better mass spectra. Carbon fiber protruding from the capillary end has been tried to provide a point for starting a Taylor cone. The Taylor cone has been found to be stable with larger bore its surface modified by roughening, resulting in altered surface energy and, additionally, less clogging.

Mode of electrical contact. The alternatives have been insertion of a wire into the emitter, or applying a conducting coating outside of the capillary. The simplest method is application of a voltage at a liquid junction provided by an electrode in contact with the solution upstream; it works even when the junction is placed upstream of a 100 μm internal diameter tube — pulled to 2 μm at exit — 10 cm long packed column. Alternatively, the outside of a traditional silicate emitter rendered conductive by coating with a conductive material — such as vacuum deposition of gold onto the distal end of the emitter. These have, however, short lifetimes — 15 min– 3 h; the thin layer is subject to deterioration. Overcoating with a 10–50 nm thick layer of SiO/SiO_2 helps. A carbon fiber (gold-coated) protruding $\sim$ 20–30 μm from the tip of the capillary has been used to stabilize a small Taylor cone — stable spray obtained for 15–20 hours. Metal-coated emitters have been seen to give the strongest signal in positive ion mode, but they degrade in discharge, and in negative ion mode. Distal metal-coated emitters have been found best under some conditions.

Integrated emitters. Nanoelectrospray emitters have also been used to produce integrated microreactors, such as the wall-functionalized tip for the online protein digestion, integration of solid phase extraction and enzymatic digestion, for sequential identification of protein analytes. Frits within the tip for packing purpose are commercially available. For LC/MS, packed beads have been used. The spray is little affected unless the material is very close to the exit aperture. Constriction of the flow gives lower flow without the increased risk of clogging.

As regards materials used for chips, initially silicon was the material of choice and it still continues to be used. Then interest in glass developed because of its surface characteristics and electro-osmotic flow properties — the high electric field needed for eOF is not compatible with silicon. Silicon is used for complex microstructures and also for applications employing pressure-driven flow. Overall, a range of materials are used: glass/pyrex, quartz, PDMS polydimethyl siloxane, PMMA polymethyl methacrylate, polycarbonate, polyimide, silicon, silicon dioxide, etc. Advion's ESI chip is made of silicon; Agilent's chip is made of polyimide by laser ablation, Phoenix S&T has an ESI chip made of polypropylene by injection moulding. Processes involved are both cleanroom intensive such as HF etching, DRIE deep reactive ion etching, UV exposure SU-8 epoxy-based negative photoresist, x-ray photolithography, and non-cleanroom methods such as replica moulding, embossing/hot embossing, laser ablation, injection moulding. Initially the technique used to be spraying from the edge of the microfluidic chip. In later developments, standard fused-silica emitters are inserted into the device, or, during chip fabrication, emitters are made integral part of the device.[81]

Chips with spray orifices directly on the edge — though they are relatively simple to fabricate and truly monolithic in nature — have several disadvantages. Hydrophilic liquids tend to spread along the edge of the glass devices. Hydrophobic liquids wet hydrophobic chips made of plastics and thus prediction of the exact position from where the spray is generated becomes difficult. Also, the total volume of liquid wetting the edge of the chip is relatively large and may act as a dead volume to negatively influence performance. Closing the exit of the microfluidic channel with a porous PTFE membrane is a possibility, from which small Taylor cones are then generated.

Coupling of (nano)-electrospray emitters with microfluidic channels improves ESI-MS performance significantly in comparison to spraying from the edge of a chip. However, dead volumes, usually introduced upon coupling, have a negative effect on the separation performance. This can be avoided using complex drilling approaches using gluing (the latter may interfere with MS) but is not the best choice if a microfluidic device needs to be cheap to fabricate and disposable. Manually inserting spray needles is not cost-effective.

Integrated emitters are the best. The microfluidic channel and the emitter can be formed simultaneously during the same fabrication process and can eliminate dead volumes. In one approach, via laser machining and

---

[81]Koster, S. and Verpoorte, E. (2007) A decade of microfluidic analysis coupled with electrospray mass spectrometry: an overview. *Lab Chip*, **7**, 1394–1412.

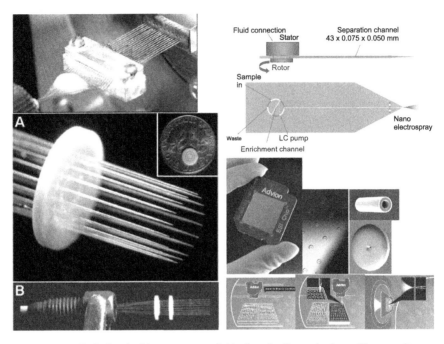

Figure 2.11: Left (top): Linear array of 19 chemically etched capillary emitters (Kelly, R. T., Page, J. S., Zhao, R., Qian, W., Mottaz, M., Tang, K., and Smith, R. D. (2008) Capillary-based multinanoelectrospray emitters. *Anal. Chem.*, **80**, 143–149). Reproduced by permission of the American Chemical Society; (bottom): circular array of chemically etched emitters to improve electric field homogeneity (Kelly, R. T., Page, J. S., Marginean, I., Tang, K., and Smith, R. D. (2008) Nanoelectrospray emitter arrays providing interemitter electric field uniformity. *Anal. Chem.*, **80**, 5660–5665). Reproduced by permission of the American Chemical Society from *Analytical Chemistry*. Right (top): Polyimide chip sandwiched in between the stator and rotor of a two-position HPLC rotary switching valve.[82] Reproduced by permission of John Wiley & Sons, Inc. from *J. Separation Science*, Vol. 29, 499-509, 2006. Right (middle): Advion nanospray microchip; Right (bottom): TriVersa Nanomate operation. Courtesy: Advion Biosystems. Reproduced by permission.

milling, one chip is produced at a time, serially. In another approach multiple chips are formed making use of photolithographic process.

In serial fabrication of chips with integrated emitters, in an Agilent polyimide chip (Fig. 2.11 right, top),[82] the microfluidic channels and the spray tip are made using laser ablation, the chip sandwiched between the stator and the rotor of a 2-position rotary switching valve normally used for HPLC.

---

[82]Vollmer, M., Hörth, P., Rozing, G., Couté, Y., Grimm, R., Hochstrasser, D., and Sanchez, J. C. (2006) Multi-dimensional HPLC/MS of the nuclear proteome using HPLC-chip/MS. *J. Sep. Sci.*, **29**, 499–509.

Of the chip's two columns,[83] a trap column is located in the rotating part of the valve. Results obtained are comparable to standard state-of-the-art nano-LC-MS. Sample introduction rate was $4\,\mu L\,min^{-1}$ with $1\,\mu L$ sample peptides from a $20\,fmol\,\mu L^{-1}$ tryptic digest of BSA, trapped first on the enrichment bed between the stator and rotor containing $5\,\mu m$ C18 particles, then switching the valve and eluting the trapped peptides from the enrichment column, the sample injected onto the analytical column containing $3.5\,\mu m$ C18 particles and separated by applying a reversed-phase gradient at 100–300 nL/min. With the same chip, a 2-dim LC separation was performed in another work. The peptides from tryptic digests and plasma were trapped in an off-chip but on-line SCX column and peptides were sequentially eluted onto the C18 precolumn with different concentrations of ammonium acetate at pH 3.5. The trapped peptides were then eluted and separated in reversed-phase mode using a gradient of H_2O/ACN/FA.

Chips with integrated emitters with capability of high-throughput analysis have been realized via parallel fabrication. Examples are: chips with one or two functions integrated from the following: separation performed on-chip/on-chip enzymatic digestion and separation/sample preconcentration and separation/sample clean-up/free-flow electrophoresis/clean-up based on diffusion/sample clean-up and preconcentration/on-chip electrochemical labeling/ on-chip reactions/large-scale integration.

The work of Xie *et al.*[84] is outstanding using a chip with a very large number of integrated components. In a single device gradient pumps, an injector, mixer, reversed-phase separation column, electrodes and ESI nozzle were integrated, the chip made of a combination of parylene, silicon, and PDMS (Fig. 2.12, left). Pumping and injection of the sample in microfluidic channels uses gas bubbles in enclosed cavities with their pressure generated by the hydrogen and oxygen formed in electrolysis using a pair of platinum/titanium electrodes. By controlling the current between the electrodes, an LC gradient of methanol/water could be generated. It is difficult to control the initial gas/liquid volume in the cavities and this adversely affects the run-to-run reproducibility. The system performance, however, is close to those of state-of-the-art nanoflow systems.

A schematic of a microfluidic device[85] fabricated out of glass and capable of two-dimensional reversed-phase liquid chromatography–capillary electrophoresis with integrated electrospray ionization (LC-CE-ESI) is shown

[83]Yin, H., Killeen, K., Brennen, R., Sobek, D., Werlich, M., and van de Goor, T. (2005) Microfluidic chip for peptide analysis with an integrated HPLC column, sample enrichment column, and nanoelectrospray tip. *Anal. Chem.*, **77**, 527–533.

[84]Xie, J., Miao, Y., Shih, J., Tai, Y. C., and Lee, T. D. (2005) Microfluidic platform for liquid chromatography-tandem mass spectrometry analyses of complex peptide mixtures. *Anal. Chem.*, **77**, 6947–6953.

[85]Chambers, A. G., Mellors, J. S., Henley, W. H., and Ramsey, J. M. (2011) Monolithic integration of two-dimensional liquid chromatography — capillary electrospray and electrospray ionization on a microfluidic device. *Anal. Chem.*, **83**, 842–849.

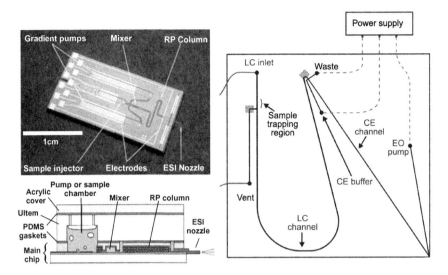

Figure 2.12: Left: Photograph of a chip, made of a combination of parylene, silicon, and PDMS, has an integrated gradient pump, injector, mixer, reversed-phase column, electrodes and ESI nozzle. Reproduced by permission of the American Chemical Society from *Analytical Chemistry*, Vol. 77, 6947–6953, 2005. Right: Schematic of a microchip-based LC-CE-MS system. The grey squares on the microchip denote the locations of the weirs (channel segments etched 6 μm deep) that were used to retain the packed particles. Electrospray was performed from the lower right corner of the microchip. Reproduced by permission of the American Chemical Society from *Analytical Chemistry*, Vol. 83, 842–849, 2011.

in Fig. 2.12 (right). It has a sample trapping region and an LC channel packed with reversed phase C18-bonded, 3.5 μm porous particles. At a cross-channel intersection rapid subnanoliter electrokinetic injections of the LC effluent into the CE dimension are performed. An integrated electrospray tip is located at one corner of the device where the CE separation channel terminated.

Volny *et al.* have described[86] an alternative interface for coupling microfluidic chips related to introduction of nanoliter plugs into a mass spectrometer. The plugs (which can be generated by T- or Y-junctions or polydimethyl siloxane PDMS chips) are separated by immiscible liquids (such as perfluorohexane). The standard Z-spray inlet of LCT Premier (ESI TOF, Waters), was replaced by a glass-lined stainless steel inlet capillary (i.d. 1.5 mm, length 300 mm) leading to an additional vacuum

---

[86]Volny, M., Rolfs, J., Hakimi, B., Schneider, T., Frycak, P., Yen, G., Liu, D., Chiu, D., and Tureček, F. (2013) Nanoliter segmented-flow sampling mass spectrometry: introducing on-line compartmentalization while avoiding sample dilution. *Proc. 61st Annual ASMS Conference on Mass Spectrometry and Allied Topics*, Minneapolis, June 9–13.

enclosure. Charged droplets generated from the plugs at an emitter needle embedded in the microfluidic chip enter the vacuum enclosure through the capillary (96% transport efficiency); droplets — separated by perfluorinated oil — evaporate by multiple interaction with the output of a 30 W, 10.6 μm laser as they move through a gold-plated tube lens (this serves both as an electrostatic lens and a multi-reflection mirror), then through a radiofrequency lens to the mass spectrometer. The oil does not ionize; the resulting set of well separated mass spectral data correspond to individual water plugs originating at the microfluidics channel. The technique, with high attomol/plug detection limits maintained, was tested with small organic molecules as well as mammalian cell lysates.

The Advion Triversa Nanomate[87] can be used for ESI spray in three modes (Fig. 2.11, right, bottom). In Mode 1, *chip-based infusion*, the instrument first picks up a pipette tip — samples and pipette tips are located inside the Nanomate — then it aspirates a sample from the 96- or 384-well plate. The sample-loaded pipette tip then seals to the back of the ESI Chip, voltage and pressure are applied, and electrospray ionization occurs. In Mode 2, *LC/MS with simultaneous fraction collection*, the Nano-Mate connects with the HPLC and mass spectrometer for LC/MS analysis; fractions are simultaneously collected during an LC/MS analysis, and are analyzed by chip-based infusion. In Mode 3, LESA[TM] (*Liquid Extraction Surface Analysis*), the NanoMate picks up a pipette tip from the rack, then takes the extraction solvent from the reservoir into the pipette tip, brings the extraction solvent from a pipette tip into contact with the surface of a sample held in the sample plate, the analyte is extracted from the surface, then the solvent is retracted back into the pipette tip and is sprayed through the ESI chip. The TriVersa NanoMate can be used, among others, to sample directly from: thin tissue sections, TLC plates and other planar separation media, dried blood spots on paper, MALDI plates for complementary information by ESI. The TriVersa NanoMate is operated using a proprietary software program called ChipSoft[TM].

Toward high transmission of ions from ESI source, Krutchinsky *et al.* have reported[88] work on a conical duct electrode — ConDuct — employing it as ion inlet to mass spectrometer. The ion channel used is very narrow — it employs a 0.1–10 μL conductive pipette tip, with the result that the ion beam emerging from ConDuct is highly focussed (angle of divergence $\sim 14 \pm 2$ mrad). ConDuct electrodes can transmit 80–95% of the total ion current into vacuum. In their experiments, with 60% methanol plus 1%

---

[87]Triversa Nanomate (Advion Biosystems, Ithaca, New York) autosampling device brochure.

[88]Krutchinksy, A. N., Padovan, J. C., Cohen, H., and Chait, B. T. (2013) A novel ConDuct interface for transmitting $\sim$ 100% ions from an ESI source into a mass spectrometer. *Proc. 61st Annual ASMS Conference on Mass Spectrometry and Allied Topics*, Minneapolis, June 9–13.

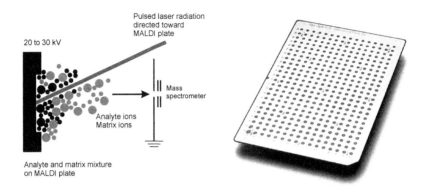

Figure 2.13: Left: MALDI schematic. Right: AB SCIEX TOF/TOF^{TM} MALDI plate. MALDI plate, Courtesy: AB SCIEX. Reproduced by permission.

acetic acid solution total ESI current of $\sim 200\,\text{nA}$ (order of several hundred nA), and with $1\,\text{M}$ ammonium acetate currents up to $1\,\mu\text{A}$ could be obtained. They report data in which the atmosphere-to-vacuum ConDuct interface transmits 1–10% of a peptide analyte in solution as detectable ion current. Measured using an instrumental construct SICE — stable isotopes comparative enhancement — in which a reference ion input line (such as through LCQ or Thermo S-lens) from one side and ConDuct from the other side are conjoined to a common mass spectrometer (LCQ DECA XP) through a T-quadrupole so that relative ion transmission efficiencies could be compared directly, SICE ConDuct transmission (tested using peptides with both light and heavy isotopes) is ~ 900 times higher than old LCQ, and ~ 4 times higher than the newest Thermo S-lens. Also, peak S/Ns were observed to be better in ConDuct transmission.

2.4.2   Matrix-assisted laser desorption ionization — MALDI

In MALDI (*M*atrix-*A*ssisted *L*aser *D*esorption *I*onization) mass spectrometry, the analyte sample is mixed with a matrix compound — which has high ability of absorbing laser radiation — and the mixture is placed on a plate and subjected to laser radiation. The matrix rapidly and efficiently absorbs laser energy, the heat generated from the absorption is shared with the analyte sample which then *desorbs* and also gets *ionized*. The ionization could be under vacuum (at $10\,\text{mTorr}$ or less) or at atmospheric pressure.

The ions generated then are guided to a mass analyzer.[89,90,91,92] Effects of spot size, laser fluence have been studied.[93] In Fig. 2.13 (left) a MALDI ionization schematic is shown, and in Fig. 2.13 (right) a commercial (AB SCIEX) MALDI plate. MALDI plates can have from hundreds to thousands of wells for the sample-matrix mixtures to be placed. In Table 2.2[94] some UV-absorbing matrix compounds and their properties are listed. Some matrices used for $2.94\,\mu$m infrared laser are: malic acid, succinic acid, glycerol (not acidic), 5-(trifluoromethyl) uracil (not acidic), urea (not acidic), TRIS (not acidic). MALDI is a batch process.

Lasers which have been used in MALDI are: Nitrogen uv laser, wavelength 337 nm, repetition rate 1–20 Hz, pulse energy 120–200 μJ; solid-state UV laser, repetition rate 0.1–5 kHz; Nd-YAG (neodymium-doped yttrium aluminium garnet; $Nd:Y_3Al_5O_{12}$) laser, wavelength 355 nm, repetition rate 0.1 –2 kHz. Infrared (IR) laser,[95] wavelength $2.94\,\mu$m, pulse duration ~ 5 ns.

Specifications of the new MALDI machine from Bruker Daltonik UltrafleXtreme: Smartbeam-II^{TM} laser; 1–2000 Hz MALDI proteomics repetition rate; laser focus diameter 10 μm — without pixel overlap; 40,000 broadband mass resolution, 1 ppm mass accuracy.

One of the advantages of MALDI is that, once set up, ions are easy to generate. At one time, for ionization conditions used, singly charged ions dominated; in latest MALDI instruments, multiply charged ion generation has been possible — Bruker DHAP matrix, efficient for Top-Down analysis, produces multiple charged ions. The sensitivity is high; earlier, for digests, it used to be better than 20 fmol. There are certain disadvantages, however; in MALDI-based analysis, one of them is the effect of matrices — results can be matrix-dependent. MALDI TOF mass spectrometry has been used in a

---

[89] Karas, M., Bachmann, D., Bahr, U., and Hillenkamp, F. (1987) Matrix-assisted ultraviolet laser desorption of non-volatile compounds. *Int. J. Mass Spectrom. Ion Proc.*, **78**, 53–68.

[90] Tanaka, K., Waki, H., Ido, Y., Akita, S., Yoshida, Y., and Yoshida, T. (1988) Protein and polymer analyses up to 100,000 by laser ionization time-of-flight mass spectrometry. *Rapid Commun. Mass Spectrom.*, **2**, 151–153.

[91] Karas, M. and Hillenkamp, F. (1988) Laser desorption ionization of proteins with molecular masses exceeding 10,000 Daltons. *Anal. Chem.*, **60**, 2299–2301.

[92] Vestal M. (2009) Modern MALDI time-of-flight mass spectrometry. *J. Mass Spectrom.* **44**, 303–317.

[93] Guenther, S., Koestler, M., Schulz, O., and Spengler, B. (2010) Laser spot size and laser power dependence of ion formation in high resolution MALDI imaging. *Int. J. Mass Spectrom.*, **294**, 7–15.

[94] https://en.wikipedia.org/wiki/matrix-assisted_laser_desorption/ionization.

[95] Zheng, W., Niu, S., and Chait, B. T. (1998) Exploring infrared wavelength matrix-assisted laser desorption/ionization of proteins with delayed extraction time-of-flight mass spectrometry. *J. Am. Soc. Mass Spectrom.*, **9**, 879–884.

wide variety of applications including for bacterial profiling at the strain level.[96]

Over the past decade there have been two major augmentations in the MALDI set-up. The first one: a much better vacuum is in use; earlier the vacuum used to be $\sim 10^{-7}$ torr, at present, $\sim 10^{-9}$ torr. The second one is in the laser. Earlier, nitrogen laser was mostly used. At present tunable (dye) laser also is used. Recent performance status[92]: Resolving power $> 30\,000$ for peptides; mass error $< 2\,\mathrm{ppm}$ RMS over the entire sample plate; detection limit $\sim 1\,\mathrm{amol}/\mu\mathrm{L}$, dynamic range $\sim 10^5$, performance independent of laser rate (to $5\,\mathrm{kHz}$). Whereas MALDI mass resolution and mass accuracy are nowhere close to the best in electrospray, it is very fast — a full spectrum is obtained per laser shot ($5\,\mathrm{kHz}$); and with no precursor selection (which implies it requires pure sample) in direct sequencing of intact proteins, the sensitivity is inferior requiring pmol or more samples. It has been observed that its performance improves when there is significant intensity of doubly charged ions, from which a possibility emerges of merged beam TOF-TOF of doubly charged positive ions from intact protein reacting with singly charged negative ions (like in ETD). Overall MALDI use however is relatively much less in sequence/structure of proteins — order of 5%; for that purpose mostly ESI is used.

Table 2.2: UV-MALDI Matrix compounds.

Compound	Other names	Solvent	Wavelength λ nm	Application
2,5-dihydroxy benzoic acid	DHB, Gentisic acid	acetonitrile, water, methanol, acetone, chloroform	337 nm, 355 nm, 266 nm	peptides, nucleotides, oligonucleotides, oligosaccharides
3,5-dimethoxy-4-hydroxy-cinnamic acid	sinapic acid, sinapinic acid, SA	acetonitrile, water, acetone, chloroform	337 nm, 355 nm, 266 nm	peptides, proteins, lipids
4-hydroxy-3-methoxy cinnamic acid	ferulic acid	acetonitrile, water, propanol	337 nm, 355 nm, 266 nm	proteins
α-cyano-4-hyrdoxy cinnamic acid	CHCA	acetonitrile, water, ethanol, acetone	337 nm, 355 nm	peptides, lipids, nucleotides
Picolinic acid	PA	ethanol	266 nm,	oligonucleotides
3-hydroxy picolinic acid	HPA	ethanol	337 nm, 355 nm	oligonucleotides

[96]Sandrin, T. R., Goldstein, J. E., and Schumaker, S. (2013) MALDI TOF MS profiling of bacteria at the strain level: a review. *Mass Spectrom. Rev.*, **32**, 188–217.

MALDI *imaging* mass spectrometry uses the basic MALDI technique, supplemented by scanning of multiple sections of tissue samples, and from that 2D and 3D images are constructed. An early report demonstrating 3D MALDI imaging mass spectrometry of mouse brain was by Crecelius *et al.*[97] in which myelin basic protein in a 3D rendered volume of the corpus callosum of a mouse brain was visualized. The brain was sectioned 20 μm thickness, resulting in 264 collected sections. Ten sections, equally spaced through the brain (400–500 μm apart), each of them manually coated with sinapinic acid using a TLC sprayer, were used for mass spectral analysis, and imaged using Applied Biosystems MALDI TOF Voyager DE-STR at 100 μm resolution.

Current workflow for 3D imaging mass spectrometry and coregistration have been reviewed.[98,99] The animal is first subjected to *in vivo* imaging (such as MRI); this is followed by sacrifice and freezing. Cryomacrotome sections of the animal are then taken, and image mass spectra and histology images are generated; in the same three-dimensional space, data from multiple imaging modalities are then coregistered.

The introduction of the Tape Transfer System (Instrumedics, Inc.) allows for more rigid reproducible sample collection and transfer to a MALDI target or histology slide without folding or tearing and macro distortions. A piece of transfer tape is placed on the surface of the tissue and a section is then transferred through the aid of a photoactive polymer coating on the target. This is important for later reconstruction, easier alignment into a 3D volume. Two types of MALDI matrices have been used: 2,5-dihydrobenzoic acid (DHB) and α-cyano-4-hydroxycinnamic acid (CHCA). Some of the software for image acquisition and viewing have been: Bruker FlexImaging, AB Sciex TissueView, Waters HDImaging, Thermo ImageQuest.

---

[97]Crecelius, A. C., Cornett, D. S., Caprioli, R. M., Williams, B., Dawant, B. M., and Bodenheimer, B. (2005) Three-dimensional visualization of protein expression in mouse brain structures using imgaging mass spectrometry. *J. Am. Soc. Mass Spectrom.*, **16**, 1093–1099.

[98]Spraggins, J. M. and Caprioli, R. M. (2011) High-speed MALDI-TOF imaging mass spectrometry: rapid ion image acquisition and considerations for next generation instrumentation. *J. Am. Soc. Mass Spectrom.*, **22**, 1022–1031.

[99]Seeley, E. H, and Caprioli, R. M. (2012) 3D-Imaging by mass spectrometry: a new frontier. *Anal. Chem.*, **84**, 2105–2110.

References to some UV-MALDI matrix compounds are: DHB[100]; SA[101,102]; ferulic acid[101,102]; CHCA[103]; PA[104]; HPA.[105]

2.4.3   Laser-induced liquid beam/bead ionization desorption — LILBID

In laser-induced liquid bead/beam ionization/desorption (LILBID)[106,107] mass spectrometry, a microscopic liquid beam — later the liquid *beam* was changed into *droplets* produced on command — is injected into high vacuum and a spot on the beam subjected to infrared laser radiation. Pre-formed ions desorb from the liquid phase of the beam, and are then analyzed by mass spectrometry. It is a soft desorption method.

In the initial set-up of Brutschy *et al.*, to generate the liquid beam, the analyte solution — an aqueous probe solution was used — is forced through a $10 \, \mu$m diameter platinum-iridium nozzle into the vacuum using a standard HPLC pump. The beam emerging from the nozzle flows through a laminar region during which it intersects the laser beam. After a few millimeters from the nozzle, the beam, while it remains macroscopically intact, breaks up into microscopically sized droplets. After a few centimeters, the beam gets disrupted and is frozen out on a liquid nitrogen cold trap. The ions generated from the beam are guided to a time-of-flight mass spectrometer.

They used an optical parametric oscillator/amplifier (OPO/OPA) system for the IR radiation source, continuously tunable between 2.5 and $4 \, \mu$m in its 'idler' component and between 1.55 and $1.85 \, \mu$m in its corresponding 'signal' component. The OPO was tuned to the absorption band ($\lambda = 3 \, \mu$m)

[100]Strupat, K., Karas, M., and Hillenkamp, F. (1991) 2,5-Dihydoxybenzoic acid: a new matrix for laser desorption ionization mass spectrometry. *Int. J. Mass Spectrom. Ion Proc.*, **72**, 89–102.

[101]Beavis, R. C. and Chait, B. T. (1989) Matrix-assisted laser-desorption mass spectrometry using 355 nm radiation. *Rapid Commun. Mass Spectrom.*, **3**, 436–439.

[102]Beavis, R. C. and Chait, B. T. (1989) Cinnamic acid derivatives as matrices for ultraviolet laser desorption mass spectrometry of proteins. *Rapid Commun. Mass Spectrom.*, **3**, 432–435.

[103]Beavis, R.C., Chaudhary, T., and Chait, B. T. (1992) α-cyano-4-hydroxycinnamic acid as a matrix for matrix-assisted laser desorption mass spectrometry. *Org. Mass Spectrom.*, **127**, 156–158.

[104]Tang, K., Taranenko, N. I., Allman, S. L., Chang, L. Y., and Chen, C. H. (1994) Detection of 500-nucleotide DNA by laser desorption mass spectrometry. *Rapid Commun. Mass Spectrom.*, **8**, 727–730.

[105]Wu, K. J., Steding, A., and Becker, C. H. (1993) Matrix-assisted laser desorption time-of-flight mass spectrometry of oligonucleotides using 3-hydroxypicolinic acid as an ultraviolet-sensitive matrix. *Rapid Commun. Mass Spectrom.*, **7**, 142–146.

[106]Kleinekofort, W., Avdiev, J., and Brutschy, B. (1996) A new method of laser desorption mass spectrometry for the study of biological macromolecules. *Int. J. Mass Spectrom. Ion Proc.*, **152**, 135–142.

[107]Wattenberg, A., Sobott, F., Barth, H., and Brutschy, B. (1999) Laser desorption mass spectrometry on liquid beams. *Eur. J. Mass Spectrom.*, **5**, 71–76.

of bulk water; the repetition rate of the laser was 10 Hz, and its output energy in the range 24–30 mJ. In their initial experiments with aqueous liquid beams they studied diglycylhistidine (GGH), lysozyme (hen egg white lysozyme at neutral pH), and DNA.

In a later version[108] of their set-up they demonstrated, using this technique, feasibility of quantitative analysis of specific biocomplexes in native solution. The continuous liquid beam used earlier was changed to an on-demand droplet source — thus generating liquid beads at 10 Hz — with duty cycle of 1 and sample solution volume in microliter range. The absolute amount of analyte that could be detected from a single droplet was in the attomolar to zeptomolar (10^{-21}M) range. To check its suitability in quantitative analysis, linearity of ion intensity in the range 10^{-7}M to 10^{-4}M was demonstrated using bovine serum albumin. Mostly lower charge state ions are produced in LILBID — which is in between the properties of ESI and MALDI. Cation spectra of hen egg lysozyme with NaCl concentration in the range 10–100 mM (corresponding ESI tolerances in the range 0.1–1 mM), low ionization efficiency at higher salt concentrations was clearly visible. Compared to nano-ESI, the LILBID technique was found tolerant against buffers such as Tris-HC, ammonium acetate, or phosphate buffers. Mass spectrum of DNA/RNA binder neomycin B obtained in LILBID showed that the ionization process is soft — no fragmentation takes place even though the neomycin B molecule possesses a number of OH groups — potential spots for absorbing IR radiation. Pictures of exploding beads show that only the ions from the upper diffuse plume are detected; the lower parts of the droplets do not contribute; they vanish unused.

The main advantage of using beads/droplets instead of a beam is use of small sample volume hence making application possible to biomolecules of low availability. In the new beam option,[109] $\phi = 50\,\mu$m microdroplets were produced on demand by a piezo-driven droplet generator and introduced into vacuum. The laser pulses used were of 6 ns duration, spot size around 300 μm in diameter, and were tuned to a wavelength of about $\lambda = 3\,\mu$m, at pulse energies of 1 to 15 mJ/pulse. Pulse energy applied was determined by the degree of fragmentation desired. The water molecules (their O-H stretch broad vibration band $\lambda = 3\,\mu$m) within the first micrometers of the liquid surface of the droplet absorb the laser energy at a threshold intensity of around 100 MW/cm^2. This causes a very fast transition, a concomitant spherical explosion, resulting in emission of ions from liquid into gas phase.

For the oligonucleotides Morgner *et al.* studied,[108] no positively charged ions were visible — indicating the potential of the method for

[108]Morgner, N., Barth, H-D., and Brutschy, B. (2006) A new way to detect noncovalently bonded complexes of biomolecules from liquid micro-droplets by laser mass spectrometry. *Austral. J. Chem.*, **59**, 109–114.

[109]Morgner, N., Kleinschroth, T., Barth, H., Ludwig, B., and Brutschy, B. (2007) A novel approach to analyze membrane proteins by laser mass spectrometry: from protein subunits to the integral complex. *J. Am. Soc. Mass Spectrom.*, **18**, 1429–1438.

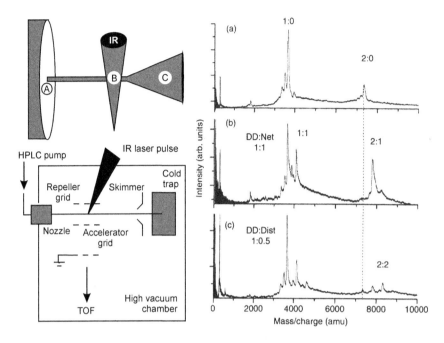

Figure 2.14: Left: LILBID ionization set-up. Reproduced by permission of IM Publications LLP from *Eur. Mass Spectrom.*, Vol. 5, 71–76, 1999. Right: Anion spectra of the Dickerson dodecamer with ligands. (a) At a concentration of 10^{-4}M, the single strand and duplex is visible. (b) At a concentration of 10^{-5}M, mixed under equimolar conditions with Net. The latter binds to the DD in a specific 2:1 complex. The duplex alone (2:0) vanishes. (c) At a concentration of 10^{-5}M, mixed with Dist in an 1:0.5 ratio. The dominance of the specific 2:2 complex is clearly visible. The additional peaks at lower masses are due to unspecific binding of the minor groove binders to the DNA single strand. For all measurements a 10 mM ammonium acetate buffer at pH 8.0 was used. The number of droplets in (a), (b), and (c) was 50, 200, and 300 respectively. Reproduced by permission of the Australian Chemical Society from *Australian Journal of Chemistry*, Vol. 59, 109–114, 2006.

detection of preformed polyanions in solution. They demonstrated usability of the method with the 12-mer oligonucleotide DNA(C) (5′-CCC CAT ATC CCC-3′) and DNA(G) (5′-GGG GAT ATG GGG-3′); there was no self-hybridization, a (CG)-DNA duplex was observed with two conjugated strands. Absence of duplices of DNA(C) or DNA(G) indicates formation of a specific duplex.

Also studied was the oligonucleotide Dickerson dodecamer DD (5′-CGC GAA TTC GCG-3′) which is self-complementary. A major purpose of their study with DD was to check specific interactions in minor groove binding of DD with netropsin (Net) and distamycin A (Dist). Both these molecules are fully crescent-shaped, interest in them is because of their anti-tumor activity with binding preferences for AT-rich regions.

Figure 2.14 (right) shows results on minor-groove binding of netropsin and distamycin with $100\,\mu$M aqueous solution of DD. Figure 2.14(a) shows peaks due to the negatively charged single strand DD$^-$ (1:0) and DD$^-$ fragments, and also a peak for (DD)$_2$ (2:0) corresponding to the DD duplex. Figure 2.14(b) shows a spectrum of 10 μM solution of DD with equimolar concentration of netropsin added; the peak owing to the duplex signal disappears and a signal due to a stochiometric (duplex):(ligand) complex appears — marked in the figure as 2:1 complex. The increase of the relative intensity of the duplex peak in Fig. 2.14(b) in comparison to Fig. 2.14(a) is possibly because of an increased melting point T_m of the duplex. The single strand shows an unspecific aggregate (1:1); the nonspecificity is concluded from the fact that the single stranded (1:0) continues to be the dominant species, whereas in contrast, the uncomplexed duplex is no longer present. This is indicative of quantitative formation of a specific 2:1 complex and nonspecific binding of the double positively charged netropsin to a negatively charged single strand. In the case of distamycin A (Dist) addition, the binding behavior is different. With Dist molarity half that of DD, at higher masses DD duplex is clearly visible (Fig. 2.14(c)) along with specific binding of a dimeric ligand — showing a maximum at 2:2 (this stoichiometry has been observed also in NMR studies), along with peaks of decreasing intensity — inverted size distribution tracking the number of attached ligand molecules, besides peaks due to nonspecific attachment to the single strand.

In their work with blue-native gels,[110] they combined blue native electrophoresis (BNE or BN-PAGE) with LILBID. Protein complexes were separated by BNE, eluted from the gel, and the sample was repeatedly diluted with 30 mM ammonium hydrogen-carbonate buffer of pH 7 containing 0.05% dodecyl-β-D-maltoside, and was concentrated. This step exchanged the buffer, also removed excess Coomassie blue G-250. The sample was then injected into vacuum as solution droplets ($\phi \sim 50\,\mu$m, V ~ 65 pL) and irradiated by pulsed mid-IR laser as described above. Mitochondrial NADH: ubiquinone reductase from *Yarrowia lipolytica* was used to validate the method. Typically $30\,\mu$g or 32 pmol of the complex was sufficient to obtain spectra and from that obtain subunit composition.

LILBID mass spectrometry has been used to demonstrate existence of intact rotary ATPases/synthases in vacuum.[111,112]

[110]Sokolova, L., Wittig, I., Barth, H. D., Schägger, H., Brutschy, B., and Brandt, U. (2010) Laser-induced liquid bead ion desorption-MS of protein complexes from blue-native gels, a sensitive top-down approach, *Proteomics*, **10**, 1401–1407.

[111] Meier, T., Morgner, N., Matthies, D., Pogoryelov, D., Keis, S., Cook, G. M., Dimroth, P., and Brutschy, B. (2007) A tridecameric c ring of the adenosine triphosphate (ATP) synthase from the thermoalkaliphilic *Bacillus* sp. strain TA2.A1 facilitates ATP synthesis at low electrochemical proton potential. *Mol. Microbiol.*, **65**, 1181–1192.

[112] Hoffmann, J., Sokolova, L., Preiss, L., Hicks, D. B., Krulwich, T. A., Morgner, N., Wittig, I., Schägger, H., Meier, Y., and Brutschy, B. (2010) ATP synthases: cellular nanomotors characterized by LILBID mass spectrometry. *Phys. Chem. Chem. Phys.*, **12**, 13375–13382.

2.4.4   Laser ablation with electrospray ionization — LAESI

LAESI (laser ablation electrospray ionization) is a laser-induced ionization method under ambient conditions. This is particularly suitable for water-containing specimens. In this method, a well-focussed infrared laser beam hits the sample surface, water molecules absorb the laser energy, sudden excitation of water results in phase explosion causing ablation of sample material. The ablated material is subjected to electrospray ionization and the resulting ions are guided into a mass spectrometer. A schematic of a LAESI experimental system is shown in Fig. 2.15. Here, we describe the basics of a LAESI set-up.[113,114,115,116]

A mid-IR Er:YAG laser at $2.94\,\mu$m at $10\,$Hz repetition rate, with laser output power of $\sim 100\,\mu$J/pulse energy, is used. By combined use of gold mirrors and a focusing lens transparent at the laser wavelength (a plano-convex CaF_2 or ZnSe lens) the laser pulse energy is deposited into the sample at normal incidence. The sample is positioned 15–20 mm below the mass spectrometer sampling cone and the spray axis, and 3–5 mm ahead of the emitter tip. The infrared beam axis is positioned 5–8 mm in front of the mass spectrometer sampling cone. By adjusting the focusing lens position and the laser pulse energy, the extent of tissue removal in the focal spot on the biological sample can be controlled. The laser spot is circular with a typical diameter of $\sim 100\,\mu$m. In imaging applications, the dimension of the ablated volume determines the pixel size (or the voxel size in 3D-imaging). The nanospray emitter is positioned in line with the inlet axis of the mass spectrometer with ~ 10 mm orifice-to-emitter distance. The spray current is monitored by a counter electrode with a 6.0 mm diameter opening in the center, placed perpendicular to the emitter axis and 10 mm from the tip. For most efficient ion generation by LAESI, the electrospray source is operated in cone-jet spraying mode. To determine the established spraying mode, the temporal behavior of the spray current is analyzed. Using a fast digital camera, the behavior of the Taylor cone and the plume of the ablated particulates can be followed. A ~ 10 ns fluorescence pulse, excited

---

[113]Nemes, P. and Vertes, A. (2007) Laser ablation electrospray ionization for atmospheric pressure, *in vivo*, and imaging mass spectrometry. *Anal. Chem.*, **79**, 8098–8106.

[114]Nemes, P. and Vertes, A. (2010) Atmospheric-pressure molecular imaging of biological tissues and biofilms by LAESI mass spectrometry. *J. Vis. Exp.*, **43**, e2097.

[115]Parsiegla, G., Shreshtha, B., Carrière, F., and Vertes, A. (2012) Direct analysis of phycobilisomal antenna proteins and metabolites in small cyanobacterial populations by laser ablation electrospray ionization spectrometry. *Anal. Chem.*, **84**, 34–38.

[116]Nemes, P. and Vertes, A. (2012) Ambient mass spectrometry for *in vivo* local analysis and *in situ* molecular tissue imaging. *Trends. Anal. Chem.*, **34**, 22–34.

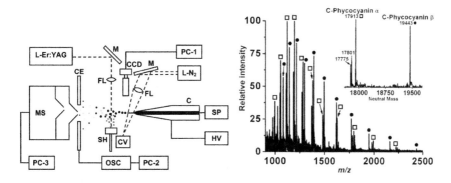

Figure 2.15: Left: Schematics of laser ablation electrospray ionization (LAESI) and fast imaging system (C, capillary; SP, syringe pump; HV, high voltage power supply; L-N$_2$, nitrogen laser; M, mirrors; FL, focusing lenses; CV, cuvette; CCD, CCD camera with short-distance microscope; CE, counter electrode; OSC, digital oscilloscope; SH, sample holder; L-Er:YAG, Er:YAG laser; MS, mass spectrometer; PC-1 to PC-3, personal computers). The cone-jet regime is maintained through monitoring the spray current on CE and adjusting the spray parameters. Black dots represent the droplets formed by the electrospray. Their interaction with the particulates and neutrals emerging from the laser ablation produces some fused particles (light dots in front of the skimmer) that are thought to be the basis of the LAESI signal. Reproduced by permission of the American Chemical Society from *Analytical Chemistry*, Vol. 79, 8098–8106, 2007. Right: LAESI mass spectrum of cyanobacteria *Anabaena* sp. PCC7120 between m/z 900 and 2500 showing multiply charged protein ions. Multiply charged ions in the $+8$ to $+18$ charge states annotated with the open squares and solid circles correspond to two proteins with molecular weights of 17,913 Da and 19,443 Da respectively. Two other protein peaks of 17,775 Da and 17,801 Da are also observed in the deconvoluted spectrum. Reproduced by permission of the American Chemical Society from *Analytical Chemistry*, Vol. 84, 34–38, 2012.

by a nitrogen laser, back-illuminates the cone and the droplets that are generated.

Ions generated are mass analyzed with an orthogonal acceleration time-of-flight mass spectrometer. The data acquisition rate is set to ∼ 1s/spectrum. The method can be used for depth profiling, with which, in combination with lateral imaging, 3D molecular imaging of a sample can be done. With current lateral and depth resolutions of ∼ 100 μm and ∼ 40 μm respectively, the method can be used for 3D imaging of biological tissues. When the technique is applied to analysis of single cells, the mid-IR ablation is done by delivering the radiation to the cell through an etched fiber; it has been applied to the analysis of a single epidermal cell from the skin of an *Allium cepa* bulb.

CHAPTER 3
INSTRUMENTS

The mass spectrometric analysis procedure has three major components: (i) ionization of the sample, (ii) mass (actually, mass-to-charge-ratio) analysis of the ions, and finally, (iii) detection of the analyzed ions. All these processes take place in gaseous, mostly rarefied environment, but this background gas pressure varies from region to region depending on the specific needs of the physical processes taking place. For example, electrospray ionization environment has a base pressure of ~ 1–10 Torr, the sampling zone $\sim 10^{-3}$ Torr, and the analyzer and detector zone $\sim 10^{-6}$–10^{-9} Torr. Besides, collisional activation process and movement of ions in an ion mobility drift tube are in the pressure domain of several Torr. The ionization methods used in peptide and protein mass spectrometry have already been discussed in the previous chapter. Detection of the analyzed ions is straightforward; secondary electron multipliers are used; when an ion impinges such a device, a huge number of electrons are produced which represent the ion signal. In this chapter we first discuss the various types of mass analyzers that are in use, that is then followed by descriptions of some complete instruments.

3.1 Mass analysis

3.1.1 Mass analyzers: time-of-flight (TOF), 2D-quadrupole (Q), Linear Trap (q,LT), 3D-ion trap (IT), Fourier transform ion cyclotron resonance (FT-ICR), Orbitrap, ion mobility (IM)

Time-of-flight (TOF)

In a time-of-flight mass spectrometer, ions which are to be mass analyzed are imparted the *same*, *fixed* kinetic energy, and in a pulse, are injected into a field-free drift region. Since the kinetic energy is $(1/2)mv^2$, where m is the mass and v the velocity of an ion, given the same kinetic energy, ions with lighter mass would possess higher velocity and ions with heavier mass lower velocity. As a bunch of ions drifts through the field-free region, the lighter ions in the bunch, because of their higher velocity, would travel faster and reach the detector-end first, with ions having heavier masses trailing them and reaching the detector later, thus resulting in separation of ions in terms of their masses. With kinetic energy in kilovots and drift path length ~ 1 m, the time taken for the ions in the drift region is typically of the order of microseconds.

Introduction to Protein Mass Spectrometry
http://dx.doi.org/10.1016/B978-0-12-805123-8.00003-5
Copyright © 2016 Elsevier Inc.
All rights reserved.

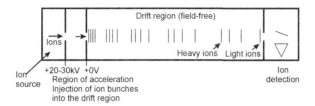

Figure 3.1: Time-of-flight mass analyzer schematic.

The basic equations determining the time-of-flight operation are:

$$\frac{1}{2}mv^2 = \frac{1}{2}m\frac{L^2}{t^2} = zeV \quad t = L\left[\frac{m}{2zeV}\right]^{1/2} \tag{3.1}$$

$$\frac{m}{z} = \frac{2eVt^2}{L^2} \tag{3.2}$$

where m/z is the mass-to-charge ratio of an ion accelerated across a potential difference of V volts, eV the ion kinetic energy ($eV = U$), L the length of the flight tube, and t the time taken by the ion to reach the end of the flight tube where a detector is placed. Mass resolution increases with increase in the length of the flight path. For a singly charged ion of mass 1000 Da accelerated across a potential difference of 20 kV, the time to travel a distance of 1.5 m is given by

$$t = \frac{L}{\sqrt{2U}}\sqrt{\frac{m}{q}} \quad t = \frac{1.5\,\text{m}}{\sqrt{2\,(20{,}000)}}\sqrt{\frac{(1000\,\text{Da})(1.66053892 \times 10^{-27}\,\text{kg Da}^{-1})}{+1.602 \times 10^{-19}\,\text{C}}}$$
$$\tag{3.3}$$
$$t = 2.415 \times 10^{-5}\,\text{s} \tag{3.4}$$

There were several early reports[117,118,119,120] on the time-of–flight mass spectrometer, but the first detailed consideration discussing the focussing action was by Wiley and McLaren.[121,122] The next significant improvement in the time-of-flight resolution was effected by extension of the flight path,

[117]Stephens, W. E. (1946) A pulsed mass spectrometer with time dispersion. *Bull. Am. Phys. Soc.*, **21**, 22.

[118]Cameron, A. E. and Eggers, Jr., D. F. (1948) An ion "Velocitron". *Rev. Sci. Instrum.*, **19**, 605–607.

[119]Wolff, M. M. and Stephens, W. E. (1953) A pulsed mass spectrometer with time dispersion. *Rev. Sci. Instrum.*, **24**, 616–617.

[120]Katzenstein, H. S. and Friedland, S. S. (1955) New time-of-flight mass spectrometer. *Rev. Sci. Instrum.*, **26**, 324–327.

[121]Wiley, W. C. (1954) U. S. Patent 2,685,035.

[122]Wiley, W. C. and McLaren, I. H. (1955) Time-of-flight mass spectrometer with improved resolution. *Rev. Sci. Instrum.*, **26**, 1150–1157.

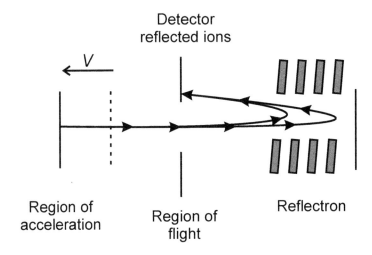

Figure 3.2: TOF-Reflectron. Mamyrin, B. A., Karataev, V. I., Shmikk, D. V., and Zagulin, V. A. (1973) The mass-reflectron, a new nonmagnetic time-of-flight mass spectrometer with high resolution. *Sov. Phys. JETP*, Vol. 37, p. 45. Mamyrin, B. A., Karataev, V. I., and Schmikk, D. V. (1978) Time-of-flight mass spectrometer, US patent 4,072,862.

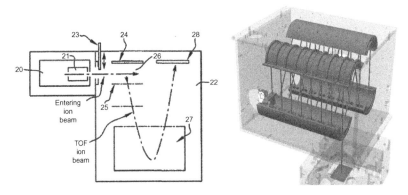

Figure 3.3: Left: Orthogonal acceleration time-of-flight mass spectrometer. Legend: 20 external continuous ion source; 21 ion transport portion; 22 Oa-TOF mass analyzer; 23 isolation valve; 24 ion repeller electrode; 25 grids; 26 ion reservoir; 27 reflector; 28 ion detector. US patent 7,230,234 T. Kobayashi, June 12, 2007. Right: Spiral ion trajectory ion optical system in Jeol SpiralTOF. Courtesy: Jeol Ltd. Reproduced by permission.

by placing an ion mirror at the point where a detector was placed in a linear analyzer. The new contraption came to be known as Reflectron.[123] After that a new modification was introduced: acceleration of ions orthogonal to their initial direction as they emerged from the ion source.[124] An early review of time-of-flight mass analyzers is by Mamyrin.[125]

A novel realization of time-of-flight ion separation is SpiralTOF. The SpiralTOF ion optical system is based on "perfect spacial and isochronous focussing"[126] and "multi-turn" principles.[127] During the flight, ion packets are focussed back in space over figure-eight trajectories (Fig. 3.3, right) — with a total length of 17 m. Ions, accelerated to 20 kV in the ion source, experience during flight four sets of layered toroidal fields — implemented by four pairs of cylindrical electrodes and nine "Matsuda plates" that are incorporated within every pair of cylindrical electrodes — and reach the detector.[128,129]

Vestal and Hayden have reported[130] success in a new approach to focussing in TOF which allows *simultaneous* space and velocity focussing that resulted in substantially improved performance. Using this focussing principle of simultaneous space and velocity focussing three new MALDI instruments have been developed. Using SIMION a gridless ion optical system was designed to overcome initial discrepancy between projected and experimentally observed performances. For imaging of proteins in tissue sections and evaluation of potential clinical application of Mass Spectrometric Immunoassay (MSIA), the linear instrument has been applied; the Reflector and TOF-TOF have been applied to imaging and identification of

---

[123]Mamyrin, B. A., Karataev, V. I., Shmikk, D. V., and Zagulin, V. A. (1973) The mass-reflectron, a new nonmagnetic time-of-flight mass spectrometer with high resolution. *Sov. Phys. JETP*, **37**, 45. Time-of-flight mass spectrometer. Mamyrin, B. A., Karataev, V. I., and Schmikk, D. V. (1978) Time-of-flight mass spectrometer, US patent 4,072,862.

[124]Kobayashi, T. (2007) Orthogonal acceleration time-of-flight mass spectrometer. US patent 7,230,234.

[125]Mamyrin, B. A. (2001) Time of flight mass spectrometry (concepts, achievements, and prospects). *Int. J. Mass Spectrom.*, **206**, 251–266.

[126]Ishihara, M., Toyoda, M., and Matsuo, T. (2000) Perfect space and time focussing ion optics for multi-turn time of flight mass spectrometers. *Int. J. Mass Spectrom.*, **197**, 179–189.

[127]Toyoda, M., Okumura, D., Ishihara, M., and Katakuse, I. (2003) Multi-turn time-of-flight mass spectrometers with electrostatic sectors. *J. Mass Spectrom.*, **38**, 1125–1142.

[128]Japanese patent application: JP2006012782, US patent 7,504,620.

[129]*J. Am. Soc. Mass Spectrom.*: May 2011 cover. SpiralTOF-Jeol. Highest resolution MALDI-TOF-TOF, > 60,000 FWHM. Monoisotopic precursor selection. True high-energy CID, no PSD artifacts. After selection: unit mass selectivity at m/z 2465 (ACTH 18-39). www.jeolusa.com/spiral.

[130]Vestal, M. and Hayden, K. (2012) A new approach to TOF mass spectrometry for high performance MS and MS-MS. *Proc. 60th ASMS Conference on Mass Spectrometry and Allied Topics*, Vancouver, May 20–24.

lipids in tissue sections. Basic quantitative relationships used in these two focussings — that is, focussing using delayed extraction and simultaneous space and velocity — are as follows.

Delayed extraction: Voltage V applied initially at origin drops to V_g at d_1 (distances measured from origin, the ion source exit point), and to 0 at d_1+d_2. The space focus takes place at the point $d_1+d_2+D_s$; velocity focus takes place at $d_1+d_2+D_v$ ($D_v > D_s$). For delayed extraction, the space focus and velocity focus relationships are: $y = V/(V - V_g)$, $D_s = 2d_1y^{3/2} - [2d_2y/(1 + y^{1/2})]$, $D_v = D_s + [(2d_1y)^2/v_n\Delta t]$, where $v_n = (2zV/m)^{1/2}$. Also, $\delta v/v = [2d_1y/(D_v - D_s)](\delta v_0/v_n)$, $D_v - D_s$ is always positive. The $\delta v/v$ relationship shows that its value is large if $D_v - D_s$ is small. This means that high mass resolution over a broad mass range is unattainable in a linear analyzer. A refocussing with a *two-field mirror* can produce high resolution over a broad mass range.

Simultaneous space and velocity focussing: At a variable distance from the origin, an accelerating pulse V_p is applied (which after a distance d_3 decays linearly to 0 amplitude). If D_s is the distance of the point of space focus from the origin, in order to adjust the velocity focus to occur at the same point — to ensure simultaneous space and velocity focussing — the amplitude V_p and the space focus D_s are varied, by adjusting the voltage ratio y. With D_v representing the distance of the point of velocity focus from the point where V_p is zero, the velocity spread is reduced if D_v is greater than d_1y. Under this condition, for a linear analyzer, high resolution is possible for a broad range of masses. This could also be the first stage of a reflectron-type analyzer or of a tandem TOF-TOF analyzer. The relevant equations in this simultaneous space and velocity focussing are: $y = V/(V - V_g)$, $D_s = 2d_1y^{3/2} - [2d_2y/(1 + y^{1/2})]$, $D_v = 2d_3V/V_p$, $\delta v/v = [d_2y/D_v](\delta v_0/v_n)$.

At SimulTOF Systems they have tried out three different configurations: (i) linear analyzer (100 Linear) where the detector is placed at the point where simultaneous space focussing and velocity focussing take place, (ii) combination analyzer (200 Combo) where an ion mirror is placed just past the point of simultaneous space and velocity focussing (along with a detector for it to possibly act as a linear analyzer) in reflectron mode with a detector at the end of the return flight paths of the ions, and (iii) a tandem analyzer (300 Tandem) where a timed ion selector is placed at the point of simultaneous space and velocity focussing, which is followed, in tandem, by a collision cell, a second ion accelerator, and an ion mirror plus a detector placed on the return flight path of the ion. And, in their instruments, they use a Photonis fast hybrid detector (typical pulse width 1 ns), efficient at low as well as high masses. They report, using gridless ion optics and the hybrid detector, high resolving power over a broad range, and for both small molecules and proteins, high ion transmission efficiency. The addition of a two-stage ion mirror provided higher sensitivity at resolving power comparable to that in delayed extraction. In 300 Tandem TOF configuration, PSD and CID fragmentation showed that it can be an efficient first stage in Tandem TOF.

They presented a number of spectra and results: (i) one of the first spectra they took was of peptides at 10 kV, resolution $\sim 8,000$, (ii) spectrum of α-cyano matrix dimer measured in SimulTOF 100 Linear at 15 kV ion energy, (linewidth 2.3 ns), (iii) mixture of peptide standards, 100 femtomole/μL, 2.5 mm spot — 1 shot — $\sim 20,000$ ions detected, resolution 3,000 or more, (iv) resolving power and peak intensity variations for 963.5 in α-cyano matrix: resolving power decreases, intensity increases with laser fluence (6–9 μJ), (v) spectrum of BSA (bovine serum albumin) in sinapinic acid matrix (Photonis fast hybrid detector with channel plate at ground potential), peak width 1500 ns, which corresponds to resolving power RP = 60, (vi) spectrum of IgG in sinapinic acid matrix, (vii) spectrum of IgG, 1 picomole/μL, 200 laser shots at 5 kHz, 0.04 s acquisition, (viii) example of angiotensin from 300 Tandem, (post-source decay), unimolecular decomposition, (PSD) 35% fragmentation (gives all the sequence ions), + CID (by adding gas) 65% fragmentation (ix) Neurotensin, 100 attomoles/μL on 2.5 mm spot, 100,000 shots, 20 s acquisition at 5 kHz. Unimolecular (PSD) pELYENKPRRPYIL. MH$^+$ = 1672.917. peaks: P, Y KP, b_3, y_5, y_7 ions. Spectra submitted to ProteinProspector (19 of the 30 peaks matching; not more, probably because the database did not include a lot of low mass peaks).

2D quadrupole (Q)

In a 2D quadrupole mass filter,[131,132,138] ions are injected into the analyzer's quadrupole field which consists of four rods having hyperbolic cross sections, as shown in Fig. 3.4 (left). The injected ions experience a field which has both DC and RF components, and as the ions move along the field, they execute oscillations whose amplitudes depend on the field conditions and on m/z values of ions. Some of the ions execute oscillations with finite amplitude, remain within the field close to the z-axis, drift along the z-direction, reach the end of the quadrupole structure, while others with higher and lower m/zs execute unstable oscillations and are lost.

[131]Paul, W. and Steinwedel, H. (1953) A new mass spectrometer without magnetic field. *Z. Naturforschung*, **A8**, 448–450.
[132]Dawson, P. H. (ed.) (1976) *Quadrupole Mass Spectrometry and its Applications.* Elsevier, Amsterdam.

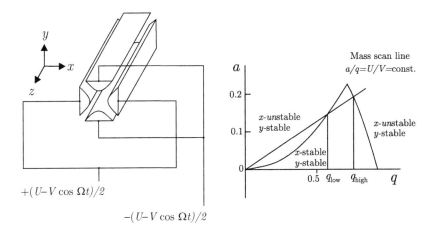

Figure 3.4: Left: 2D quadrupole mass filter structure. Right: 2D stability diagram.

The equations of motion of a charged particle in a two-dimensional quadrupole field as shown in Fig. 3.4 (left) are (the quadrupole, hexapole, and octopole potentials are given below):[133]

$$\ddot{x} + \frac{e}{mr_0^2}(U - V\cos\Omega t)x = 0 \qquad (3.5)$$

$$\ddot{y} - \frac{e}{mr_0^2}(U - V\cos\Omega t)y = 0 \qquad (3.6)$$

$$\ddot{z} = 0 \qquad (3.7)$$

---

[133]the quadrupole potential:

$$\phi_2(x,y) = \frac{(x^2 - y^2)}{r_0^2}$$

the hexapole potential:

$$\phi_3(x,y) = \frac{(x^3 - 3xy^2)}{r_0^3}$$

the octopole potential:

$$\phi_4(x,y) = \frac{(x^4 - 6x^2y^2 + y^4)}{r_0^4}$$

If the following substitutions are made,

$$\frac{4eU}{mr_0^2\Omega^2} = a; \quad \frac{2eV}{mr_0^2\Omega^2} = q; \quad \Omega t = 2\zeta \tag{3.8}$$

Then eqn 3.5 and eqn 3.6, respectively, become

$$\frac{d^2x}{d\zeta^2} + (a - 2q\cos 2\zeta)x = 0, \tag{3.9}$$

$$\frac{d^2y}{d\zeta^2} - (a - 2q\cos 2\zeta)y = 0 \tag{3.10}$$

Which are Mathieu equations, the canonical form of the equation being

$$\frac{d^2u}{d\zeta^2} + (a - 2q\cos 2\zeta)u = 0 \tag{3.11}$$

The solutions of Mathieu equations are either *stable* or *unstable* depending on the values of the parameters a and q, thus we have regions of stability or instability on the a-q plane. The lowest region of stability (Fig. 3.4 (right)) is the one mostly used for operation of the quadrupole mass filter. The segment of the stability diagram that is intersected by a constant a/q (that is, constant U/V) line creates a mass stability window, within which a range of m/z ions is stable, and ions with both higher and lower m/zs outside that range are unstable. In commonly used mass-selective stability scan, at a fixed rf frequency Ω, the ratio of a/q (that is U/V) is held constant, with U (hence simultaneously also V) scanned. As a result, at a time a group of m/z ions is brought within the stability region, ions having higher and lower m/zs remain unstable.

Linear trap (q, LT)

From a consideration of the equations of motion of the 2D quadrupole, it would be apparent that it can be operated under different sets of conditions for different purposes. In its rf-only mode (that is, DC $= 0$, $a = 0$ — see the a–q stability diagram) it has no mass filtering action and it can act essentially as an ion guide. In a configuration where a quadrupole mass filter structure has at its inlet and outlet two quadrupole structures to which by applying appropriate DC potentials ion transmission can be controlled — restricted or stopped, then after ions are inside the middle quadrupole if such stopping potential is applied, the 2D quadrupole would act as a confinement device, a trap, a *linear ion trap*. This device is frequently utilized in such a manner. For collision-induced dissociation, the linear trap is

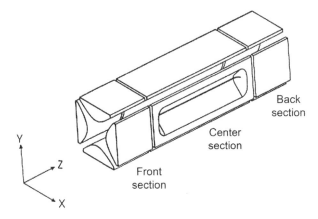

Figure 3.5: 2D Linear Ion trap. Reproduced by permission of Wiley Periodicals, Inc. from *Mass Spectrometry Reviews*, Vol. 24, 1–29, 2005.

often used, with higher gas pressure inside the middle 2D structure so that fragmentation may take place there (Fig. 3.5).[134,135,136]

3D ion trap (IT)

In a 3D quadrupole ion trap,[137,138] an ion inside the field structure — the latter consists of a ring electrode and two end-cap electrodes (Fig. 3.6 (left)) — experiences a quadrupole field in all the three x-, y-, and z-directions. As in a 2D quadrupole mass filter, the ion trajectories here again are either stable with finite amplitude or unstable with monotonically increasing amplitude. The ions having unstable trajectories go beyond the field or strike the electrodes and are lost. Unlike the stable ions in a 2D quadrupole mass filter, those ions which have stable trajectories, cannot escape along *any* direction, therefore remain confined, or are trapped.

The potential in a 3D quadrupole field is given by:

$$\phi = \phi_0 \left(\frac{x^2 + y^2 - 2z^2}{4z_0^2} \right) = \phi_0 \left(\frac{x^2 + y^2 - 2z^2}{2r_0^2} \right) \qquad (3.12)$$

[134]Prestage, J. D., Dick., G. J., and Maleki, L. (1989) Linear ion trap for frequency standard applications. *J. Appl. Phys.*, **66**, 1013–1017. First suggested by Dehmelt in connection with stored-ion frequency standards applications. See *Frequency standards and metrology*, ed. A. De Marchi, Springer-Verlag, Berlin, 1989, p. 286.

[135]Douglas, D. J., Frank, A. J., and Mao, D. (2005) Linear ion traps in mass spectrometry. *Mass Spectrom. Rev.*, **24**, 1–29.

[136]Douglas, D. J. (2009) Linear quadrupoles in mass spectrometry. *Mass Spectrom. Rev.*, **28**, 930–960.

[137]Paul, W., Osberghaus, O. and Fischer, E. (1958) Ion cage. *Forschungsberichte des Wirtschaft und Verkehrministeriums Nordrhein Westfalen*, No. 415, Westdeutscher Verlag, Köln.

[138]Ghosh, P. K. (1995) *Ion Traps*, Clarendon Press, Oxford.

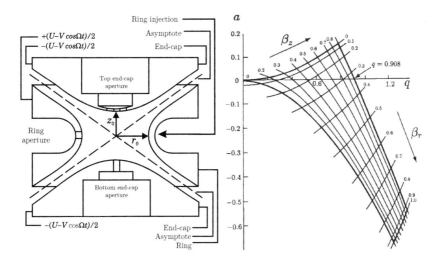

Figure 3.6: Left: 3D ion trap; Right: 3D stability diagram.

The equations of motion are given by:

$$\ddot{r} + \left(\frac{e}{mr_0^2}\right)(U - V\cos\Omega t)r = 0 \tag{3.13}$$

$$\ddot{z} - \left(\frac{2e}{mr_0^2}\right)(U - V\cos\Omega t)z = 0 \tag{3.14}$$

where r represents either x or y.

If the following substitutions are made,

$$a_z = -2a_r = -\frac{8eU}{mr_0^2\Omega^2}; \; q_z = -2q_r = -\frac{4eV}{mr_0^2\Omega^2}; \; \zeta = \frac{\Omega t}{2} \tag{3.15}$$

one obtains

$$\frac{d^2r}{d\zeta^2} + (a_r - 2q_r\cos 2\zeta)r = 0 \tag{3.16}$$

$$\frac{d^2z}{d\zeta^2} - (a_z - 2q_z\cos 2\zeta)z = 0 \tag{3.17}$$

The three Mathieu equations — eqn 3.16, eqn 3.17 — yield characteristic stability regions on the a-q plane. For simultaneous ion stability in the x-, y-, and z-directions, the operating point on the a-q plane *must* reside in the region of common stability. The common stability zones are obtained by superimposing the r- (that is, x- and y-) and z-stability diagrams. Because of the factor 2 difference between the parameters a, a_z in eqns 3.8 and 3.15, *unlike* the stability zones for the quadrupole mass filter, the stability zones for the three-dimensional quadrupole field are *not* symmetrical with respect to the q-axis. The lowest stability zone is shown in Fig. 3.6 (right).

The operation of the device using the three-dimensional quadrupole field is similar to that of the quadrupole mass filter. By using a fixed a/q = U/V ratio, an operating line is set which cuts the stability diagram (here the lower part of the stability diagram is more commonly used). For a given DC (or RF) voltage, ions within only a small range of m/z have stable trajectories in all three directions — x, y, and z — simultaneously. The main difference from the quadrupole mass filter is that whereas in the mass filter the ion amplitudes remain bound in the x- and y-directions and the ions continue to move with constant velocity in the z-direction and ultimately out of the field along the z-direction, in the three-dimensional quadrupole field, the stable ions have bound trajectories in *all* the three directions; they cannot escape the electrode structure, therefore are *trapped*. This is the fundamental confinement process in the quadrupole ion trap. Like the quadrupole mass filter, here also the resolution of the stable ions depends on the slope of the operating line, that is, the steeper the slope the higher is the resolution and hence a narrower range of ions is confined in the trap.[138]

A novel high-capacity ion trap has been tried out by Krutchinsky *et al.*,[139] and they have demonstrated[140] its efficacy in comprehensive *linked-scan* identifications of precursors and fragment ions from the trap.

---

[139]Krutchinsky, A. N., Cohen, H., and Chait, B. T. (2007) A novel high-capacity ion trap-quadrupole mass spectrometer. *Int. J. Mass Spectrom.*, **268**, 93–105. This ion trap is composed of two non-linear quadrupole arms each composed of four tapered rods, joined at their widest ends by a short linear quadrupole. Called the *dual tapered ion trap*, the purpose of this configuration was to create in each non-linear quadrupole arm an electric potential that closely approximates

$$U(x,y,z) = U_0 \left(\frac{x^2 - y^2}{R^2} \right) + C, \ R = \frac{r_0}{\sqrt{1 + kz/L}} \qquad (3.18)$$

This potential satisfies the Laplace's equation ($\Delta U = 0$), where the constants U_0, r_0, L, k, and C are determined from particular initial and boundary conditions. The key element of the mass spectrometer is high ion capacity in a TrapqQ configuration. It has been shown that the trap can store $> 10^6$ ions without significant degradation of its performance. At the time of its report the mass resolution of the trap was 100–450 full width at half maximum for ions in the range 800–4000 m/z, yielding a 10–20 m/z selection window for ions ejected at any given time into the collision cell. Ions are stored and scanned out from the trap as a function of m/z, collisionaly fragmented and analyzed by a TOF mass spectrometer. Accurate mass analysis was achieved on both the precursor and fragment ions of all species ejected from the trap. The approach has been demonstrated[140] for comprehensive linked-scan identification of phosphopeptides in mixtures with their corresponding unphosphorylated peptides.

[140]Myung, S., Cohen, H., Fenyö, D., Padovan, J. C., Krutchinsky, A. N., and Chait, B. T. (2011) High-capacity ion trap coupled to a time-of-flight mass spectrometer for comprehensive linked scans with no scanning losses. *Int. J. Mass Spectrom.*, **301**, 211–219.

Fourier transform ion cyclotron resonance (FT-ICR)

In the Fourier transform ion cyclotron resonance (FT-ICR) method of obtaining mass spectra, the frequency of ion motion in a spatially uniform static magnetic field is measured through scanning for any frequency resonance by an externally applied uniform radiofrequency field. When resonance is detected, then, through Fourier transform of the resonance signal — which transforms a time-domain signal to a frequency domain signal — the mass-to-charge ratio of the ion is obtained. The ion motion that is of most interest is the ion cyclotron motion; in this the ion mass is inversely proportional to the resonance frequency. Usually ions are generated outside the magnetic field of the FT-ICR trap and then are guided into the confinement region. Collisions inside the trap can be utilized for cooling and compressing an ion packet; this improves detection efficiency.

This mass spectrometric technique was invented by Comisaraw and Marshall.[141] The ICR ion trap used in FT-ICR mass spectrometry can have a number of configurations (some are shown in Fig. 3.7); the trap sits inside a uniform static magnetic field. Some of the basic principles of operation of the FT-ICR trap and important relationships are presented in the following sections from the review of Marshall *et al.*[142]

Ion cyclotron motion. In a spatially uniform magnetic field ($\mathbf{B} = -B_0\mathbf{k}$ — along the z direction — the z-axis is defined as the direction opposite to B), an ion traverses a circular path of radius r on the xy-plane, with an angular frequency ω_c (the *cyclotron frequency*) and orbital frequency ν_c ($\nu_c = \omega_c/2\pi$). The cyclotron frequency is the *same* for *all ions of a given m/z*, and is *independent* of the *ion velocities*; therefore it provides a convenient way of determining ionic mass from ω_c, no translational energy focussing is necessary. From the equation of motion of an ion subjected to the Lorentz force, assuming that the ion maintains constant speed — true in the absence of collisions, the ion cyclotron frequency and the orbital frequency are given by

$$\omega_c = \frac{qB_0}{m} \text{ (S. I. units)} \quad \nu_c = \frac{\omega_c}{2\pi} = \frac{1.535611 \times 10^7 B_0}{m/z} \qquad (3.19)$$

where ω_c is in radians/s, ν_c is in Hz, B_0 is in tesla, m in dalton, z is in multiples of elementary charge, and r is in m. Typically, for $\sim 15 \leq m/z \leq 10{,}000$, and in the magnetic field range 1.0–9.4 Tesla, the ICR frequencies lie between a few kHz and a few MHz. Relationships between the radius (r) and velocity (v_{xy}, in m/s) of the ion on the xy-plane, and also kinetic

[141]Comisaraw, M. B. and Marshall, A. G. (1974) Frequency-sweep Fourier transform ion cyclotron resonance spectroscopy. *Chem. Phys. Lett.*, **26**, 489–490.

[142]Marshall, A. G., Hendrickson, C. L., and Jackson, G. S. (1998) Fourier transform ion cyclotron resonance mass spectrometry: a primer. *Mass Spectrom. Rev.*, **17**, 1–35.

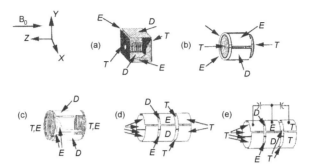

Figure 3.7: ICR ion trap configurations. E = Excitation; D = Detection; T = End Cap "Trapping." (a) cubic; (b) cylindrical; (c) end-cap segmented to linearize excitation potential ("infinity trap"); (d) and (e) open-ended, without or with capacitive rf coupling between the three sections. Reproduced by permission of John Wiley & Sons, Inc. from *Mass Spectrometry Reviews*, Vol. 17, 1–35, 1998.

energy (K. E.) of a trapped ion in the field in terms of q, B_0, r, and m are given below:

$$r = \frac{mv_{xy}}{qB_0} \text{ (S. I. units) } \quad r = \frac{1.036427 \times 10^{-8}(m/z)v_{xy}}{B_0} \tag{3.20}$$

$$\frac{m\langle v_{xy}^2 \rangle}{2} \approx kT \quad r = \frac{1}{qB_0}\sqrt{2mkT} \text{ (S. I. units) } \quad r = \frac{1.336510 \times 10^{-6}}{zB_0}\sqrt{mT} \tag{3.21}$$

$$\text{Kinetic energy} = \frac{mv_{xy}^2}{2} = \frac{q^2B_0^2r^2}{2m}\text{(S. I. units)}$$

$$\text{K. E.} = \frac{4.824265 \times 10^7 z^2 B_0^2 r^2}{m} \tag{3.22}$$

Azimuthal dipolar single-frequency excitation, detection. In order to detect the ion cyclotron motion and from that obtain ionic mass-to-charge ratio, various techniques are used. One is by application of an excitation of the trapped ion by a spatially uniform oscillating (or rotating) electric field at a frequency close to the cyclotron frequencies of ions of interest. Such an excitation can coherently accelerate the ions to larger orbital radii which can be detected; also, orbital radii reaching beyond the trap size can result in ejection which too can be detected (Fig. 3.8). Excitation can result in an increase in kinetic energy causing ion-neutral reaction or it can result in ion dissociation. The radius r of the orbital motion and the kinetic energy of the ion depend on the applied field strength and on the time duration the external field is on. The radius and the post-excitation kinetic energy are:

$$r = \frac{E_0 T_{\text{excite}}}{2B_0} \quad r = \frac{V_{p-p} T_{\text{excite}}}{2dB_0} \text{ (S. I. Units)} \tag{3.23}$$

$$\text{K. E.}_{\text{post-excitation}} = \frac{m\omega_c^2 r^2}{2} = \frac{q^2 E_0^2 (T_{\text{excite}})^2}{8m} = \frac{q^2 V_{p-p}^2 (T_{\text{excite}})^2}{8d^2 m} \tag{3.24}$$

$$\text{K. E.}_{\text{post-excitation}} = \frac{1.20607 \times 10^7 z^2 V_{p-p}^2 (T_{\text{excite}})^2}{d^2 m} \tag{3.25}$$

where the applied field is $\mathbf{E}(t) = E_0 \cos \omega_c t \mathbf{j}$, $E_0 = 2V_0/d = V_{p-p}/d$ — $V_0 = V_{p-p}/2$ applied across two infinitely extended parallel conductive flat electrodes located at $y = \pm d/2$ with interelectrode spacing of d, and the time duration of excitation is T_{excite}.

The *azimuthal dipolar single-frequency detection* works as follows. Resonant excitation increases ion cyclotron radius, creates spatial coherence, and moves the ion packet significantly off-axis. An ion of charge q induces a difference in image charge ΔQ between two opposed infinitely extended parallel flat conductive plates:

$$\Delta Q = -\frac{2qy}{d} \tag{3.26}$$

The ICR signal, proportional to the induced current $d\Delta Q/dt = -2q(dy/dt)/d$, is *independent* of the *magnetic field strength*. It increases linearly with ion cyclotron post-excitation radius because dy/dt increases linearly with r (eqn 3.20). It is directly proportional to ion charge, so the ICR signal strength is *higher* with *multiply charged ions*. Signals from ions having any m/z contribute to the total time-dependent signal, the Fourier transform of which can then yield the frequency spectrum, hence the mass spectrum.

Broadband excitation of an ICR signal and its detection. For a single-frequency excitation at $\omega_c = 2\pi\nu_c$ applied for $T_{\text{excitation}}$ seconds, the frequency domain spectrum $E(\nu)$ is given by

$$E(\nu) = E_0 \frac{\sin(2\pi\nu_c T_{\text{excitation}})}{2\pi\nu_c} \tag{3.27}$$

The duration of $T_{\text{excitation}}$ has effect on spectral broadening. If the duration of the excitation is $T_{\text{excitation}}$, the spectral range is broadened to a bandwidth of $\sim 1/T_{\text{excitation}}$ rad s^{-1}. This means that shorter (longer) the time duration of excitation, broader (narrower) is the spectral range in the frequency domain. One consequence of this is: that *even if* the excitation *frequency* is distant from the cyclotron *frequency* of an ion (that is, the ion is in a state of off-resonance), it can still excite the ion if the excitation *period* is *sufficiently short* to cause a broadening of the frequency spectral range resulting in covering the cyclotron frequency of the ion; for this the criterion to satisfy is: $(1/T_{\text{excitation}}) \geq 2\pi |\nu - \nu_c|$. This also indicates the ultimate

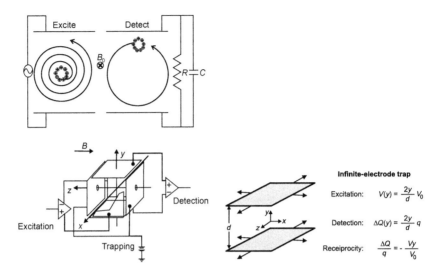

Figure 3.8: Left: Incoherent ion cyclotron orbital motion (top left) is converted to coherent (and, therefore, detectable) motion (top right) by the application of a rotating electric field, which rotates in the same sense and at the ICR frequency of the ions of a given m/z value. The electronic circuitry is shown in the bottom (left) diagram. Right: Excitation voltage difference and detection-induced charge difference in FT-ICR MS, demonstrated for two infinitely extended parallel flat electrodes located at $y = \pm d/2$ m away from the z-axis. Reproduced by permission of John Wiley & Sons, Inc. from *Mass Spectrometry Reviews*, Vol. 17, 1–35, 1998.

precision achievable; since ICR frequencies are high, high-mass selectivity can still be achieved even with a short duration pulse. (For 1s excitation pulse, this criterion amounts to $\sim 10\,\text{ppm}$ mass-selectivity for ions of m/z 1000 at 7.0 tesla.)

What is desired is a flat broadband frequency domain excitation signal, so that ions of a wide range are excited to nearly uniform cyclotron radii. Since a very large ($\sim \pm\ 10{,}000$ volt) rf excitation voltage within $\sim 0.1\,\mu\text{s}$ would be required for single-frequency resonant excitation that could excite ions within a bandwidth of $\sim 1\,\text{MHz}$, something which is not experimentally practical, one alternative in broadband ion-cyclotron excitation is to use a broadband frequency sweep (a range of frequencies at the excitation signal — "chirp"), which involves applying a relatively low excitation voltage and achieving relatively flat excitation over a broad frequency range. However, there are disadvantages of chirp excitation — nonuniform excitation amplitudes occur across the spectrum and cause limited mass selectivity. An efficient alternative is to define the desired excitation profile in the mass domain and then by performing an inverse Fourier transform gener ate the desired time-domain excitation waveform. Imagine an absolutely flat (sort of 'rectangular') frequency response curve with "start" and "end" frequencies; then through inverse Fourier transform generate the corresponding time domain signal; store the latter — which would be, the stored waveform inverse Fourier transform (SWIFT). Now apply this SWIFT excitation to the trap, a flat frequency domain excitation will result. Following this method mass spectral windows in the frequency domain can be generated at will and thus assisting in mass selectivity for MS/MS. For a given time-domain excitation period, the SWIFT method provides the flattest and the most frequency-selective excitation magnitude spectrum theoretically possible.

Broadband detection; detection limit. A possible ion detection circuit has been shown in Fig. 3.8. The differential current induced between two opposed detection plates by the coherently orbiting ion packet is equivalent to a current source. The receiver plates and connecting wires to the ion trap detection preamplifier provide some inherent resistance and capacitance in parallel. Since at typical ICR frequencies ($> 10\,\text{kHz}$) excited in FT-ICR MS experiments, such as in chirp or SWIFT, the S/N ratio is independent of cyclotron frequency (only at very low frequency, $< 10\,\text{kHz}$, the signal varies directly with frequency) the detected S/N ratio can be taken as representative of the current differential induced on the detection plates.

For an undamped signal in a single 1 s acquisition period with a requirement to yield a S/N ratio of 3:1, the minimum number of ions N that can be detected — can be termed the *detection limit* — can be obtained from the following equation:

$$N = \frac{CV_{d(p-p)}}{qA_1(r)} \tag{3.28}$$

where C is the capacitance of the detection circuit, $V_{d(p-p)}$ is the peak-to-peak amplitude of the detected voltage, and $A_1(r)$ is a coefficient approximately proportional to r and is amenable to graphical determination. For $C = 50\,\text{pF}$, $V_{d(p-p)}$ 3×10^{-7}, $A_1(r) = 0.5$, $q = 1.6 \times 10^{-19}$ coulomb, one obtains $N \sim 187$.

Effect of ion-neutral collisions. Ions inside the FT-ICR trap undergo collisions with the background neutral molecules, and this influences the peak shape of the resonance signal. In the absence of collisions, that is, at the limit of zero pressure, the time domain signal is a *sinc* function. This is one limit. The other would be the high-pressure limit, when there are many collisions. A simple way to model the effect of ion-neutral collisions is to incorporate in the overall equation of motion a frictional damping force. This results in an exponential damping of the time domain ICR signal, and the FT-ICR spectral line shape turns out to be of Lorentzian shape — signal damped nearly to zero — in the "high-pressure" limit. The line shape depends on the number of ion-collisions per second, which using the Langevin

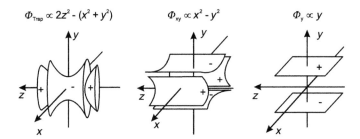

Figure 3.9: Plots of isopotential surfaces for trapping (left), 2D-quadrupolar (middle), and dipolar (right). Reproduced by permission of John Wiley & Sons, Inc. from *Mass Spectrometry Reviews*, Vol. 17, 1–35, 1998.

model is ≈0.2 collisions per second per ion. The hard-sphere collision model is however much more appropriate — which results in a frequency-domain line shape of narrow width at half-maximum peak height, but very broad at the base.

Trap of finite size: effects of axial confinement. In a trap of finite size, in order to *prevent ions from escaping* along the axial direction, it is *necessary* to provide a small (~ 1 volt) *electrostatic trapping* potential to the two "end-cap" electrodes positioned at $z = \pm a/2$ from the center of the "cell." The resulting electric field — from this trapping potential (shown below, in eqn 3.29) on finite-sized electrodes and the excitation (or detection) potentials used — is nonlinear, and this perturbation alters the ion oscillation frequencies.

Axial ion oscillation due to a z-component of electrostatic trapping potential. A three-dimensional axial quadrupolar electrostatic trapping potential

$$\Phi(x, y, z) = V_{\text{trap}} \left(\gamma + \frac{\alpha}{2a^2}(2z^2 - x^2 - y^2) \right)$$

$$\Phi(r, z) = V_{\text{trap}} \left(\gamma + \frac{\alpha}{2a^2}(2z^2 - r^2) \right) \qquad (3.29)$$

that satisfies Laplace's equation $\nabla^2\Phi(x, y, z) = 0$ and can be used is of the above form, where $r = \sqrt{x^2 + r^2}$ is the radial position of the ion in the x-y plane, a is a measure of trap size, and γ and α are constants that depend on the trap shape. There are methods for computing γ and α for orthorhombic, tetragonal, and cylindrical traps of arbitrary aspect ratio. For the cubic shape, $\gamma = 0.33333$ and $\alpha = 2.77373$; for the cylinder shape, $\gamma = 0.2787$ and $\alpha = 2.8404$. (The shape of the potential — Fig. 3.9 — can be seen from a plot of its isopotential surfaces.)

For cubic and cylindrical (unit aspect ratio) traps, the dipolar excitation fields are essentially identical; if the excitation voltage is ~ 1.4 times higher for the cylindrical trap (with respect to that of an inscribed cubic trap), the detected signal from the cylindrical trap is ~ 0.7 of that for the cubic trap. In terms of effectiveness of dipolar excitation, the traps differ considerably. For the hyperbolic trap, it is very nonlinear for dipolar excitation, although for trapping potential it is almost ideal. Dipolar excitation linearity improves slightly for ions in the $z = 0$ midplane for the three-cylinder open trap, but the capacitively coupled design results in near-linear excitation from one end of the trap to the other; as a result "z-ejection" is virtually eliminated. Whereas dipolar excitation improves in the "infinity" trap, but not dipolar detection, unless an additional switching network is used; further, it requires that quadrupolar excitation be performed with two-plate rather than four-plate excitation.

The z-component of the electrostatic trapping potential results in oscillatory z-motion, and the resulting axial frequency is

$$z(t) = z(0)\cos(2\pi\nu_z t) \quad \nu_z = \frac{1}{2\pi}\sqrt{\frac{2qV_{\text{trap}}\alpha}{ma^2}} \text{ (S. I. units)} \tag{3.30}$$

$$\nu_z = 2.21088 \times 10^3 \sqrt{\frac{zV_{\text{trap}}\alpha}{ma^2}} \tag{3.31}$$

From this, in a cubic trap of side length $a = 2.54$ cm, for $V_{\text{trap}} = 1$ volt, an ion of $m/z = 1000$ will oscillate at a "trapping" frequency of 4,580 Hz.

Radial ion "magnetron" rotation due to a combination of B-field and r-component of electrostatic trapping potential. From the DC "trapping" potential of eqn 3.29 the radial force is

$$\text{Radial force} = qE(r) = \frac{qV_{\text{trap}}\alpha}{a^2}r \tag{3.32}$$

This radial electric field acts on the ion, producing an outward-directed electric force; acting simultaneously on the ion is the inward-directed Lorentz magnetic force originating from the applied magnetic field; the latter opposes the former, thus, has the opposite sign. The equation of motion in the static magnetic field and the three-dimensional axial quadrupolar electrostatic potential can be obtained by combining eqn 3.32 with eqn 3.19:

$$\text{Force} = m\omega^2 r = qB_0\omega r - \frac{qV_{\text{trap}}\alpha}{a^2}r \tag{3.33}$$

$$\omega^2 - \frac{qB_0\omega}{m} + \frac{qV_{\text{trap}}\alpha}{ma^2} = 0 \tag{3.34}$$

We note that the quadratic equation (eqn 3.34) in ω is independent of r. Application of the three-dimensional axial quadrupolar and the DC trapping potential results in ion motional frequency (see below) that is independent of ion position inside the trap. Two natural rotational frequencies — in place

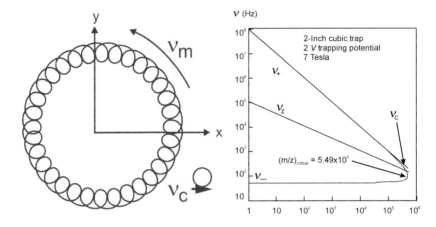

Figure 3.10: Left: The cyclotron rotation and magnetic rotation in a homogeneous magnetic field. Right: Mass-to-charge ratio (m/z) dependence of the frequency of each of the three ion natural motions. Reproduced by permission of John Wiley & Sons, Inc. from *Mass Spectrometry Reviews*, Vol. 17, 1–35, 1998.

of the original unperturbed cyclotron frequency observed in the absence of a DC trapping potential — are obtained by solving eqn 3.34 for ω:

$$\omega_+ = \frac{\omega_c}{2} + \sqrt{\left(\frac{\omega_c}{2}\right)^2 - \frac{\omega_z^2}{2}} \quad (\text{"Reduced" cyclotron frequency}) \qquad (3.35)$$

$$\omega_- = \frac{\omega_c}{2} - \sqrt{\left(\frac{\omega_c}{2}\right)^2 - \frac{\omega_z^2}{2}} \quad (\text{"Magnetron" frequency}) \qquad (3.36)$$

$$\omega_z = \sqrt{\frac{2qV_{\text{trap}}\alpha}{ma^2}} \quad (\text{"Trapping" oscillation frequency, in S. I. units}) \quad (3.37)$$

$$\omega_c = \frac{qB_0}{m} \quad (\text{"Unperturbed" cyclotron frequency, S. I. units}) \qquad (3.38)$$

The cyclotron rotation and magnetron rotation are shown in Fig. 3.10 (left) and the relative frequencies of ν_+, ν_c, ν_z, and ν_- as a function of ion m/z in Fig. 3.10 (right). The magnetron and trapping frequencies are usually much less than the cyclotron frequency, and generally are not detected — they appear as small sidebands when the ion trap is misaligned with the magnet axis and/or the amplitudes of ion trajectories approach trap dimensions.

Calibration of masses of trapped ions. In the absence of trapping potentials and electric space charge, measurement of the orbital frequency of one known mass followed by application of eqn 3.19 could be used for obtaining masses of all other ions. In the presence of the trapping potential, the following relationship, obtained from eqn 3.35, can be used for mass calibration:

$$\frac{m}{z} = \frac{A}{\nu_+} + \frac{B}{\nu_+^2}$$ (3.39)

where ν_+ is the reduced ion cyclotron orbital frequency, and A and B are constants which can be evaluated by fitting at least two ions of known m/z from a series of mass spectral peaks. Though the relationship is strictly valid only in the single-ion limit — that is, no Coulomb interaction present between ions — it can be used in an assembly of ions where the space charge perturbation is insignificant. Since the perturbation affects equally the calibrant ion and the analyte ions, the internal calibration method works out particularly well where, in the sample, the calibrant ion is present along with analyte ions.

Quadrupolar excitation, axialization. The two-dimensional azimuthal quadrupolar rf potential (shown in Fig. 3.9 (middle)), if excited at $\omega_c = qB_0/m$, periodically *interconverts* magnetron and cyclotron motions. When collisions are present, ion magnetron radius increases *slowly* with time, ion cyclotron radius decreases *rapidly*. In the limiting case, the cyclotron radius will be damped to *zero*, with ions relaxing toward the central axis of the ion trap. This conversion of magnetron to cyclotron motion, followed by collisional damping of the cyclotron radius is known as the *quadrupolar axialization.* Such shrinkage of the ion packet improves practically all aspects of FT-ICR performance. For example, axialization assists in cooling of ion internal energy, improves mass resolving power and accuracy, collision-induced dissociation efficiency, and mass selectivity for MS^n.

The frequency of interconversion is directly proportional to the voltage applied to the electrodes and inversely proportional to the product of mass-to-charge ratio of the ion, square of the electrode separation distance, and the difference of the reduced cyclotron frequency and the magnetron frequency. To give a numerical example: the application of two-dimensional azimuthal quadrupolar excitation of ± 1 volt to electrodes separated by 2.54 cm, at the unperturbed ion cyclotron frequency of 107,500 Hz at 7.0 tesla will interconvert magnetron and cyclotron motion at a frequency of ~ 141 Hz, which means that if at time zero those ions have zero initial cyclotron radius, then in ~ 3.5 ms (at $\frac{1}{2}\nu_{\text{interconvert}}$) their magnetron motion will be completely converted to cyclotron motion.

Effect of trap size and shape on dipolar and 2D-quadrupolar excitation. Relationships for ion cyclotron radius and post-excitation kinetic energy of the ion following a single-frequency resonant azimuthal dipolar excitation by a spatially uniform rf electric field were presented earlier (eqns 3.23 and 3.24). Those relationships, to a good approximation, are applicable to finite-size traps by incorporating a scale factor β_{dipolar}. Eqns 3.23 and 3.24 then become

$$r = \frac{\beta_{\text{dipolar}} E_0 T_{\text{excite}}}{2B_0} \quad r = \frac{\beta_{\text{dipolar}} V_{p-p} T_{\text{excite}}}{2dB_0} \text{ (S. I. Units)} \qquad (3.40)$$

$$\text{K. E.}_{\text{post-excitation}} = \frac{\beta_{\text{dipolar}}^2 q^2 E_0^2 (T_{\text{excite}})^2}{8m} = \frac{\beta_{\text{dipolar}}^2 q^2 V_{p-p}^2 (T_{\text{excite}})^2}{8d^2 m}$$
$$\qquad (3.41)$$

$$\text{K. E.}_{\text{post-excitation}} = \frac{1.20607 \times 10^7 \beta_{\text{dipolar}}^2 z^2 V_{p-p}^2 (T_{\text{excite}})^2}{d^2 m} \qquad (3.42)$$

For *broadband frequency-sweep* dipolar excitation, corresponding relationships for the ion cyclotron radius r and post-excitation kinetic energy can be obtained by replacing T_{excite} by $\sqrt{1/(\text{Sweep Rate})}$ in eqns 3.40 and 3.41. In the magnitude mode, spectral peak height is proportional to $1/\sqrt{\text{Sweep Rate}}$.

In the above equations energy is in eV; V_{p-p} in volts; T_{excite} in s; d in m; m in Da; and z is in multiples of elementary charge. In a cubic Penning trap with excitation plates 2 cm apart, immersed in a 7 T magnetic field, a single-frequency resonant excitation at 10 V V_{p-p} for 400 μs results in post-excitation cyclotron radius of 1.03 cm. The same radius can be obtained by spanning the ICR frequencies of interest through a frequency-sweep excitation using $V_{p-p} = 123$ V and a sweep rate of 1.00×10^9 Hz/s.

For the magnetron-cyclotron interconversion frequency for azimuthal two-dimensional *quadrupolar* excitation in a *finite-size* trap a scale factor β_{quad} needs to be introduced. With other operational parameters the same, the magnetron-cyclotron conversion frequency in finite-size trap changes by the factor $(2.30342 \times 10^7 / 9.77601 \times 10^6) \beta_{\text{quad}}$. Some scale factor values: β_{dipolar} for "ideal" trap is 1.00000, "cube": 0.72167, "cylinder": 0.80818, "infinity": $\approx$0.900, "open uncoupled": 0.86738, "open coupled": 0.89699. β_{quad} for "ideal" trap is 2.66667, "cube": 2.77373, "cylinder": 3.25522, "open uncoupled": 3.30527, "open coupled": 3.34800. A methodology for computing β_{dipolar} and β_{quad} for traps of arbitrary shape is available.

There is a "matrix-shimmed" trap[142] — a cubic trap, in which each of its six sides is segmented into a 5 × 5 grid for a total of 150 electrodes. In this shimmed trap's construction a large amount of capacitance gets added causing its excitation and detection efficiencies unacceptably low; but this trap is near-ideal in trapping, dipolar excitation, and quadrupolar excitation potential *shape*.

Mass resolution and mass resolving power. In Fourier transform spectroscopy, *resolution* is defined as the full width of a spectral peak at half-maximum peak height, that is, $\Delta\omega_{50\%}$ (in frequency domain) or $\Delta m_{50\%}$ (in mass domain). *Resolving power* is defined as $\omega/\Delta\omega_{50\%}$ or $m/\Delta m_{50\%}$.

From the first derivative of eqn 3.19 with respect to m, we obtain

$$\frac{d\omega_c}{dm} = \frac{-qB_0}{m^2} = -\frac{\omega_c}{m} \qquad \frac{\omega_c}{d\omega_c} = -\frac{m}{dm} \qquad (3.43)$$

Eqn 3.43 shows that frequency resolving power and mass resolving power in ICR mass spectroscopy are the same, differing only by a minus sign. Since the frequency of an FT-ICR mass spectral peak is approximately qB_0/m, experimental mass resolution in eqn 3.43 can be expressed as

$$\frac{m}{\Delta m_{50\%}} = -\frac{qB_0}{m\Delta\omega_{50\%}} \quad \text{S. I. units} \qquad (3.44)$$

The ICR mass resolution is best expressed stating whether it is evaluated at the low-pressure limit, $T_{\text{acq'n}} \ll \tau$, or, at the high-pressure limit, $T_{\text{acq'n}} \gg \tau$, where $T_{\text{acq'n}}$ is the time-domain acquisition period and τ is the ion-neutral collisional damping constant; τ is expressed as: $1/\tau = [m_{\text{neutral}}/(m_{\text{ion}} + m_{\text{neutral}})]\nu_{\text{collisions}}$. The FT-ICR mass spectral width is independent of m/z at the low-pressure limit. At low-pressure $(T_{\text{acq'n}} \ll \tau)$, $(m/\Delta m_{50\%})$ is $1.274 \times 10^7 (z/m)B_0 T_{\text{acq'n}}$, and at high-pressure $(T_{\text{acq'n}} \gg \tau)$, $(m/\Delta m_{50\%})$ is $2.785 \times 10^7 (z/m)B_0\tau$, m/z in Da per elementary charge.

In FT-ICR the highest mass resolution is achieved by operating at the maximum magnetic field strength.

Upper mass limits. Estimates of the *upper* mass limit of ions that can be kept confined in a FT-ICR trap are of interest. One estimate based on thermal ions can be from consideration of trap dimension — when the mass at which the ion cyclotron radius of a *thermal* ion reaches the radius of the trap. Calculation shows that in the absence of an electric field, in a trap of 2.54 cm cross-sectional radius the upper mass limit for a singly charged room-temperature ion at 7.0 tesla would be ~ 5.89 megaDalton. However, since for a detectable coherent signal the ion must *originate* at an ICR orbital radius much *smaller* than the trap (and be detected through an excitation leading the orbital radius to be higher), the actual upper mass limit would be smaller.

Electrostatic trapping potential or Coulomb repulsions between ions also affect the mass limit. The magnetron frequency ω_z and the reduced cyclotron frequency ω_c converge to a common value for a mass m/z at the so-called 'critical mass' $(m/z)_{\text{critical}}$. Beyond $(m/z)_{\text{critical}}$ the ion cyclotron motion is no longer stable and the ion spirals outward and exits the trap. For $V_{\text{trap}} = 1$ volt applied to a cubic trap $(\alpha = 2.77373)$ of $a = 2.54$ cm, m_{critical} at 7.0 tesla is 274,000 z. The upper mass limit rapidly (exponentially) falls with increasing trapping potential, and that mass limit is lower with lower trapping potential. At $V_T = 1$ V, the limit is ~ 138 kDa for 7.0 T and ~ 2.8 kDa for 1.0 T. This indicates desirability of cooling the ions; relatively cold ions mean that the trapping potential may be lowered, and as a result the upper mass limit would be increased.

When more than $\sim 10,000$ ions are present in the trap, the static and dynamic effects of ion-ion repulsions need to be considered. It has been shown that whereas in a spatially uniform static electromagnetic field, Coulomb repulsions between ions do *not* affect the ICR orbital frequency of the ion packet, in the spatially *non*uniform electromagnetic field of a *typical* ion trap, Coulomb repulsions by pushing like-charge ions apart into regions of a different applied external electric or magnetic field can shift and broaden FT-ICR mass spectral peaks. Peak coalescence — observation of a single FT-ICR mass spectral peak whose width is narrower than the separation between the cyclotron frequencies of ions of two different but very similar m/z ions — occurs when two clouds of such ions have sufficiently large ion populations. The maximum molecular weight m_{max} at which coalescence first occurs have been calculated — though from a highly idealized theoretical model; m_{max} depends directly on the magnetic field induction, directly on the square root of the product of length of the cylinder, single-ion cyclotron radius, ion cloud radii, and mass difference between the two ions, and inversely on the square root of the average number of ions in the two ion clouds. Based on this relationship the highest mass at which ions whose masses differ by one Da begin to coalesce is $\sim 10,000\ B_0$ Da (B_0 in T).

Ion excitation in ICR trap for CAD/CID. In SORI — sustained off-resonance irradiation — ions of a selected m/z ratio are alternately excited and de-excited owing to the difference between the excitation frequency and the ion cyclotron frequency; in VLE — very low energy excitation — ions are alternately excited and de-excited by resonant excitation whose phase alternates bimodally between $0°$ and $180°$, and in multiple excitation for collisional activation — MECA — excitation by resonance, relaxation of ions by collisions.

Advantages of high magnetic field. There are numerous advantages of using high magnetic field. At a fixed ion-neutral collision frequency, the mass resolving power and acquisition speed increase *linearly*; the upper mass limit, maximum ion kinetic energy, maximum number of trapped ions, maximum ion trapping duration, and two-dimensional FT-ICR mass resolving power increase *quadratically*, whereas peak coalescence tendency varies *inverse quadratically* with the field strength.[143] As a result, other FT-ICR performance parameters, such as signal-to-noise ratio, dynamic range, mass accuracy, ion remeasurement efficiency, and mass selectivity for MS/MS, improve. A timeline on employment of high magnetic field in FT-ICR mass spectrometers along with some key figures of merit that the high magnetic field augments is shown in Fig. 3.11 (top). Progress report on the construc-

---

[143]Marshall, A. G. and Guan, S. (1996) Advantages of high magnetic field for Fourier transform ion cyclotron resonance mass spectrometry, *Rapid Commun. Mass Spectrom.*, **10**, 1819–1823.

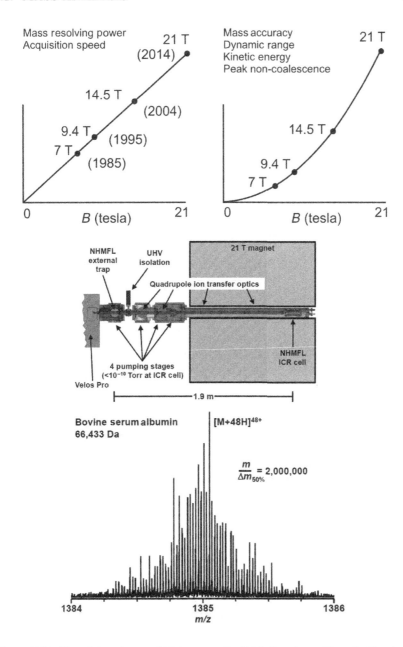

Figure 3.11: Top: Advantages of high magnetic field. Courtesy: Alan G. Marshall, National High Field Magnet Laboratory and Department of Chemistry and Biochemistry, Florida State University, Tallahassee. Reproduced by permission. Middle and bottom: NHFML 21T FT-ICR instrument and BSA spectrum. Reproduced by kind permission of Springer Science and Business Media from *Journal of the American Society for Mass Spectrometry*, DOI: 10.0007/s13361-015-1182-2.

tion of a 21T FT-ICR mass spectrometer appeared earlier;[144] recently the design and *initial performance* of the first 21 tesla FT-ICR mass spectrometer have been published.[145]

The 21 tesla magnet is the *highest field* superconducting magnet *ever used* for FT-ICR; it features high spatial homogeneity, high temporal stability, and negligible liquid helium consumption. The FT-ICR instrument hardware includes a modified dual linear quadrupole trap front end Velos Pro (Thermo Fisher Scientific); a third linear quadrupole trap offers high ion capacity and ejection efficiency, and RF quadrupole ion injection optics deliver ions to a novel dynamically harmonized ICR cell.

The 21 tesla magnet (Bruker Daltonics) has room temperature bore diameter 123 mm, the distance to field center is 1,047 mm, and the overall length is 2,272 mm (a section of the ICR mass spectrometer is shown in Fig. 3.11 (middle)). A set of eight cryoshims was used to achieve magnet spatial inhomogeneity less than 5 ppm over a 60 mm diameter by 100 mm long cylinder, which closely matches the working volume of the cell. The measured magnet drift was ~ 4 ppb/h two months after energization and settled to less than 2 ppb/h after eight months of operation.

The Velos Pro front end offers high sensitivity, efficient ion isolation, precise control of ion number, and both collisional dissociation and *front-end* electron transfer dissociation. A set of two quadrupoles facilitates transfer of ions from the Velos Pro ion trap to the NHMFL external quadrupole trap. The external quadrupole trap allows more ions to be delivered to the ICR cell by use of multiple fills from the Velos trap. CAD is implemented by acceleration of mass-selected ions as they travel from the Velos trap into the external quadrupole trap and collide with nitrogen (or other heavier gas). At the ICR cell the pressure is less than 10^{-10} Torr.

The dynamic harmonization is similar in concept to the design of Boldin and Nikolaev (see, the subsection on dynamic harmonization and also Nikolaev *et al.*[150]) but with several modifications. Here, first, the side electrodes are segmented into 12 biconcave and 12 biconvex electrodes (instead of eight), which allows convenient application of excitation and detection onto $120°$ cell segments for improved excitation electric field, detection sensitivity, and *minimization of third harmonic* signals. Excitation and detection on the same $120°$ segments is accomplished[146] with a novel electronic

[144]Kaiser, N., Weisbrod, C., Quinn, J., Blakney, G., Beu, S., Chen, T., Smith, D., Hendrickson, C., and Marshall, A. (2014) Development of an FT-ICR mass spectrometer in preparation for 21 tesla. *Proc. 62nd Annual ASMS Conference on Mass Spectrometry and Allied Topics*, Baltimore, June 15–19.

[145]Hendrickson, C. L., Quinn, J. P., Kaiser, N. K., Smith, D. F., Blakney, G. T., Chen, T., Marshall, A. G., Weisbrod, C. R., and Beu, S. C. (2015) 21 tesla Fourier transform ion cyclotron resonance mass spectrometer: A national resource for ultrahigh resolution mass analysis. *J. Am. Soc. Mass Spectrom.*, DOI: 10.1007/s13361-015-1182-2.

[146]Chen, T., Beu, S. C., Kaiser, N. K., and Hendrickson, C. L. (2014) Optimized circuit for excitation and detection with one pair of electrodes for improved Fourier transform ion cyclotron resonance mass spectrometry. *Rev. Sci. Instrum.*, **85**, 066107.

circuit. Second, the trap endcaps are segmented similarly to the "Infinity" cell (but with only four segments) and RF excitation voltage is capacitively coupled onto the end-cap segments, which minimizes axial excitation attributable to excitation electric field inhomogeneity.

A *single-scan* electrospray FT-ICR mass spectrum of the isolated 48+ charge state of bovine serum albumin (66 kDa) following a 12 s detection period is shown in Fig. 3.11 (bottom). Mass resolving power is approximately 2,000,000, and the signal-to-noise ratio of the most abundant peak is greater than 500:1. The ion accumulation period was 250 ms and the ion target was 5,000,000. Mass resolving power for 0.38 s detection period is approximately 150,000, and the signal-to-noise ratio of the most abundant peak is greater than 150:1. Externally calibrated broadband mass measurement accuracy is typically less than 150 ppb rms, with resolving power greater than 300,000 at m/z 400 for a 0.76 s detection period. Variation of measured masses is limited to 140 pp rms over thousands of measurements (taken over the course of ~ 1 h). Combined analysis of electron transfer and collisional dissociation spectra results in 68% sequence coverage for carbonic anhydrase.

Resolving power for higher charge states is pressure-limited and lower base pressure is required for higher resolving power of higher charge states. High signal magnitude was evident at the end of the detection period, so that future improvement in the memory depth of the digitizer or implementation of heterodyne detection should extend the resolving power further.

FT spectroscopy aspects. Since frequency can be measured more accurately than any other experimental parameter, FT-ICR MS — where, from frequency, m/z is obtained — offers inherently higher resolution/mass accuracy. It utilizes the Fellgett advantage — multichannel, opening the exit slit — which provides increased speed (factor of 10,000), increased sensitivity (factor of 100), increased mass resolution (factor of 10,000) and increased mass range (factor of 500), besides the advantages of fixed magnetic field. Also, it contributes to more efficient MS/MS, and lower detection limit. In FT ICR MS, spectra are usually reported in magnitude-mode rather than absorption-mode. This results in lower resolving power; to address this, new peak-fitting algorithms have been developed. FT-ICR MS, unlike FT NMR, is often performed by sampling the ICR signal directly without heterodyning; in analysis massive data sets are involved to take advantage of potentially ultrahigh mass resolving power over a wide mass range. The ICR time-domain signal typically exhibits nonexponential damping; hence, prior to FT, it is common to apodize — preferentially damp the initial portion of the time-domain signal.

Detection limit for biological applications. A fundamental limit in FT-ICR broadband image current detection is, to induce a measurable signal, typically ~ 100 ions of a given m/z are required. Therefore, destructive detection through ion counting is inherently more sensitive. However, the

FT-ICR detection limit can still be extremely low. With CE coupling, hemoglobin mass spectra have been obtained from a single red blood cell that contained ~ 450 amol of hemoglobin. FT-ICR mass spectra have been observed also from sub-attomole protein samples; the same researchers, from a CAD spectrum of 9 amol of carbonic anhydrase, unambiguously identified the protein present in spite of N-terminal acetylation. For biological sample analysis, the very low detection limit with non-destructive detection, makes FT-ICR extremely attractive.

ESI with FT-ICR detection has been generally successful in the analysis of large ions. Accurate mass analysis at unit mass resolution and at 67 kDa has been routine; the highest-mass molecules for which unit mass resolution was achieved were two 112 kDa proteins (the masses of two chondroitinase enzymes), determined with 3 Da accuracy. Also has been possible the trapping and detection of a single 100 megadalton DNA ion with $\sim 30,000$ charges.

Dynamic harmonization of an FT-ICR cell. For ions in a hyperbolic trap and in a spatially homogeneous magnetic field, in the absence of ion-neutral collisions and ion-ion interactions and under the condition of complete decoupling of cyclotron, magnetron, and axial oscillations, synchronized during cyclotron excitation, will continue in synchronous motion indefinitely. However, in a real ICR trap the cyclotron and magnetron frequencies are perturbed by the axial oscillation amplitude, which results in dephasing of the ion cloud and making the cloud assume comet-like shape.[147] For any ion trap, this can be quantitatively evaluated by calculating the dependence of the radial component of the electric field on the axial z-coordinate. Further, after some time, when the phase difference across the original cloud approaches 2π, the head of the comet reaches toward the tail, there is resulting loss of signal from the cloud. Since FT-ICR MS resolving power depends directly on the duration of the synchronous motion, the time needed for dephasing to take place limits it, with the dephasing time depending on ion trap geometry and magnetic field strength and homogeneity.

Ion motion inside the cell also is adversely affected by any inhomogeneity in the magnetic field. For ions having different axial oscillation amplitudes, magnetic field homogeneity causes differences in their cyclotron frequencies. Ions in different positions of the ion cloud experience different conditions. This also contributes toward comet formation.

To avoid comet formation it is necessary to make electric potential distribution inside the cell hyperbolic. Whereas the simplest way to effect

---

[147]Nikolaev, E. N., Miluchihin, N. V., and Inoue, M. (1995) Evolution of an ion cloud in a Fourier transform ion cyclotron resonance mass spectrometer during signal detection: its influence on spectral line shape and position. *Int. J. Mass Spectrom. Ion Proc.*, **148**, 145–147.

this is just construct a cell with hyperbolic electrodes, in such a cell the usable volume can be very small. Small operational volume does not permit working with large number of ions in the cell; and this limits its dynamic range.[148]

Recently a new method of harmonization, called *dynamic harmonization*, has been introduced. In this, the field of cylindrical cell becomes effectively hyperbolic in the whole volume. This cell design[149] is a cylinder segmented by curves along the axial direction. The slits are parabolic and they are defined by

$$\alpha = \frac{2\pi}{N} n \pm \alpha_0 \left(1 - \left(\frac{z}{a}\right)^2\right); n = 0, 1, ..., N-1 \qquad (3.45)$$

where N is the number of electrodes of each kind, α_0 is an arbitrary coefficient, and a is *half* of the cell length $a = L/2$. Then the boundary condition is

$$\hat{\varphi}(z, R) = V_0 \left(1 - \frac{\alpha_0}{(\pi/N)} \left(1 - \left(\frac{z}{a}\right)^2\right)\right) \qquad (3.46)$$

In order to obtain the hyperbolic-averaged potential in a finite-length cell, end-cap boundary condition is to be added. According to boundary condition, eqn 3.46, average potential should be of this form

$$\hat{\varphi}(z, R) = V_0 \left(1 + \frac{\alpha_0}{2a^2(\pi/N)} \left(2(z^2 - a^2) - (r^2 - R^2)\right)\right) \qquad (3.47)$$

Therefore, a possible variant of the end-cap boundary condition is — hyperbolic electrodes on both sides of the cell set at V_0 voltage. According to eqn 3.47, their shape is defined by

$$2(z^2 - a^2) - (r^2 - R^2) = 0 \qquad (3.48)$$

For end-cap electrodes, there is no need to resort to segmentation; it satisfies boundary conditions of hyperbolic field statically without averaging. Some actual details of the design of their cell follow.

The original experimentally tested ion trap with dynamic harmonization had eight segments with width decreasing to the center of the cell and eight grounded electrodes with width increasing to the center, four of which are divided into two segments, each belonging to either excitation

[148]Nikolaev, E. N., Kostyukevich, Y. I., and Vladimirov, G. N. (2014) Fourier transform ion cyclotron resonance (FTICR) mass spectrometry: theory and simulations. *Mass Spectrom. Rev.*, doi: 10.1002/mas.21422.

[149]Boldin, I. A. and Nikolaev, E. N. (2011) Fourier transform ion cyclotron resonance cell with dynamic harmonization of the electric field in the whole volume by shaping of the excitation and detection electrode assembly. *Rapid Commun. Mass Spectrom.*, **25**, 122–126.

or detection groups of electrodes. The trapping potential V is applied to
the first group of electrodes and to trapping electrodes. Other electrodes
are grounded. RF voltages are applied via capacitors to excitation group of
electrodes and detection electrodes are connected with each other and with
preamplifier by capacitors of appropriate value of capacity.

A picture of the cell that Nikolaev *et al.* used[150] in protein mass spectrometry is shown in Fig. 3.12 (bottom left). The length L ($L = 2a$) of the
cell is 150 mm, the inner diameter 56 mm. The curved slits are 1 mm wide,
defined by eqn 3.45 where α is the azimuthal angle, N = 8, $\alpha_0 = \pi/8$–$\pi/60$.
Ideally, α_0 should be equal to $\pi/8$ so that adjacent slits are tangential to
the mid-plane of the cell. However, from machining consideration the straits
were left at approximately 2 mm thickness. The inner surface of the end-
cap electrodes was made spherical (rather than hyperbolic), for machining
convenience. (A spherical surface differs insignificantly from the hyperbol-
ical one by 0.02 mm at maximum.) The end-cap electrodes have orifices of
6 mm diameter at their centers. An additional electrode at the entrance of
the cell focussed incoming ions.

During the excitation and detection sequences, typically a DC voltage
of 1.5 V is applied on both end-cap electrodes as well as the entrance lens.
The same DC voltage is applied to the *convex* electrodes of the cylindrical
surface; the *concave* electrodes remain grounded.

In Fig. 3.12 (right bottom), electric connections of the cylinder-surface
electrodes, in two-dimensional representation, are shown. All *concave-shaped*
electrodes are connected to the *trapping voltage*; all *convex-shaped* electrodes are *grounded*. The complete electrode set is divided into four sections each forming a 90° angle. The two *excitation* sections consist of one
full and two half convex electrode segments plus two concave electrodes.
The excitation chirp is applied to all electrodes of the detection section
via a 33 nF/100 kΩ high-pass filter. The two opposite *detection* sections
are *identically* structured as the *excitation* sections. The *preamplifier* is
connected to the *convex electrode* and the two *half-convex segments*. The
concave electrodes are *not* used for detection; the trapping voltage is applied via 100 MΩ resistors.

This ICR ion trap with dynamic harmonization has shown highest resolving power. The time of transient duration reached 300 sec and seemed
limited only by the level of vacuum inside the FT-ICR cell and magnetic
field inhomogeneity. It has demonstrated a mass resolving power of more
than twenty millions of reserpine (m/z 609) and more than one million
of highly charged BSA molecular ions (m/z 1357) — such performance has
been reached on moderate 7 T magnetic field solenoids. Figure 3.12 (top

---

[150]Nikolaev, E. N., Boldin, I. A., Jertz, R., and Baykut, G. (2011) Initial experimental
characterization of a new ultra high resolution FTICR cell with dynamic harmonization.
J. Am. Soc. Mass Spectrom., **22**, 1125–1133.

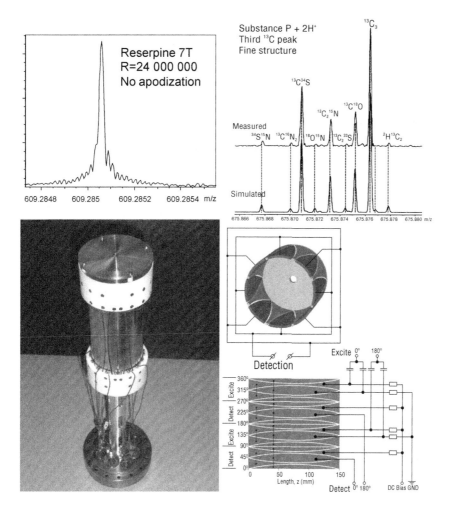

Figure 3.12: Bottom left: Dynamically harmonized FT-ICR cell. Bottom right: Electrode connections of the cylinder surface electrodes in two-dimensional representation. Top left: Frequency spectrum of the monoisotopic peak of singly charged reserpine (m/z 609.28066). Reproduced with kind permission from Springer Science and Business Media from *Journal of the American Society for Mass Spectrometry*, Vol. 22, 1125–1133, 2011. Top right: Experimental Fourier transform mass spectrum of doubly protonated Substance P isotope cluster 3 ^{13}C group. Reproduced by permission of the American Chemical Society from *Analytical Chemistry*, Vol. 84, 2275–2283, 2012. Right middle: A view of the electrodes from inside (half of the cell) looking toward the ion inlet aperture and trapping electrode of spherical form. Reproduced by permission of John Wiley & Sons, Ltd. from *Rapid Communications in Mass Spectrometry*, Vol. 25, 122–126, 2011.

left) shows the frequency spectrum (not calibrated) of the monoisotopic peak of singly charged, protonated reserpine (m/z 609.28066) with a resolving power of 24,000,000. They have also studied[151] the fine structure in isotopic peak distributions in mass spectra of reserpine and substance P. The resolved peaks in the fine structure consisted of ^{13}C, ^{15}N, ^{17}O, ^{18}O, ^{2}H, ^{33}S, ^{34}S, and combinations of them. The positions of the experimentally obtained fine structure peaks on the mass scale agreed with the isotopic distribution simulations with ≤ 200 ppb error.

In a recent report[152] from the same group a time-domain transient detection signal of about 180 s in duration and resolution — 12,000,000 at m/z 675, was experimentally achieved for the doubly protonated Substance P peptide (Fig. 3.12, top right), using an FT-ICR instrument (Bruker Apex Qe equipped with an unshielded low-Tesla 4.7 T magnet; along with a laboratory prototype of the dynamically harmonized ICR cell, cylindrical, 56 mm i.d., and 150 mm long). They were measuring the resolving power as full width at half maximum (FWHM) of individual peak. The theoretical resolution corresponding to a 180 s duration of an undamped ICR signal is about 18,000,000 in accordance with the expression for FT-ICR resolution as a function of the observation period. The small difference between the theoretical and measured resolution values for the current signal seemed to result from close peaks owing to relatively low magnetic field. The magnitude of the frequency drift during detection was about $\Delta\nu_+ = +0.0085$ Hz and this corresponds to the measured resolution value at the current cyclotron frequency of 104.2 kHz.

Orbitrap

The Orbitrap mass analyzer consists of two coaxial electrodes: an outer barrel-shaped electrode, and an inner spindle-like electrode (Fig. 3.13). Due to the special shapes of the electrodes and voltages on them, an ion in the interelectrode space experiences an electrostatic field which has one component similar to that present in the 3D ion traps, and another, similar to that inside a cylindrical capacitor. The field generates a potential well along the axial direction which can trap an ion within the well if the kinetic energy of the ion is not too high for escape. In the field, the bottom of the potential

---

[151]Nikolaev, E. N., Jertz, R., Grigoryev, A., and Baykut, G. (2012) Fine structure in isotopic peak distributions measured using a dynamically harmonized Fourier transform ion cyclotron resonance cell at 7 T. *Anal. Chem.*, **84**, 2275–2283.

[152]Popov, I. A., Nagornov, K., Vladimirov, G. N., Kostyukevich, Y. I., and Nikolaev, E. N. (2014) Twelve million resolving power on 4.7 T Fourier transform ion cyclotron resonance with dynamically harmonized cell — observation of fine structure in peptide mass spectra. *J. Am. Soc. Mass Spectrom.*, **25**, 790–799.

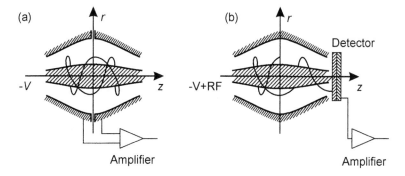

Figure 3.13: Modes of mass analysis in the Orbitrap: (a) Fourier transform mass spectrometry (image current detection); (b) mass selective instability (detection using secondary electron multiplier). Reproduced by permission of the American Chemical Society from *Analytical Chemistry*, Vol. 72, 1156–1162, 2000.

in the radial direction lies along the longitudinal axis; an ion enters a spiral trajectory around the inner cylindrical electrode. The ion motion has three characteristic frequencies: one is of the harmonic motion in the axial direction, the second one of the oscillation in the radial direction, and the third is the frequency of angular rotation. While the ions are moving in the interelectrode space these frequencies can be measured. Among these, the frequency of the harmonic motion in the axial direction, which is independent of the energy and spatial spread of the ions, is the most appropriate one to provide the required high performance as a mass analyzer. From a measurement of the frequency of the ions' harmonic oscillations (of the orbitally trapped ions) along the axis of the electric field, mass/charge ratios of the ions can be determined. In this non-destructive image current method of detection the ions are not lost. A Fourier transform (in practice, fast Fourier transform FFT) of the transient image current signal then yields the mass spectrum. An alternative method to obtain mass spectra — based on mass selective instability of the ions — is ejection of the trapped ions to an external secondary electron multiplier detector. High mass accuracy, dynamic range, and upper mass limit is possible in Orbitrap. High mass resolution (up to 100,000 to 200,000) can be obtained because the field may be defined with very high accuracy, and also because of *larger* trapping volume and *increased* space charge capacity at higher masses — in contrast to 3D ion trap and FT-ICR cell — arising out of the trapping potential being independent of mass-to-charge ratio.

In some detail,[153,154,155,156] the shapes of the two electrodes in cylindrical coordinates are given by

$$z_{1,2}(r) = \sqrt{\frac{r^2}{2} - \frac{(R_{1,2})^2}{2} + (R_m)^2 \ln\left[\frac{R_{1,2}}{r}\right]} \qquad (3.49)$$

where subscripts 1 and 2 stand for the central electrode and the outer electrode, respectively, $R_{1,2}$ are the maximum radii of the corresponding electrodes, and $z = 0$ is the plane of symmetry. The potential distribution in the electrostatic field is given by

$$U(r,z) = \frac{k}{2}\left(z^2 - \frac{r^2}{2}\right) + \frac{k}{2}(R_m)^2 \ln\left[\frac{r}{R_m}\right] + C \qquad (3.50)$$

where k is field curvature, R_m is the characteristic radius, and C is a constant. It will be noticed that the field is the sum of a quadrupole field of the ion trap and a logarithmic field of a cylindrical capacitor. Equations of motion for an ion with mass to charge ratio m/q in this quadro-logarithmic (hyper-logarithmic) field in polar coordinates (r, ϕ, z) are given by[157]

$$\ddot{r} - r\dot{\varphi}^2 = -\frac{q}{m}\frac{k}{2}\left[\frac{(R_m)^2}{r} - r\right] \qquad (3.51)$$

$$\frac{\mathrm{d}}{\mathrm{d}t}(r^2\dot{\varphi}) = 0 \qquad (3.52)$$

$$\ddot{z} = -\frac{q}{m}kz \qquad (3.53)$$

with initial conditions at $t = 0$ being

$$r(0) = r_0 \quad \dot{r}(0) = \dot{r}_0 \qquad (3.54)$$

$$\varphi(0) = \varphi_0 \quad \dot{\varphi}(0) = \dot{\varphi}_0 \qquad (3.55)$$

$$z(0) = z_0 \quad \dot{z}(0) = \dot{z}_0 \qquad (3.56)$$

For $r < R_m$, ions are attracted toward the field axis, and for $r > R_m$ they are repelled; therefore, only radii below R_m are useful for trapping purposes.

[153] Makarov, A. A., HD Technologies Limited, Manchester UK. United States Patent, Patent Number: 5,886,346. Date of Patent: March 23, 1999.

[154] Makarov, A. (2000) Electrostatic axially harmonic orbital trapping: a high-performance technique of mass analysis. *Anal. Chem.*, **72**, 1156–1162.

[155] Hardman, M. and Makarov, A. (2003) Interfacing the Orbitrap mass analyzer to an electrospray ion source. *Anal. Chem.*, **75**, 1699–1705.

[156] Qizhi, H., Noll, R. J., Hongyan, L., Makarov, A., Hardman, M., and Cooks, R. G. (2005) The Orbitrap: a new mass spectrometer. *J. Mass Spectrom.*, **40**, 430–443.

[157] In this book, we generally use m/z to represent mass-to-charge ratio of ions; only here we have used m/q to avoid conflict with expressions involving the z-coordinate.

Since the motion along z is *independent* of r and ϕ, it is convenient to define separate energy characteristics for each direction as

$$qE_r = (m/2)(\dot{r}_0^2) \tag{3.57}$$
$$qE_\varphi = (m/2)(r_0\dot{\varphi}_o)^2 \tag{3.58}$$
$$qE_z = (m/2)(\dot{z}_0)^2 \tag{3.59}$$

From eqn 3.53 the z-motion can be obtained as a simple harmonic oscillation given by

$$z(t) = z_0\cos(\omega t) + \sqrt{(2E_z/k)}\sin(\omega t) \tag{3.60}$$

where, ω, the frequency of axial oscillation (in rad/s) is:

$$\omega = \sqrt{(q/m)k} \tag{3.61}$$

This frequency is independent of the energy and the positions of the ions and can be used for mass analysis.

Unlike the z-equation of motion, the r, φ equations (eqn 3.51, eqn 3.52) cannot be integrated analytically. The general trajectory in the polar plane resembles a rotating ellipse. The influence of field imperfections is reduced when this ellipse is close to a circle, which is achieved when $E_r = 0$ and $E_\varphi = \tilde{E}_\varphi$, $\tilde{E}_\varphi$ given by

$$\tilde{E}_\varphi = (k/4)((R_m)^2 - R^2) \tag{3.62}$$

R being the radius of the circle, eqn 3.62 is obtained by adopting $\dot{r} = \ddot{r} = 0$. The r and φ equations have been solved approximately;[158] in the first approximation, the solutions for $r(t)$ and $\varphi(t)$ are

$$r(t) = R + (r_0 - R)[\eta + (1-\eta)\cos(\omega_r t)] + \sqrt{\tfrac{\eta}{2}\tfrac{E_r}{\tilde{E}_\varphi}}R\sin(\omega_r t)$$
$$+ \tfrac{E_\varphi - \tilde{E}_\varphi}{\tilde{E}_\varphi}\tfrac{\eta}{2}R(1 - \cos(\omega_r t)) \tag{3.63}$$

$$\varphi(t) = \omega_\varphi t\left[1 + \tfrac{r_0 - R}{R}(1 - 2\eta)\right] + \tfrac{E_\varphi - \tilde{E}_\varphi}{\tilde{E}_\varphi}\left(\tfrac{1}{2} - \eta\right)t + \varphi_0 + \vartheta \tag{3.64}$$

where ϑ represents the sum of all oscillating terms (it is negligible for $t \gg 1/\omega_r$), and η is given by

$$\eta = 1 + \frac{1}{(R_m/R)^2 - 2} \tag{3.65}$$

ω_r is the frequency of radial oscillations:

$$\omega_r = \sqrt{\tfrac{q}{m}k}\sqrt{\left(\tfrac{R_m}{R}\right)^2 - 2} = \omega\sqrt{\left(\tfrac{R_m}{R}\right)^2 - 2} \tag{3.66}$$

---

[158] Kolesnikov, S. V. and Monastyrskii, M. A. (1988) General theory of spatial and temporal aberrations in cathode lenses with weakly violated symmetry. *Sov. Phys. Technol. Phys.*, **33**, 1–11.

and ω_φ is the frequency of rotation:

$$\omega_\varphi = \frac{1}{R}\sqrt{\frac{2q}{m}\tilde{E}_\varphi} = \omega\sqrt{\frac{\left(\frac{R_m}{R}\right)^2 - 1}{2}} \qquad (3.67)$$

For $R < R_m$, the electric field attracts ions to the axis, and this condition is usually sufficient to trap ions with small E_φ: $E_\varphi \ll \tilde{E}_\varphi$. For ions with E_φ close to $\tilde{E}_\varphi$, however, $(E_\varphi - \tilde{E}_\varphi) \ll \tilde{E}$; eqn 3.66 indicates that only at radii $R < R_m/2^{1/2}$ attractive electric force starts to exceed repulsive centrifugal force caused by circular rotation, otherwise radial oscillations become unstable; for $E_\varphi > \tilde{E}_\varphi$, for ion stability, even lower R are required.

The *image current* is detected on split outer electrodes and amplified by a differential amplifier. The image current Q induced[159] on the electrodes by a charge q at a point (r, φ, z), and potential change $\Delta\Phi(r, \varphi, z)$ induced in the same point by a voltage ΔU applied to the same electrodes in the absence of any charge (because of axial symmetry of electrodes, potential $\Delta\Phi$ does not depend on angle φ) are related by

$$Q/q = -(\Delta\Phi(r, \varphi, z)/\Delta U) \qquad (3.68)$$

The differential image current on the detection electrodes then is

$$\frac{dQ}{dt} = -\frac{q}{\Delta U}\left[\frac{\partial\Delta\Phi(r, z)}{\partial r}\dot{r} + \frac{\partial\Delta\Phi(r, z)}{\partial z}\dot{z}\right] \qquad (3.69)$$

The total image current for a cloud of N ions excited to amplitude Δz and rotating around the central electrode at *average radius* ρ, is approximately

$$I(t) \approx -\frac{q}{\Delta U}N\frac{\partial\Delta\Phi(r, 0)}{\partial z}\dot{z} \approx -qN\omega\frac{\Delta z}{\lambda}\sin(\omega t) \qquad (3.70)$$

where λ is *effective gap* between detection electrodes, which, because of the complex geometry of the trap, depends on ρ. If the central electrode is also split and used for ion detection, this dependence may be made much weaker.

The sensitivity and signal-to-noise ratio in Orbitrap image current signal is similar to those of the FT-ICR cell. However, the ion frequency decrease with mass is much slower in Orbitrap. In this regard, the Orbitrap has some similiarity to a time-of-flight mass spectrometer (square root dependence). In Orbitrap, the mass resolution in FT-MS mode is twice lower than the frequency resolution: $M/\delta M = (1/2)(\omega/\delta\omega)$.

---

[159]Marto, J. A., Marshall, A. G., and Schweikhard, L. (1994) A two-electrode ion trap for Fourier transform ion cyclotron resonance mass spectrometry. *Int. J. Mass Spectrom. Ion Processes*, **137**, 9–30.

In the *mass selective instability* mode, an RF voltage is added to its regular DC voltage on the central electrode. On the right sides of eqns 3.51 and 3.53 this causes modulation with RF frequency. The z-motion is then governed by the canonical Mathieu equation. The equation for radial motion becomes even more complex and nonlinear. Introducing Mathieu parameters $a_u = 4(\omega^2/\Omega^2)$ and $q_u = 2\mu(\omega^2/\Omega^2)$ for the axial motion, where Ω is the angular frequency (in rad/s) of the RF field and μ is the ratio of the RF amplitude (0–peak) to the average voltage between the central and the outer electrode, the nonlinearity of the radial oscillations in the Orbitrap allows it to operate it in a unique region of the stability diagram — with a scan line cutting across several stable regions on the a_u-q_u plane. At low RF voltages (low q_u), for some ions, parametric resonance along r and the excitation of the radial oscillations may develop. But, due to their very strong nonlinearity, the radial oscillations appear to be self-quenching. For axial oscillations, parametric resonance develops when, for ions with specific m/q, $a_u \to 1$ (that is, $\Omega = 2\omega$, $q_u = \mu/2$). The axial coordinate of the ions then increases rapidly until they are ejected to the detector placed along the axis of the orbitrap. Here, unlike the 3D quadrupole ion trap, for any mass, just several volts of RF are sufficient.

In the following paragraphs we present a few studies on Orbitrap properties reported by the Orbitrap inventing group. Lange *et al.*[160] studied accelerating spectral acquisition rate in Orbitrap mass spectrometry (FT-MS requires relatively long detection times for high resolving powers). Using LTQ Orbitrap XL and Velos mass spectrometers, they compared several Orbitrap constructions: a standard commercial, a high-field, a compact, and an overtone design. The overtone design employed four detection electrodes for image current pick-up and allowed detection of third harmonics of axial oscillations. In parallel, to enable them use magnitude spectra instead of absorption spectra for obtaining mass spectra, they modified their processing software to provide broadband phase correction of ion oscillations in the acquired spectra. The compact design was found to provide 1.8-fold resolution enhancement (the increase in resolving power for the same duration of acquisition (~ 0.5 sec) and the same voltage on the central electrode); the overtone design provided an enhancement factor of 3 but with unacceptably high harmonics in spectra, compared to the standard design. In all the Orbitraps, space-charge dependent mass shifts — substantially m/z-independent — proportional to total ion signal — in the compact were relatively lower than in the standard or overtone design. Although *excitation by injection* in Orbitrap analyzers makes phase correction more straightforward than for FT-ICR, its routine use required better

---

[160]Lange, O., Makarov, A., Balschun, W., and Denisov, E. (2010) Accelerating spectral acquisition rate of Orbitrap mass spectrometry. *Proc. 58th Annual ASMS Conference on Mass Spectrometry and Allied Topics*, Salt Lake City, May 23–27.

synchronization of ion pulsing/detection as well as at start of detection soon after ion injection; resolution enhancement of up to a factor of 1.8–2.1 was confirmed. The resolution enhancement factor improvement was found up to a factor of 3.5 for the compact design leading to much higher resolving powers at short acquisition times.

Hauschild *et al.*[161] reported results of performance test on Orbitrap with an added quadrupole mass filter, enhanced Fourier transform (eFT) for Orbitrap data processing, predictive automatic gain control (pAGC) and parallel filling and detection, with possibility of multiple fills for spectrum multiplexing, stacked S-lens at the front end for higher transmission and the C-trap directly interfaced to HCD (both as in LTQ Orbitrap Velos instrument). Whereas in full MS, total C-trap charge capacity is shared between multiple signals of different intensity, thus S/N ratio becomes dependent on the ratio of compound of interest to other analytes — this is much less so in Selected Ion (Reaction) Monitoring SIM (SRM). In Orbitrap instruments SIM (SRM) could become MRM without any additional overhead. They discussed spectrum multiplexing; they also reported no ion loss in C-trap over a broad range of storage times.

Grinfeld *et al.*[162] used perturbation theory calculations and reported effects of space charge and tolerances in Orbitrap mass spectrometry, and classified the observed effects. Routine mass accuracy in low ppm and sub-ppm range over orders of magnitude of ion intensities become possible by tight control of tolerances and space charge effects within the analyzer. The purpose of the work was quantitatively to assess the role of the two latter causes and evolve corrective steps.

The perturbations of the electric field were calculated with finite-element method; as the field perturbations are relatively small, equations of motion were averaged over fast-oscillating phases. The total effect on the ion frequencies come from three sources: static perturbations owing to imperfections of trap construction, non-resonant space-charge interactions between ions with fairly separated frequencies, and resonant space-charge interactions between ion populations with close frequencies.

Results of the calculations show that static perturbations do not change oscillation amplitudes but induce dispersion of amplitude- and radius-dependent frequency shifts that tend to dephase the ion bunch after multiple oscillations. Small displacements and tilts of the central electrode with respect to the detection electrodes are not critical to the

[161]Hauschild, J., Fröhlich, U., Lange, O., Makaraov, A., Damoc, E., Kanngiesser, S., Crone, C., Xuan, Y., Kellman, M., and Wieghaus, A. (2011) Performance investigation of an Orbitrap mass analyzer with a quadrupole mass filter. *Proc. 59th Annual ASMS Conference on Mass Spectrometry and Allied Topics*, Denver, June 5–9.

[162]Grinfeld, D., Makarov, A., Denisov, E., Skoblin, M., and Monastyrskiy, M. (2012) "Crowd control" of ions in Orbitrap mass spectrometry. *Proc. 60th Annual ASMS Conference on Mass Spectrometry and Allied Topics*, Vancouver, May 20–24.

analyzer's performance. Another conclusion is: that it is mostly possible to compensate for important static perturbations by both electrical and mechanical means.

The frequency shift effect of non-resonant space-charge interactions is proportional to the total number of ions; this can be compensated by using automatic gain control. The resonant interaction causes coalescence of closely separated mass peaks, with the threshold number of ions decreasing for closer peaks. The coalescence effect cannot be eliminated, but its threshold can be significantly shifted toward larger numbers of ions by proper shaping of initial phase volume of injected ion packets. By tuning the analyzer and optimizing its geometry, coalescence and self-bunching effects can be brought under control.

Ion mobility spectrometry — Traveling Wave Ion Mobility Spectrometry (TWIMS)

Relationships for obtaining ion mass from drift-time measurements:
The mobility K $(= L/t_D E)$ is given by[163]

$$K = \frac{1}{\Omega} \frac{(18\pi)^{1/2}}{16} \frac{ze}{(k_b T)^{1/2}} \left[\frac{1}{m_I} + \frac{1}{m_B} \right]^{1/2} \frac{760}{P} \frac{T}{273.2} \frac{1}{N} \qquad (3.71)$$

where K is the mobility, Ω the cross section, E the electric field strength, t_D the drift time, and L the length of the drift region, z the ionic charge, e the elementary charge, m_I the ionic mass, m_B the mass of the neutral gas, P the pressure, T the temperature, and N the neutral number density. From eqn 3.71, we obtain

$$m_I = \frac{\left[\frac{1}{K}\right]^2 \left[\frac{(18\pi)^{1/2}}{16}\right]^2 \left[\frac{ze}{(k_b T)^{1/2}}\right]^2 \left[\frac{760}{P} \frac{T}{273.2} \frac{1}{N}\right]^2 m_B}{m_B \Omega^2 - \left[\frac{1}{K}\right]^2 \left[\frac{(18\pi)^{1/2}}{16}\right]^2 \left[\frac{ze}{(k_b T)^{1/2}}\right]^2 \left[\frac{760}{P} \frac{T}{273.2} \frac{1}{N}\right]^2} \qquad (3.72)$$

If Ω for two ions with masses m_{I1} and m_{I2} are about the same, then equating the Ωs, one can write

$$\frac{Et_{D1}}{L} \frac{(18\pi)^{1/2}}{16} \frac{ze}{(k_b T)^{1/2}} \left[\frac{1}{m_{I1}} + \frac{1}{m_B} \right]^{1/2} \frac{760}{P} \frac{T}{273.2} \frac{1}{N} = \qquad (3.73)$$

$$\frac{Et_{D2}}{L} \frac{(18\pi)^{1/2}}{16} \frac{ze}{(k_b T)^{1/2}} \left[\frac{1}{m_{I2}} + \frac{1}{m_B} \right]^{1/2} \frac{760}{P} \frac{T}{273.2} \frac{1}{N}$$

---

[163]Ruotolo, B. T., Benesch, J. L. P., Sandercock, A. M., Hyung, S. and Robinson, C. V. (2008) Ion mobility-mass spectrometry analysis of large protein complexes. *Nature Protocols*, **3**, 1139–1152.

With two ions having a mass difference of 1

$$m_{I2} = m_{I1} + 1 \qquad (3.74)$$

We then have two equations, two unknowns (m_{I1}, m_{I2}); solving, we obtain

$$\frac{t_{D2}^2}{t_{D1}^2} \approx \frac{\frac{1}{m_I} + \frac{1}{m_B}}{\frac{1}{m_I+1} + \frac{1}{m_B}} \qquad (3.75)$$

From which m_I can be obtained as

$$
\begin{aligned}
m_I &\approx \frac{-m_B t_{D1}^2 - t_{D1}^2 + m_B t_{D2}^2 + t_{D2}^2}{2(t_{D1}^2 - t_{D2}^2)} \\
&\pm \frac{\sqrt{(m_B t_{D1}^2 + t_{D1}^2 - m_B t_{D2}^2 - t_{D2}^2)^2 - 4 m_B t_{D1}^2 (t_{D1}^2 - t_{D2}^2)}}{2(t_{D1}^2 - t_{D2}^2)}
\end{aligned}
$$
$$(3.76)$$

The basic mechanical structure of the traveling wave ion mobility mass spectrometer (TWIMS)[164] is a stacked ring ion guide (SRIG) which consists of a series of ring electrodes on which opposite phases of RF voltage are applied to adjacent rings. The effective potential V* in the SRIG is given by

$$\mathrm{V}^*(r,z) = V_{\max} \frac{[I_1^2(r/\delta)\cos^2(z/\delta) + I_0^2(r/\delta)\sin^2(z/\delta)]}{I_0^2(\rho/\delta)} \qquad (3.77)$$

with $V_{\max}$ the maximum effective potential at $r = \rho$, $z = d(i + \frac{1}{2})$, $i = 0, 1, \ldots$

$$V_{\max} = \frac{q V_{\mathrm{RF}}^2}{4 m \omega^2 \delta^2} \qquad (3.78)$$

where I_0 and I_1 are, respectively, the zero and first-order Bessel functions, r is the radial coordinate and z is the axial coordinate, d is the center-to-center spacing of the electrodes, $\delta = d/\pi$, ρ is the radius of the aperture, V_{RF} is the base-to-peak RF voltage with RF angular frequency ω, and q and m in eqn 3.78 are, respectively, the charge and mass of the ion. An expanded view of the SRIG effective potential (for aperture radius $\rho = 2.5$ mm, ring spacing $d = 1.5$ mm, $V_{\mathrm{RF}} = 100$ V at 1 MHz frequency for an ion mass of $m = 500$ Da and charge $q = +1\,e$) in Fig. 3.14 (left bottom) shows depressions — "traps" — formed along the z-axis of the device; the depth of these traps determine the extent to which the ions would be retained there. The potential variation from the axis to the periphery shows that

[164]Giles, K., Pringle, S. D., Worthington, K. R., Little, D., Wildgoose, J. L., and Bateman, R. H. (2004) Applications of a travelling wave-based radio-frequency-only stacked ring ion guide. *Rapid Commun. Mass Spectrom.*, **18**, 2401–2414.

the potential gradient is shallow for most of the diameter; it rises steeply near the walls.

Ions are propelled through the SRIG by superposing a DC potential on the RF of an electrode and then, after a given time, the DC potential is switched to an adjacent electrode — this provides a moving electric field — so as to guide the ion along that direction. On the moving electric field — *the traveling wave* — ions can move, "surf," toward the exit end of the drift region. Figure 3.14 (right) shows the results of ion movement simulation by using SIMION 3D (version 7.0) ion optics package. This particular simulation used electrodes of 9.5 mm thickness, and a drag model for collisions was used which, for enabling clear visualization of ion axial motion due to the traveling wave, did not include scattering effects. It will be noticed that the ion motion amounts to a series of impulses; the ion rolls down the potential gradient to the low field region, the pulse moves forward, and the process is repeated. The pulse height, the pulse velocity, and the gas pressure are the main factors that determine the ion transport; by varying these parameters different modes of operation can be achieved. A photograph of the traveling wave ion guide (TWIG) is shown in Fig. 3.14 (left). This TWIG cell with 122 electrodes is supported between two printed-circuit boards. The smaller picture shows the sealed unit with the 2 mm diameter ion entrance aperture.

Beyond what has been stated above, the theory of TWIMS has been developed using both derivations and simulations.[165] Some major observations have been expressed in terms of the key parameter, the ratio (c) of ion drift velocity at the steepest wave slope to wave speed. At low c, the ion transit velocity is proportional to the *squares* of mobility (K) and electric field intensity (E), as opposed to *linear* scaling in drift tube (DT) IMS and differential mobility analyzers. The scaling deviates from quadratic with increase of c; depending on the waveform profile, first steeper, then more gradual for realistic profiles with variable E, turning to more gradual with the ideal triangular profile. The transit velocity asymptotically approaches the wave speed at highest c. The resolving power in TWIMS, unlike with DT IMS, depends on mobility; it scales as $K^{1/2}$ in the low-c limit and less at higher c. The major predicted trends were found to be in agreement with TWIMS measurements for peptide ions.

Results of a new hybrid quadrupole/traveling wave ion mobility separator/orthogonal acceleration time-of-flight instrument[166] and comparison of the mobility data (mobility separation of lysozyme and cytochrome c,

[165]Shvartsburg, A. A. and Smith, R. D. (2008) Fundamentals of traveling wave ion mobility spectrometry. *Anal. Chem.*, **80**, 9689–9699.

[166]Pringle, S. D., Giles, K., Wildgoose, J. L., Williams, J. P., Slade, S. E., Thalassinos, K., Bateman, R. H., Bowers, M. T., and Scrivens, J. H. (2007) An investigation of the mobility separation of some peptide and protein ions using a new hybrid quadrupole/travelling wave IMS/oa-ToF instrument. *Int. J. Mass Spectrom.*, **261**, 1–12.

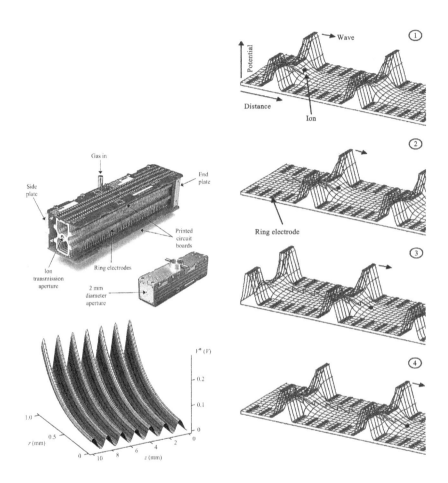

Figure 3.14: Traveling wave propelling ions in stacked ring ion guide (SRIG) system. Left, bottom: The SRIG effective potential illustrating the "traps" along the z-axis of the device. Left, top: The TWIG cell with 122 electrodes. Right: Series of SIMION plots showing an ion surfing on the front of the traveling wave as it passes along the SRIG. Parameters used: wave velocity = 300 m/s, pulse height 2 V, gas pressure = 0.03 mbar (argon), and ion/neutral interaction cross section = 500 Å^2. Reproduced by permission of John Wiley & Sons, Ltd. from *Rapid Commun. Mass Spectrom.*, Vol. 18, 2401–2414, 2004.

lysozyme and α-lactalbumin) showed that the instrument provides mobility separation without compromising the base sensitivity (at analytically significant levels) of the mass spectrometer.

Test results on the second-generation TW IM separator, which show up to a four-fold increase in mobility resolution have been reported.[167] This improvement has been demonstrated using two reverse peptides (mw 490 Da), small ruthenium-containing anticancer drugs (mw 427 Da), a cisplatin-modified protein (mw 8776 Da), and the noncovalent tetradecameric chaperone complex GroEL (mw 802 kDa). Collision cross-sections determined using this mobility separator have been found to correlate well with earlier measurements using the previous-generation separator and theoretically derived values.

Assessments have also been made of the performance characteristics of a second-generation traveling-wave ion mobility separator, focussing on those figures of merit that lead to making measurements of collision cross-section having both high precision and high accuracy.[168] Through a comprehensive survey of instrument parameters and settings, instrument conditions for optimized drift time resolution, cross-section resolution, and cross-section accuracy for a range of peptide, protein, and multi-protein complex ions have been determined. The conditions for high-accuracy IM results are significantly different from those optimized for separation resolution; a balance between these two metrics must be attained to yield optimized instrument conditions for traveling wave IM separation of biomolecules.

3.1.2 Theoretical ion mobility cross-sections using PA, PSA

There have been attempts to theoretically derive ion mobility cross-sections. Bowers *et al.*,[169] via a Monte Carlo technique (*projection approximation*, PA: in this approximation a polyatomic ion is taken as a collection of spheres; the orientation-averaged area of the union of these spheres projected on a plane perpendicular to the axis of orientation leads to a measure of the collision integral), attempted to obtain theoretical cross-sections and from that theoretical mobilities of carbon clusters. Their later *projec-*

---

[167]Giles, K., Williams, J. P., and Campuzano, I. (2011) Enhancements in travelling wave ion mobility resolution. *Rapid Commun. Mass Spectrom.*, **25**, 1559–1566.

[168]Zhong, Y. Y., Hyung, S. J., and Ruotolo, B. T. (2011) Characterizing the resolution and accuracy of a second-generation traveling-wave ion mobility separator for biomolecular ins. *Analyst*, **136**, 3534–3541.

[169]von Helden, G., Hsu, M. T., Gotts, N., and Bowers, M. T. (1993) Carbon cluster cations with up to 84 atoms: structures, formation mechanism, and reactivity. *J. Phys. Chem.*, **97**, 8182–8192.

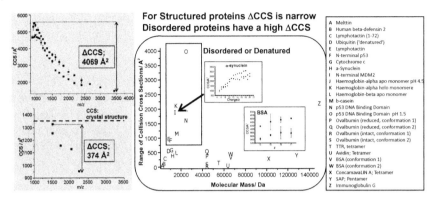

Figure 3.15: Range of collision cross-sections (CCS) vs. Molecular mass of some proteins. Left (top): The rotationally averaged CCSs of each charge state of β-casein; left (bottom): the rotationally averaged CCSs of each charge state of Cytochrome C. Courtesy: Perdita Barran, Manchester Institute of Biotechnology, University of Manchester. Data published in *Anal. Chem.*, Vol. 86, 10979–10991, 2014. Permission from the American Chemical Society taken for reproduction.

tion superposition approximation PSA[170],[171] is a development over PA. There is also the trajectory method TJM[172] for calculation of collision cross-section in which deflection angle, momentum transfer over many *trajectories* are averaged using the exact hard-sphere scattering EHSS[173] or the 12-6-4 (Lennard-Jones and ion-induced dipole) potential. TJM, PSA both are satisfactory for large systems (PA satisfactory for small systems); however, PSA, in comparison to TJM, takes much *less* computational time. Results of study of molecular dynamics using combination of ion mobility spectrometry and mass spectrometry have been reviewed.[174] Here we present some results obtained using the PSA.

[170]Bleiholder, C., Wyttenbach, T., and Bowers, M. T. (2011) A novel projection approximation algorithm for the fast and accurate computation of molecular cross sections (I). Method. *Int. J. Mass Spectrom.*, **308**, 1–10.

[171]Wyttenbach, T., Bleiholder, C., and Bowers, M. T. (2013) Factors contributing to the collision cross section of polyatomic ions in the kilodalton to gigadalton range: application to ion mobility measurements. *Anal. Chem.*, **85**, 2191–2199.

[172]Mesleh, M. F., Hunter, J. M., Shvartsburg, A. A., Schatz, G. C., and Jarrold, M. F. (1996) Structural information from ion mobility measurements: effect of long-range potential. *J. Phys. Chem.*, **100**, 16082–16086.

[173]Shvartsburg, A. A., and Jarrold, M. F. (1996) An exact hard-sphere scattering method. *Chem. Phys. Lett.*, **261**, 86–91.

[174]Wyttenbach, T., Pierson, N. A., Clemmer, D. E., and Bowers, M. T. (2014) Ion mobility analysis of molecular dynamics. *Annu. Rev. Phys. Chem.*, **65**, 175–196.

Collision cross-sections [Å²]

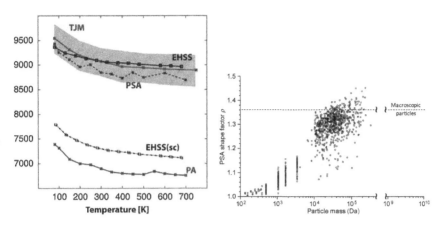

Figure 3.16: Left: Temperature-dependent cross sections computed by the various methods. The PSA and EHSS models agree well with the TJM data; the PA method underestimates the TJM data by approximately 30%. (EHSS(sc) model: the effect of multiple collisions on the EHSS collision cross section removed.) Reproduced by permission of Elsevier B.V. from *International Journal of Mass Spectrometry*, Vol. 308, 1–10, 2011. Right: PSA shape factor ρ (300 K, helium) for a range of molecular structures. Data below 500 Da are based on model structures for protonated and sodiated oligoglycines Gly_2–Gly_6. Data in the 500-1400 Da range include calculations for six peptides with dozens of model structures for each peptide giving rise to the spread in the y-direction at a given mass (some peptides overlap with each other in mass). The remaining data capture ~ 400 protein and protein complex structures downloaded from the PDB data bank. The limit for spherical macroscopic particles of 1.36 is noted. Reproduced by permission of the American Chemical Society from *Analytical Chemistry*, Vol. 85, 2191–2199, 2013.

The relationship between the reduced ion mobility K_0 (normalized to 760 Torr and 273.15 K) to molecular momentum transfer integral Ω is given by

$$K_0 = \frac{3e}{16N_0} \sqrt{\frac{2\pi}{\mu kT}} \frac{1}{\Omega} \tag{3.79}$$

where e is the charge on the analyte, N_0 the number density at 760 Torr and 273.15 K, and T the temperature of the buffer gas. Ω, which contains information of the ion geometry, is difficult to obtain rigorously; it depends on highly non-local and correlated van der Waals interaction between the buffer gas and the analyte. In the PSA model, the overall concept of the PA was retained (that is, relatively simple and high computational efficiency) while approximating the molecular momentum transfer integral $\Omega^{(1,1)}(T)$ as ($\Omega_{\text{PSA}}^{(1,1)}(T)$) an orientation-averaged projection area $\langle\Omega_{\text{PSA}}\rangle$

(which accounts for the *size effect*) multiplied by a *shape factor* ρ (which corrects for any deviation of the molecular shape from full convexity):

$$\Omega_{\text{CALC}} \equiv \Omega^{(1,1)}(T) \equiv \Omega_{\text{PSA}}^{(1,1)}(T) = \rho(T)\langle\Omega_{\text{PSA}}\rangle(T) \quad \langle\Omega_{\text{PSA}}\rangle = f(T, \Sigma P_j)$$

$$(3.80)$$

Whereas the computation of ρ is complicated, the basic concept is simple: it is essentially the ratio of the actual molecular surface area A_{mol} of a molecule to the surface area A_{ce} of the convex envelope of the molecule (for smooth convex surface $\rho = 1$, for objects with concave features ρ is >1):

$$\rho = \frac{A_{\text{mol}}}{A_{\text{ce}}} = \frac{\text{Actual molecular surface area}}{\text{Surface area of the convex envelope of the molecule}} \quad (3.81)$$

They used the PSA method to theoretically determine contribution of factors to the cross-section Ω_{IMS} as a function of ion size — from diglycine to a $1\,\mu$m oil droplet. In Fig. 3.16, temperature-dependent cross-sections computed by various methods (such as the TJM, EHSS, PA, PSA) as well as PSA-determined shape factor ρ-values (scatter plot) for a wide range of molecular structures, are given. Small molecules in the 100–1000 mass range tend to have $\rho \approx 1$, that is a near-convex surface. ρ values increase with increasing molecule size, then tend to settle, for most molecules above 10 kDa, in the 1.15–1.45 region (that is, concaveness $\sim 30\%$ on average).

Despite the trends, the shape contribution to Ω cannot be predicted dependably based on molecular mass in the 10,000 to 150,000 Da range (true probably for higher masses as well). A protein with numerous dents and crevasses on the surface leads to a larger shape factor, owing to higher resisting forces.

Bowers *et al.* report that for ~ 400 different molecules, thousands of PSA calculations carried out in the temperature range from 80 to 700 K showed that even for small peptides, a major component of Ω_{IMS} is the molecular framework made up of atomic hard spheres. For very small peptides, the ion-buffer gas interaction is almost equally important; and, whereas its relative significance decreases with increasing ion size, it is still a factor for megadalton ions, and it only tends to become negligible for ions in the gigadalton range.

Frictional resisting force is minimal for fully convex ions drifting in a buffer gas; with ion surface concaveness, the resisting force increases; for proteins it results in increase in the ion mobility cross-section from 0 to greater than 40% — with the highest degree of concaveness (which is reflected in PSA ρ values in the 1.15–1.45 range) the effect of shape-effected friction can be as large or even larger than that of macroscopic particles such as oil droplets. Above 1 MDa, it is expected that ρ will eventually begin to converge on the macroscopic limit of 1.36 for protein-based assemblies.

That *macroscopic* spherical assemblies (liquid oil droplets and solid glass spheres (with radius $a \approx 1$ μm)) experience a factor of 1.36 larger resisting force than expected for hard sphere is interesting. It is predominantly because of diffuse reflections of the buffer gas on sphere surface that the momentum transfer cross section (Ω_{IMS}) is larger than the size of the sphere ($\Omega \approx 1.36\pi a^2$ (large spheres)). Since for the case of a hard sphere, both the quantities πa^2 and $\langle \Omega_{PSA} \rangle$ correspond to the sphere projection cross-section, a comparison of eqn 3.80 and $\Omega \approx 1.36\pi a^2$ (large sphere) suggests an expected value of $\rho \approx 1.36$ for macroscopic particles. The larger ρ values seen for molecules in the 10 kDa to 1 MDa range in Fig. 3.16 are of the same order as the values 1.39 and 1.36 given in $\Omega_{IMS} = 1.39\sigma$ and $\Omega \approx 1.36\pi a^2$ (even though there is a spread of $\pm 20\%$ in the values). This is taken as an indication that the PSA algorithm in essence captures the physics of the collision process correctly, and that (diffuse) scattering from a protein surface begins at some point above the megadalton range. Applicability of this approximation, for systems including such as the 800 kDa GroEL complex, has recently been suggested in the literature. In a recent GroEL study, applying this approximation on diffuse scattering contribution, IMS data (20,700 Å^2) has been scaled down by a factor of 1.36. Shown experimentally, this approximation holds only in the gigadalton range where both Ω and σ can be measured unambiguously in independent experiments (such as Ω by IMS; σ by microscopy).

Since the ion-buffer gas interaction remains still a factor for megadalton ions, their "size" is still a poorly defined quantity — especially for buffer gases other than helium (nitrogen interaction effect appears to be four times that of helium, larger effect for larger buffer gas). This suggests that no attempts be made to correct experimental Ω_{IMS} values — using *any* value of shape factor — to arrive at the "true" size of the ion.

Their results indicate that the transition from nanoparticle (with Lennard-Jones-like interaction with the buffer gas) to macroscopic particle (with hard sphere-like interaction) occurs at ~ 1 GDa (10^6Å^2) corresponding roughly to a $0.1\,\mu$m diameter oil droplet.

In an update Bowers *et al.*[175] have reported that their (theoretical/experimental) data show that no simple relationship exists between the cross-sections in He and N$_2$. Molecular shape, drift gas temperature etc. affect cross-sections differently in He and N$_2$. They discuss the impact of their results on current protocols that are in use for elucidating molecular structure by IM/MS.

---

[175]Bleiholder, C., Wyttenbach, T., and Bowers, M. T. (2014) Collision cross sections: the effect of size, shape, and bath gas. *Proc. 62nd Annual ASMS Conference on Mass Spectrometry and Allied Topics*, Baltimore, June 15–19.

3.1.3   MS/MS instrumental supplements: CAD/CID, ECD, ETD, EDD, HCD, IRMPD

A few instrumental supplements to the basic mass spectrometer are briefly described here.

Ion funnel

The ion funnel — a device for focussing of ions emerging from the ion source — was originally implemented by Smith *et al.*[176] to efficiently capture ions, losslessly, from an ESI source. The basic ion funnel (Fig. 3.17, top), serving to radially confine ions, consists of a series of closely spaced ring electrodes whose inner diameters gradually decrease. To adjacent electrodes out-of-phase RF potentials are applied besides a DC gradient along the axis of the funnel to drive the ions through the device. Whereas the gain depends upon details of the platform being compared with and the ion funnel implementation being used, a straightforward use of a single ion funnel will generally result in 2–10 fold gain, while (concatenated) dual funnel designs (a couple of examples will be seen in the next section presenting some commercial instruments) with multi-inlets incorporating increased pumping etc. can have gains as much as — or even more than — hundred-fold.

The ion funnel has also been adapted to trap, store, and release ions from continuous beam ESI sources along with pulsed IMS operation. Such ion funnel traps (IFT) are of broad applicability. The IFT design of Ibrahim *et al.*[177] — hourglass ion funnel structure which included a final section that refocussed the ions released from the trap to a conductance limiting orifice — improved by Clowers *et al.*,[178] is shown in Fig. 3.17 (bottom). This latter design has an additional second grid at the exit of the trapping region providing more uniform radial potential distribution and thus allowing ions to accumulate closer to the exit grids. The ion funnel technique has been reviewed.[179]

---

[176] Shaffer, S. A., Tang, K., Anderson, G. A., Prior, D. C., Udseth, H.R., and Smith, R. D. (1997) A novel ion funnel for focussing ions at elevated pressure using electrospray ionization mass spectrometry. *Rapid Commun. Mass Spectrom.*, **11**, 1813–1817.

[177] Ibrahim, Y., Belov, M. E., Tolmachev, A. V., Prior, D. C., and Smith, R. D. (2007) Ion funnel trap interface for orthogonal time-of-flight mass spectrometry. *Anal. Chem.*, **79**, 7845–7852.

[178] Clowers, B. H., Ibrahim, Y. M., Prior, D. C., Danielson, W. F., Belov, M. E., and Smith, R. D. (2008) Enhanced ion utilization efficiency using an electrodynamic ion funnel trap as an injection mechanism for ion mobility spectrometry. *Anal. Chem.*, **80**, 612–623.

[179] Kelly, R. T., Tolmachev, A. V., Page, J. S., Tang, K., and Smith, R. D. (2010) The ion funnel: theory, implementations, and applications. *Mass Spectrom. Rev.*, **29**, 294–312.

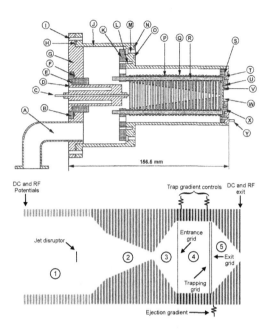

Figure 3.17: Top: The prototype ion funnel interface: (C) stainless steel capillary tubing, (P) ceramic insulating washer, (Q) nickel-coated brass ion funnel electrode, (R) ceramic rod, (W) nickel-coated brass final orifice electrode (conductance lmit). Reproduced by permission of the American Chemical Society, from *Analytical Chemistry*, Vol. 70, 4111–4119, 1998. Bottom: Schematic of the ion funnel trap consisting of five distinct regions that: (1) accept the rapidly expanding free jet of ions, (2) focus ions entering the chamber that houses the next three regions that (3) guide ions into (4) the ion trapping region and (5) the final converging region and funnel exit. Regions 3, 4, and 5 are separated by a series of DC-only grids. Reproduced by permission of the American Chemical Society from *Analytical Chemistry*, Vol. 80, 612–623, 2008.

C-trap

Figure 3.18 (left) shows a schematic of the C-trap, as used with Orbitrap; Fig. 3.18 (right) shows some steps in ETD operation, discussed later in this Section.

Collisionally Activated Dissociation, Collision-Induced Dissociation (CAD/CID)

In *high*-energy CAD/CID, the energy in the laboratory frame E_{lab} is quite high, in keV range. In terms of reactivity, however, it is the relative energy

Protein	MW	$E_{lab} = 1$ keV	5 keV	20 keV
Substance P	1,348	$E_{rel} = 2.96$ eV	14.79 eV	59.17 eV
Ubiquitin	8,565	0.47 eV	2.33 eV	9.34 eV
Cytochrome C (human)	12,233	0.33 eV	1.63 eV	6.54 eV
Ovalbumin	45,000	0.09 eV	0.44 eV	1.78 eV
Bovine serum albumin	66,433	0.06 eV	0.30 eV	1.20 eV

E_{rel} of the collision that matters; that depends upon the molecular and target masses, and can be calculated using the following equation:

$$E_{rel} = \frac{m_n}{m_n + m_M} E_{lab} \tag{3.82}$$

The table above gives E_{rel} for a few E_{lab} values using $m_n = 4$ for He. The fragmentation generally takes place as a result of a single collision.

In *low*-energy CAD/CID, the kinetic energy is 20–100 eV in the laboratory frame. There are two basic alternatives for carrying out CID/CAD: one (CC-CAD) in which there is a collision chamber (CC) — in this, effective energetic collisions are few (1–5) with target gas nitrogen or argon, here no effort is made to produce many collisions. The MS/MS product ions are extracted and then analyzed by any of the common analyzers such as: 2D quadrupole, time-of-flight, ICR, or Orbitrap. The other alternative (IT-CAD) is to confine the precursor ions in an ion trap (IT) such as the quadrupole ion trap — the trapping permits many (100s) weak collisions, then the fragments could be analyzed by any of the usual analyzers. CAD mechanism has been discussed by Mayer *et al.*[180]

Ultraviolet, Infrared Multiphoton Dissociation (IRMPD)

Figure 3.19 (left) shows an *ultraviolet* radiation arrangement of Zubarev *et al.*[181] with an ion trap. Figure 3.19 (right) shows an ion trap arrangement for *infrared* multiphoton dissociation.[182] Mass-selected ions are injected into the 3D trap; at less than 10^{-5} mbar the ion in the trap is exposed to infrared radiation from a parametric oscillator; the precursor and the

[180]Mayer, P. M. and Poon, C. (2009) The mechanisms of collisional activation of ions in mass spectrometry. *Mass Spectrom. Rev.*, **28**, 608–639.

[181]Zubarev, R., Kjeldsen, F., Ivonin, I., and Silivra, O. — US Patent 6,800,851 B1 October 5, 2004.

[182]Gulyuz, K., Stedwell, C. N., Wang, D., and Polfer, N. C. (2011) Hybrid quadrupole mass filter/quadrupole ion trap/time-of-flight-mass spectrometer for infrared multiple photon dissociation spectroscopy of mass-selected ions. *Rev. Sci. Instrum.*, **82**, 054101.

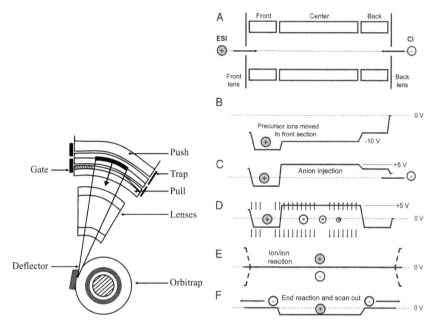

Figure 3.18: Left: C-Trap to push ions into Orbitrap. Courtesy: Thermo Fisher Scientific. Reproduced by permission. Right: Schematic of steps involved in the operation of an LTQ mass spectrometer used for peptide sequence analysis by ETD. See text for details. Reproduced by permission of the National Academy of Science, U.S.A. from *Proc. Nat. Acad. Sci., U.S.A.*, Vol. 101, 9528–9533, 2004.

photofragmented ions are analyzed by a time-of-flight mass spectrometer IRMPD.[183,184,185]

Electron Transfer Dissociation (ETD)

A schematic of an ETD instrument along with operational steps is shown in Fig. 3.18 (right).[186] In a commercial (Thermo Electron) LTQ mass spectrometer a Finnigan 4500 CI source was placed at the rear of the instrument (A). In the first step (B), a DC offset moves precursor ions (from the ESI

---

[183]Brodbelt, J. S. and Wilson, J. J. (2009) Infrared multiphoton dissociation in quadrupole ion traps. *Mass Spectrom. Rev.*, **28**, 390–424.

[184]Eyler, J. R. (2009) Infrared multiple photon dissociation spectroscopy of ions in Penning traps. *Mass Spectrom. Rev.* **28**, 448–467.

[185]Polfer, N. C. (2011) Infrared multiple photon dissociation spectroscopy of trapped ions. *Chem. Soc. Rev.*, **40**, 2211–2221.

[186]Syka, J. E. P., Coon, J. J., Schroeder, M. J., Shabanowitz, J., and Hunt, D. F. (2004) Peptide and protein sequence analysis by electron transfer dissociation mass spectrometry. *Proc. Nat. Acad. Sci., U.S.A.*, **101**, 9528–9533.

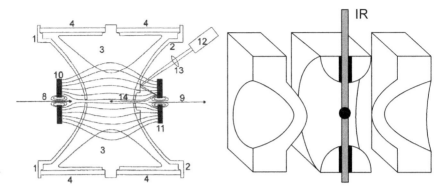

Figure 3.19: Left: 3D quadrupole ion trap with a nanosecond *ultraviolet* pulsed laser (12) for electron generation and two washer-shaped magnets (10,11) for the guidance of electrons along path (14) into the ion trap (Zubarev *et al.*, US Patent 6,800,851 B1 October 5, 2004.). Right: IRMPD using ion trap. *Infrared* radiation is from a parametric oscillator. Reproduced by permission of the Royal Society of Chemistry from *Chemical Society Reviews*, Vol. 40, 2211–2221, 2011.

source) to the front section of the linear trap. In the next step (C), negatively charged reagent ions from the CI sources are injected into the center section. A supplementary broadband ac field is then applied which ejects all ions except those within three mass-unit windows centered around the (positively charged) precursor ions and the (negatively charged) electron-donor reagent ions (D). Positive and negative ion populations are then allowed to mix and react by removing the DC potential well and application of a secondary RF voltage (100 V V_{0-p}, 600 kHz) to the end lens plates of the linear trap (E). The ion/ion reaction is terminated by creating a potential well for the positive ions at the center of the trap and axial ejection of the negative ions (F). This is followed by mass-selective radial ejection of positively charged fragment ions and recording of MS/MS spectrum.

3.2   Some examples of instruments

3.2.1   Agilent 6560 IM QTOF

A schematic of the Agilent 6560 IM QTOF mass spectrometer which combines the strengths of ion mobility separation, quadrupole mass filter selection of precursor ions for MS/MS, and time of flight analysis of the MS/MS products is shown in Fig. 3.20. The chief attributes of the instrument are: system sensitivity optimization using electrodynamic ion funnels to focus and transmit ions; ion mobility resolution optimization while maintaining QTOF performance (mass resolution and accuracy); ion fragmentation

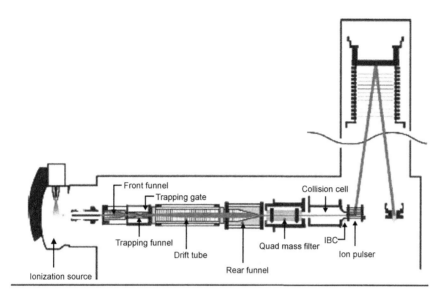

Figure 3.20: Agilent 6560 IM QTOF mass spectrometer. Courtesy: Agilent Technologies. Reproduced by permission.

selection using standard QTOF collision cell (CID); and increase of bandwidth of QTOF data acquisition and processing channel by tenfold to match the ion mobility data rates.[187]

As shown in Fig. 3.20, the ion mobility system consists of a front funnel, trapping funnel, trapping gate, drift tube, and a rear funnel that couples through a hexapole to the Q-TOF mass analyzer. Ions generated in the source region are carried into the front ion funnel through a single bore capillary. The front ion funnel improves the sensitivity by efficiently transferring gas phase ions into the trapping funnel while pumping away excess gas and neutral molecules. The front funnel operates at high pressure where funnel DC and RF voltages propel ions toward the trapping funnel. The key function of the front ion funnel is to enrich the sample ions and remove excess gas.

The trapping funnel accumulates and releases ions into the drift tube. The drift cell is ∼ 80 cm long and generally operated at 20 V/cm drift field. Ions exiting the drift tube enter the rear ion funnel that efficiently refocuses and transfers ions to the mass analyzer. The continuous ion beam from the electrospray process has to be converted into a pulsed ion beam prior to ion mobility separation. The trapping funnel operates by first storing and then releasing discrete packets of ions into the drift cell.

---

[187]Agilent 6560 IM QTOF mass spectrometer brochure.

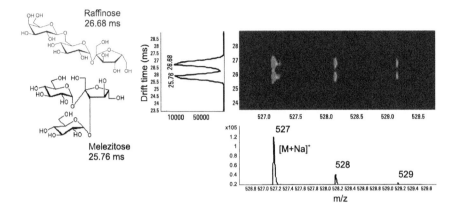

Figure 3.21: Separation of structural sugar isomers $C_{18}H_{32}O_{16}$. The folded structure for melezitose has a higher mobility, and shorter drift time, than the elongated structure for raffinose. Courtesy: Agilent Technologies. Reproduced by permission.

Ions are separated as they pass through the ion mobility cell based on their size and charge. The drift cell is operated under low field limit conditions allowing the instrument to generate accurate structural information for compounds. Under the low electric field conditions the mobility is not dependent on the electric field but rather on the structure of the molecule and its interaction with the drift gas. Ions exiting the drift cell are refocussed by the rear ion funnel before entering the hexapole ion guide.

The separation of two structural sugar isomers $C_{18}H_{32}O_{16}$ using this instrument is shown in Fig. 3.21. For separation of isobaric tri-saccharides using the IM-QTOF, an 1:1 mixture of melezitose and raffinose was infused using a syringe pump. These two carbohydrates can be baseline separated using the ion mobility drift cell and detected using the Q-TOF mass analyzer as sodium adducts. The ion mobility resolving power for this separation — based on FWHM definition — was 60.

3.2.2   Bruker Daltonik — Impact II, maXis II™

A schematic of Bruker Daltonik Impact II mass spectrometer is shown Fig. 3.22. Ultra-high resolution QTOF technology is used for superior hardware performance: (i) enhanced dynamic range, (ii) 50-Gbit/s data sampling technology, (iii) 10-bit ADC technology (high dynamic range detection system), (iv) new TOF with improved resolving power. It uses patented dual ion funnel; new broad mass transfer quadrupole CID cell with DC gradient results in in-spectrum dynamic range (1.7×10^5) at real LC speed.

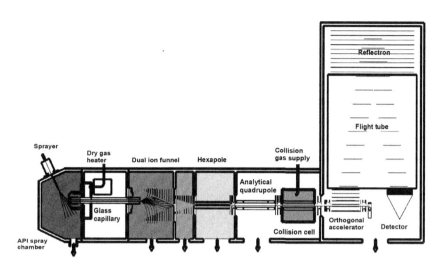

Figure 3.22: Schematic of Bruker Daltonik Impact II mass spectrometer. Courtesy: Bruker Daltonics, Inc. Reproduced by permission.

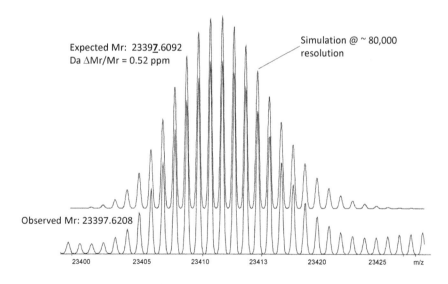

Figure 3.23: Deconvoluted Adalimumab light chain 23 kDa data, an application to true isotope patterns to proteins. Spectrum obtained with Bruker maXis II^{TM}. High resolution 85,000 FWHM. Courtesy: Bruker Daltonics, Inc. Reproduced by permission.

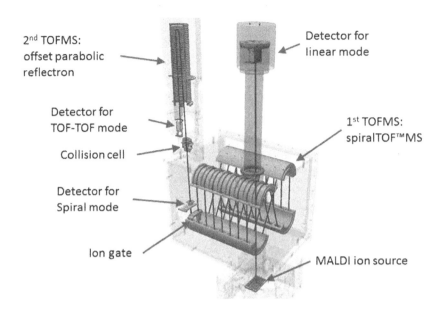

Figure 3.24: Ion optical system of JMS-S3000 Spiral TOF-TOF. Courtesy: Jeol Ltd. Reproduced by permission.

A spectrum collected from a maXis II^{TM}/maXis II^{TM} ETD, high resolution (85,000 resolution FWHM) QTOF mass spectrometer is shown in Fig. 3.23. Specifications of maXis are: full sensitivity resolution (FSR) of > 80,000; robust ETD capabilities with novel nCI source and gas enrichment apparatus (optional), native mass spectrometry (optional) with enhanced desolvation of super molecular complexes, high definition 5 orders of magnitude dynamic range, DC gradient collision cell for high sensitivity, and fast MS/MS.

3.2.3   JEOL — JMS-S3000 SpiralTOF

JEOL's JMS-S3000 SpiralTOF is available in four configurations: (i) JMS-S3000, MALDI-TOFMS with high mass resolution, high mass accuracy, and high throughput, (ii) JMS-S3000 with TOF-TOF option (Fig. 3.24, left side), in which precursor ions can be selected and they undergo high-energy CID; product ions thus generated are analyzed with an offset parabolic reflectron TOFMS — spectra yielding detailed structural information of analyte molecules, (iii) JMS-S3000 with Linear TOF option (Fig. 3.24, right side) in which voltages are not applied to electric sectors, ions travel linearly from the ion source to the detector; molecular ions of high molecular weight compounds such as proteins and synthetic polymers can be analyzed, complementary needs of a wide range of compounds are addressed, and (iv) JMS-S3000 with TOF-TOF plus Linear TOF option, in which both the linear TOF and TOF-TOF options are present.

In the TOF-TOF option, ion groups accelerated to 20 kV in the ion source travel through 15 m (this distance is more than one order of magnitude longer than that in conventional MALDI TOF-TOF instruments); in this mode electric sectors of ion optics remove ions from post-source decay (PSD) which permits monoisotopic selection of precursor ions. The latter, without deceleration, enter the collision cell, produce product ions through high-energy collisions — high ion transmission achieved by placing two sets of deflectors before and after the collision cell. The second TOFMS effects re-acceleration and then MS analysis using an offset parabolic reflectron — resulting in seamless observation of product ions, the latter monoisotopic originating from monoisotopic precursor ions, ranging from low m/z up to that of the precursor ion. Absence of effects of PSD ions enables clear observation of fragmentation pathways characteristic of high-energy CID. With monoisotopic precursor ions two product ions from two fragmentation pathways can be discerned relatively easily.

Specifications: *Standard configuration*: Mass resolving power^a: $> 60,000$ (FWHM), Resolving power at wide mass range^b: m/z 1046.5 : $> 30,000$, m/z 2093.1 : $> 60,000$, m/z 2465.2 : $> 50,000$, Sensitivity^c: < 500 amol, Mass accuracy (internal calibration): < 1 ppm (average error), Mass accuracy (external calibration): < 10 ppm (average error), Mass range: m/z 10–30,000. For *TOF-TOF*: mass resolving power: $> 2,000$ (FWHM), Sensitivity^c: < 5 fmol, Mass range: ± 0.1 Da, Precursor ion selectivity^d: $> 2,500$, Precursor ion selectivity range: m/z 100–4,000. For *Linear TOF*: Mass resolving power^e: $> 2,000$ (FWHM), Sensitivity^c: < 500 amol, Mass accuracy (internal calibration): ± 50 ppm (peptide region), Mass range: m/z 10–200,000. ^{a,b,c,d,e} stand for: ^{a}: Positive ion detection mode: ACTH fragment 1–17 [M+H]$^+$: m/z 2093.1 mass resolving power; ^{b}: Mass resolving power of AngiotensinII, ACTH fragment 1–17, ACTH fragment 18–39 [M+H]$^+$; ^{c}: AngiotensinII [M+H]$^+$: m/z 1046.5 sensitivity; ^{d}: Selection of mono-isotopic ions is possible at ACTH fragment 18–39 [M+H]$^+$; ^{e}: AngiotensinII [M+H]$^+$: m/z 1046.5 mass resolving power. (From Jeol SpiralTOF JMS-S3000 Brochure.)

3.2.4 SCIEX TripleTOF® 6600

One of the present high-end mass spectrometric instruments from SCIEX is their Triple-TOF® 6600.[188] The TripleTOF® 6600 extends the power and

---

[188]http://sciex.com/products/mass-spectrometers/tripletof-systems/tripletof-6600-system?country= United%20States (Accessed 13 April 2015).

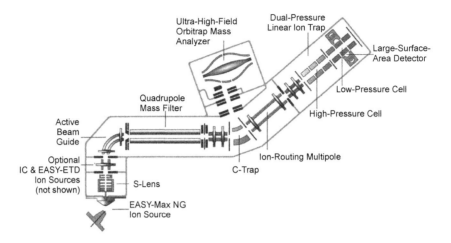

Figure 3.25: Thermo Fisher Orbitrap Fusion instrument. The hardware, in order (from left to right): Electrospray ion source, S-lens, Quadrupole mass filter, C-trap and Orbitrap, Dual pressure linear ion trap, and Detector. From the C-trap, ions can be injected into the Orbitrap. Courtesy: Thermo Fisher Scientific, Inc. Reproduced by permission.

features of SCIEX TripleTOF® technology, delivering high performance and reproducibility that can detect virtually every peptide in every run when executing SWATH^{TM} Acquisition 2.0. The TripleTOF® 6600 offers: (i) broader linear dynamic range: greater than five orders, (ii) wider Q1 mass range: isolation for ions up to 2,250 m/z, (iii) enhanced mass accuracy stability, (iv) higher quality MS/MS with superior mass resolution and speed: 35,000 TOF MS mode, 20,000 high-sensitivity MS/MS mode; 30,000 high-resolution TOF MS/MS mode; up to 100 MS/MS per second. The TripleTOF® 6600 maximizes productivity with its real-time management of MS/MS acquisition time, based on the intensity of selected precursors.

3.2.5   Thermo Fisher Scientific — Orbitrap Fusion

In Thermo Fisher's Orbitrap Fusion instrument (shown in Fig. 3.25) not only are three different mass analyzers combined into a single instrument, a new concept is introduced: parallel acquisition of mass spectra by an Orbitrap and a linear ion trap mass analyzers, thus significantly increasing the instrument's capability. The description below includes some parts of the Fusion instrument which are configurationally similar to Elite and Velos (2009). Fusion provides low detection limits, and with respect to their earlier instruments, higher sensitivity, resolving power, dynamic range of detection, mass accuracy, and quantitation in the presence of a matrix. It also

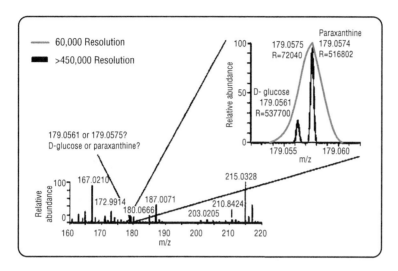

Figure 3.26: D-Glucose, Paraxanthine in spectrum from Fusion. Courtesy: Thermo Fisher Scientific, Inc. Reproduced by permission.

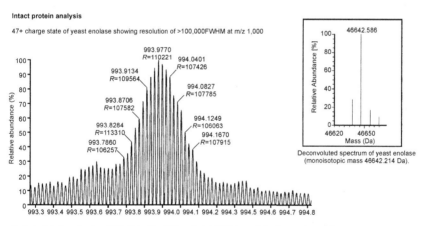

Figure 3.27: Analysis of intact Yeast Enolase (48.64 kDa). Left: Zoom-in on charge state 47+ ions, and right: deconvoluted spectrum of yeast enolase (monoisotopic mass 46642.214 Da). Courtesy: Thermo Fisher Scientific, Inc. Reproduced by permission.

provides multiple fragmentation technologies for MS^n including HCD, CID, and ETD for high reliability of identification of a large number of precursors and fragments. The order of the hardware in Fig. 3.25 is: first an S-lens for high transmission at high pressures, followed by a quadrupole mass filter with precursor ion isolation window widths from 0.4 Da to 20 Da, a nitrogen-filled C-trap which feeds ions into an Orbitrap mass analyzer with resolving power up to 450,000 FWHM and isotopic fidelity up to 240,000 FWHM at m/z 200 with scan rates up to 15 Hz, followed by a dual-pressure linear ion trap containing a high-pressure cell for high efficiency ion injection, ion isolation with precursor ion isolation from 0.2 Da to full mass range, and ETD and CID activation techniques, and a low pressure cell for fast scan rates at nominal resolutions.

The S-lens at the front end of the instrument was part of the previous Orbitrap Elite version of instrument and provides RF focussing of ions in 2 mbar region. In the dual ID variable-spacing stacked lens system, the larger ID of the stacked lens enables capture of the entire expansion from the transfer tube whereas the smaller ID section helps in focussing the ion beam through the exit lens and downstream optics; the increased spacing results in increased field penetration which causes radial focussing of the ion beam, all with a small number of electrodes. The ion transfer tube is 2x shorter than previous versions and a neutral beam blocker was added which provides interference protection from droplets and neutrals.

RF focussing worked well even in the simple design (Fig. 3.27 shows that S-lens spectrum is much cleaner). The High Pressure Cell (from LTQ Orbitrap Velos technology) provides high trapping efficiency; high isolation and dissociation efficiency; faster trapping and activation; 5.0×10^{-3} Torr He. The Low Pressure Cell provides higher resolution at a higher scan rate; 3.5×10^{-4} Torr He. The HCD cell with axial field was directly attached to the C-trap; there is no gas supply to the C-trap any more; as gas leaks from the HCD cell (3–5 mbar) and is used for ion trapping and cooling in the C-trap; a dedicated pump for the HCD cell protects the low pressure Orbitrap cell from gas overload. As regards ion transfer into collision cell, initially the HCD potential is low — therefore precursor and/or product ions are initially trapped there — then the potential in the HCD cell is raised, allowing the fragments to be transferred to the C-trap, which subsequently transfers them to the Orbitrap.

The capability of a high-temperature bake-out of the Orbitrap results in improved ultimate vacuum (10^{-11} Torr) for analysis of small and medium-size proteins with isotopic resolution. A water-cooled pre-amplifier helps provide low noise levels allowing data with high S/N.

In Fig. 3.26, a spectrum from Fusion is shown, and in Fig. 3.27 a spectrum obtained from Elite.

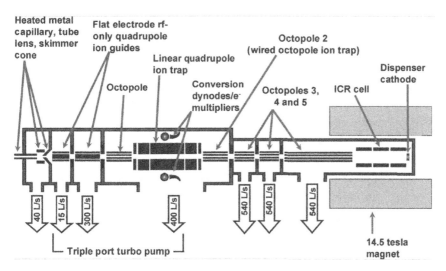

Figure 3.28: Schematic diagram (not to scale) of the 14.5 T hybrid linear quadrupole ion trap/FT-ICR mass spectrometer. The wired octopole enables a second mode of mass-selective ion accumulation whereby ions are mass selected in the LTQ and transferred to the adjacent storage octopole. That process can be repeated for numerous cycles before the entire octopole ion population is transferred to the ICR cell. Reproduced by permission of the American Chemical Society from *Analytical Chemistry* Vol. 80, 3985–3990, 2008.

3.2.6 Thermo Fisher Scientific – Hybrid FT-ICR at 14.5 tesla

Here we describe the design and performance of a 14.5 T hybrid linear quadrupole ion trap and Fourier transform ion cyclotron resonance mass spectrometer.[189] The instrument (Fig. 3.28) is a hybrid, linear quadrupole ion trap coupled with FT-ICR MS (LTQ-FT, Thermo Fisher Corp., Bremen) which has been adapted to operate in an actively shielded 14.5 T superconducting magnet. Its fringe magnetic field substantially higher than most (shielded) FT-ICR magnets — approximately 75 gauss at a distance of 1.5 m from field center in any direction. In very high tesla magnetic field systems, an ultrahigh vacuum is most essential to keep down collisional fragmentation, particularly of low-mass ions. Because kinetic energy and collision rate scale inversely with m/z, the former scales as the square of the magnetic field, and also with the square of the orbital radius. Their early measurements suggested m/z-dependent collisional dissociation during ICR

---

[189]Schaub, T. M., Hendrickson, C. L., Horning, S., Quinn, J. P., Senko, M. W., and Marshall, A. G. (2008) High-performance mass spectrometry: Fourier transform ion cyclotron resonance at 14.5 Tesla. *Anal. Chem.*, **80**, 3985–3990.

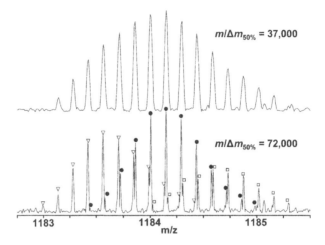

Figure 3.29: Mass scale expanded segments from a positive-ion electrospray FT-ICR mass spectrum of insect neuropeptides collected at 1 scan/s and 200,000 resolving power (m/z 400). Bottom: resolved overlapping isotopic distributions for three 8 kDa proteins. Top: The same m/z segment from FT of the same time-domain data truncated by half (i.e., equivalent to halving the resolving power), for which the three isotopic distributions are no longer resolved. Reproduced by permission of the American Chemical Society from *Analytical Chemistry*, Vol. 80, 3985–3990, 2008.

detection; here, unwanted collisional dissociation, particularly for low m/z ions, were later reduced by equipping the three lowest-pressure stages with high-speed, high-compression turbomolecular pumps, decreasing the base pressure to 2×10^{-10} Torr. The ion optics between the linear trap and the ICR cell were modified while retaining the full functionality (such as the AGC, data-dependent MS and MS^n, parallel linear trap/ICR detection etc.) of the standard commercial system. For a second mode of external ion accumulation and for improved ion ejection efficiency, the octopole directly behind the LTQ was modified to incorporate tilted wire extraction electrodes. For the Octopole 2 drive, a high-power RF supply coupled through a center-tapped transformer was used; voltages (RF and DC offset) to the 1, 3, and 4 Octopoles were supplied by the tuned circuit oscillators of the standard system.

The ICR cell, supplied by vendor, rewired for external capacitive coupling, is open cylindrical with a diameter of 55 mm and an axial trapping dimension of 82 mm. For dipolar excitation, software-specified waveforms were used; the instrument control PCB supplied those to a high power RF supply — coupled through a center-tapped transformer — to a custom-built circuit for distribution of voltages that facilitated capacitive coupling.

At the front end, microelectrospray was performed with standard hardware, and a commercial ion source was used for atmospheric pressure potoionization (APPI). Collisionally activated dissociation (CAD) could be carried out in the linear trap, and electron-capture dissociation (ECD) in the ICR cell. A part of a mass spectrum of insect neuropeptides taken using this instrument is shown in Fig. 3.29.

A comparison of the present 14.5 T with 7 T LTQ FT-ICR instruments shows that mass accuracy scales as the square of magnetic field strength. Mass accuracy specification for the commercial 7 T LTQ-FT is 2 ppm rms (m/z 400, 1 s duty cycle) for broadband analyses, and 1–2 ppm has been reported. In the 14.5 T system this is 200–300 ppb. Further, very high selected-ion mass accuracy (34 ppb rms error) has been observed for 800 measurements of substance P (500 fmol/μL, not present in the calibration solution), selected from a mixture of five standard peptides.

FT-ICR data acquisition speed increases linearly with magnetic field strength. At 14.5 T, mass resolving power $m/\Delta m_{50\%} = 213,000$ (at m/z 400) is observed for time-domain acquisition period of 767 ms. With ~ 100 ms (or less) of external ion accumulation, spectra with high mass resolving power could be collected at a rate greater than 1 spectrum per second, which is the benchmark for chromatographic sampling. Mass resolving powers of 50,000 and 25,000 (m/z 400) have been observed for chromatographic sampling at 4 and ~ 6 mass spectra/s; by simultaneous ion accumulation and detection the full theoretical limit may be achieved.

Figure 3.29 illustrates how a factor of two improvement in mass resolving power may mean obtaining or not obtaining an answer in biological analyses. The lower panel shows an m/z scale-expanded segment from an FT-ICR mass spectrum collected at 1 spectrum/s ($m/\Delta m_{50\%}$) 207,000 at m/z 400 and 70,000 at m/z 1183 for a complex mixture of neuropeptides and intact proteins extracted from *Periplaneta americana* (cockroach) brain. The upper trace was simulated by truncating the time domain by half to see what would be observed if this sample were analyzed at the same acquisition rate but with a 7 T FT-ICR mass spectrometer. The three overlapping isotopic distributions in the lower trace each originate from 7^+ ions of similar mass (8281.012, 8284.121, and 8288.189 Da, measured) — likely from amino acid substitution and/or post-translational modification of similar structures. The three peptides are clearly resolved at 14.5 T, but would not be resolved at 7 T. In the absence of high-resolution mass analysis, the identity and number of species present would be incorrectly determined; they are not always separated by chromatography.

A major advantage of high-field instruments over lower field instruments is reduced sensitivity to space charge effects. The resolving power achievable with the 14.5 T instrument for a mixture of high complexity likely exceeds 1,000,000 (m/z 400).

Unlike commercial 7 T LTQ FT systems, where ions are stored in and transferred from the linear quadrupole trap to the ICR cell, in the 14.5 T instrument a second mode of external ion accumulation is provided. In this, ions (mass selected or not) in the LTQ are transferred to an adjacent octopole equipped with wire ion extraction electrodes. This accumulation process in the octopole may be repeated for numerous cycles before the entire octopole ion population is transferred to the ICR cell. This mode of operation, which permits a two- to three-fold improvement in ion storage capacity, is particularly useful for in-cell MS^n (ECD or IRMPD) in comparison to those selected, stored, and transferred from the LTQ directly. This octopole ion storage is invoked only if a large ion population is needed for ICR analysis (such as structural characterization which requires internal ECD, IRMPD, or external CAD with FT-ICR product ion detection) and the duty cycle is of less concern.

3.2.7 IMS-IMS-MS

The mobility spectrometer[190] used by Clemmer *et al.* in their multiplexed IMS-IMS-MS measurements (Fig. 3.30) has two drift regions in tandem — this enables two-dimensional IMS — interspersed by a region utilized for collisional activation where ions can be activated to alter their shape/ structure and hence mobility. The second drift region is followed by a TOF mass spectrometer. At the front end of the spectrometer there is an ion funnel and a gate between the ion source and the first drift region; at the end of this region a second gate connects to the interface region. Ion packets extracted from the source undergo the first stage of mobility-based separation in the first drift region; after a predetermined delay using the second ion gate, ions of a selected mobility are activated in the activation region where the conditions present alter primarily the structures of +2 and +3 states while limiting fragmentation.

In some more detail, ions are periodically extracted from the source into the first drift region for mobility-based separation. After a predetermined delay time, the data acquisition system records the arrival time of ions — which is the sum of the drift and the TOF times. The gate to the intermediate activation region opens after a second predetermined delay time, set equal to or greater than the first, and collisional activation takes place resulting in new structures; following this, there is a second stage of mobility-based separation and mass spectrometric detection.

---

[190]Valentine, S. J., Kurulugama, R. T., Bohrer, B. C., Merenbloom, S. I., Sowell, R. A., Mechref, Y., and Clemmer, D. E. (2009) Developing IMS-IMS-MS for rapid characterization of abundant proteins in human plasma. *Int. J. Mass Spectrom.*, **283**, 149–160.

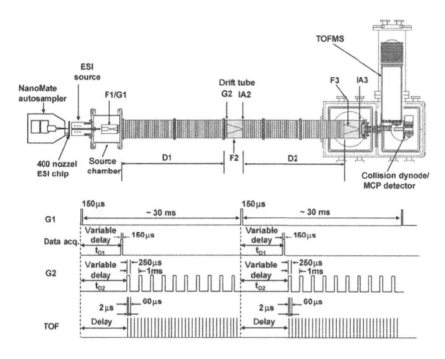

Figure 3.30: Top: Schematic of the IMS-IMS-MS instrument.[190] Bottom: Timing pulse sequence used in the IMS-IMS-MS experiment. Shown are the timing pulses applied to the instrument gates (G1 and G2), the data acquisition system, and the TOF high-voltage pulse. The delay times t_{D1} and t_{D2} have the relationship $t_{D2} \geqq t_{D1}$. Reproduced by permission of Elsevier B. V. from *International Journal of Mass Spectrometry*, Vol. 283, 149–160, 2009.

In a single experimental sequence, the data acquisition with multiple mobility selections takes place in the following multiplexed manner. The gate to the activation region opens for a duration of $250\,\mu s$ — when mobility-selected ions enter the activation region in sequence — at intervals spaced $1\,ms$ apart (within the time span of $\sim 30\,ms$ — the time difference set between two ion drawout pulses). After activation, these then undergo the second mobility-based separation and mass spectrometric detection. Together, this constitutes the first data frame. In three additional sets of data collection — meaning, a total of four data frames — the gate openings shifted by $250\,\mu s$ each time, all possible drift times are covered. The time required for each data frame is $45\,s$ for a total of $3\,min$ with additional time of $1\,min$ for saving the data and sample switching time; the total analysis time amounts to $4\,min$. The mass spectrometer resolving power is $\sim 5,000 - \sim 7,000$ at m/z of 500. For data manipulation (peak picking,

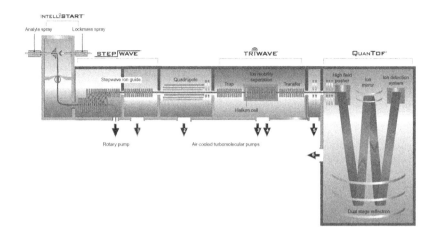

Figure 3.31: Synapt G2-S schematic. Courtesy: Waters Corporation. Reproduced by permission.

normalization, and peak alignment) it is operated at a somewhat lower resolution $\sim 2{,}500 - \sim 3{,}000$ at the same m/z. The mass accuracy for the measurement is $\sim 70 - \sim 100$ ppm ($m/z \approx 1060$).

3.2.8   Waters — Synapt G2-S

A schematic of the Waters Synapt G2-S is shown in Fig. 3.31. The principal sections of the instrument are: the spray section (IntelliStart), input ion guide (StepWave), quadrupole, ion mobility separation section (TriWave), and Tof-Reflectron m/z analyzer section (QuanTof). The Triwave part has three constituents: Trap, IMS, and Transfer. Ion-mobility-based separation takes place in the IMS section; the Trap and Transfer sections facilitate, respectively, ion store and movement to and from the IMS section. They can also operate as collision cells. Ions continually enter the Trap region; a potential barrier applied at the gate prevents ions exiting unless desired; they accumulate for the entire period during which the previous cycle of ion mobility separation is taking place. During ion accumulation, no traveling wave is applied in the Trap T-Wave region. At the *end* of the previous mobility separation cycle, the gate is opened by removing the potential barrier, a packet of ions enters the IMS T-Wave, and mobility-based separation in the presence of buffer gas takes place. (Because of high gas pressure — typically 2.5–3 mbar N_2 — multiple collisions take place; high-mobility ions, because of their speed, surf more at the wave front. To improve separation, other gases like CO_2 can be used.) At the end of the mobility-based separation when the ions exit, they, with the help of a traveling wave, are transported through the Transfer cell to the Tof analyzer. The gating of the next ion packet occurs once *all* of the ions are detected in the Tof analyzer. Presence of the helium cell at the entrance of the T-wave — where there is transition from high-vacuum to high-pressure environment — minimizes scatter and fragmentation of ions.

Whereas in the classic form of IMS, for separation of ions, a uniform static electric field is used, in T-wave *non-uniform, moving* electric fields/ voltage pulses are used. The T-Wave ion guide consists of an even number of ring electrodes. For high transmission and radial ion confinement, opposite phases of RF voltage are applied to adjacent electrodes. To propel the ions along the device, the traveling wave DC voltages are applied to electrode pairs in repeating sequence along the device, and step to the next adjacent electrode pair at fixed intervals. The voltage profile driving the ions is not sharp-edged; it has a smoothly varying profile due to relaxation of the DC voltage in the IMS T-Wave. The repeat pattern optimizes the speed and degree of separation; the separation timeframe is typically tens of milliseconds.

A significant difference between many conventional drift tube IMS operation and Synapt is the presence of trapping capability in the latter. With limited ion confinement and without trapping, the duty cycle can be low (as low as 1%) — in Triwave the mobility separations are at near maximal transmission.

Synapt G2-S specifications (Synapt Brochure, June 2011): z-spray[TM] source, StepWave[TM] ion optics, Quantitative Tof (QuanTof)[TM]. FWHM mass resolving power $> 40,000$, exact mass ($< 1\,$ppm RMS), in-spectrum dynamic range and linearity in detector response of up to 10^5, up to 30 spectra(s). QuanTof's high-field pusher and dual stage reflectron, incorporating high-transmission parallel wire grids, reduce ion transmission times due to pre-push kinetic energy spread and improved focussing of high energy ions, respectively. High ion mobility resolving power $> 40(\Omega/\Delta\Omega)$ FWHM.

In a recent application by Merenbloom *et al.*,[191] dissociation of singly protonated leucine enkephalin dimer into monomer, as well as subsequent dissociation of monomer into a-, b-, and y- ions, besides effective ion temperatures, have been measured in TWIMS analysis using a Synapt G2. The extent of dissociation, at fixed helium cell and TWIMS cell gas flow rates, was found not to vary significantly with either wave velocity or wave height, except at low ($< 500\,$m/s) wave velocities. Greater dissociation resulted when the flow rate of nitrogen gas into the TWIMS cell was increased

---

[191]Merenbloom, S. I., Flick, T. G., and Williams, E. R. (2012) How hot are your ions in T-wave ion mobility spectrometry? *J. Am. Soc. Mass Spectrom.*, **23**, 553–562.

and the flow rate of helium gas into the helium cell was decreased. The mobility distributions of the fragment ions formed by dissociation of the dimer upon injection into the TWIMS cell and those of fragment ions formed in the collision cell prior to TWIMS analysis were found to be about the same. The results indicated that heating and dissociation occur as ions are injected into the TWIMS cell, and during the TWIMS analysis the effective temperature decreases to a point at which no further dissociation takes place. At a helium flow rate of 180 mL/min, TWIMS flow rate of 80 mL/min, and traveling wave height of 40 V, an upper limit to the effective ion temperature was obtained as 449 K.

They observed that the heating can be minimized by increasing the helium cell flow rate and/or decreasing the TWIMS cell flow rate. They obtained an upper limit to the effective temperatures of 449 and 470 K for the protonated leucine enkephalin dimer and the 10+ and 11+ ions of myoglobin, respectively. These values are well below the upper estimate of Shvartsburg and Smith,[165] and are also lower than those reported by Morsa *et al.*[192] The extent of ion activation is expected to depend on many factors, including ion charge state, mass, conformation, and instrumental parameters. For example, the cross section of a spherical polymer increases by $\sim (\mathrm{mass})^{2/3}$. Thus, the extent of collisional activation per degree of freedom should decrease with ion size for large proteins and protein complexes formed from aqueous solutions.

---

[192]Morsa, D., Gabelica, V., and De Pauw, E. (2011) Effective temperature of ions in traveling wave ion mobility spectrometry. *Anal. Chem.*, **83**, 5775–5782.

CHAPTER 4
MASS SPECTROMETRY OF PEPTIDES AND PROTEINS

In this chapter, beginning with brief basics of protein structure, we discuss mass spectrometric determination of molecular mass of *intact* proteins. This is followed by a consideration of alternate methods SRM, MRM for obtaining mass spectra and *quantitative* analytical methods such as *"label-free,"* ICAT, SILAC, iTRAQ, TMT, and AQUA. Thereafter CAD/CID, ETD/ECD, IRMPD, and other MS/MS *peptide* fragmentation methods are presented and also a summary of empirical information available on fragmentation products of peptides and their statistical treatment. Then, we discuss mechanisms of peptide fragmentation. This is followed by an overview of mass spectrometric studies of post translational modifications and the chemical cross-linking method for obtaining structural information of proteins. In the last part of the chapter, we consider manual *de novo* sequencing of peptides using mass spectral data and present a few examples.

4.1 Proteins

4.1.1 Primary, secondary, tertiary, and quaternary structures

Primary structure: The basic molecular building blocks of proteins are *amino acids*, which are characterized by terminal -COOH and -NH$_2$ groups. The *primary* structure of proteins is concatenated amino acids, which are bound to one another by peptide bonds (a peptide bond —-C(O)-N(H)-— is formed from bonding of -C(O)OH from the C-terminus of one amino acid with NH$_2$ at the N-terminus of another amino acid (Fig. 4.1)). There are 20 naturally occurring amino acids (Fig. 4.5, Table 4.1). A molecule with a series of amino acids/peptide bonds is a *polypeptide*; a protein molecule is a *huge* polypeptide.[193]

Secondary structure: There are two manifestations of the secondary structure. One consists of the peptide backbone progressing *helically* (torsional rotation about the axis of the bonds makes the helical form possible), leading to a structural pattern called α-helix; the other, containing two peptide strands stretched next to each other with hydrogen bonding possible *between* the two strands (between a carbonyl O of one strand

---

[193]In *Homo sapiens*, median protein length 375 amino acids. See also, Molecular machinery: a tour of the Protein Data Bank. (poster: by Goodsell, D. S.; Graphic design: Bamber, G. W. 2002. http://www.rcsb.org/pdb/education_discussion/molecule_of_the_month/poster_full.pdf).

Introduction to Protein Mass Spectrometry
http://dx.doi.org/10.1016/B978-0-12-805123-8.00004-7

Copyright © 2016 Elsevier Inc.
All rights reserved.

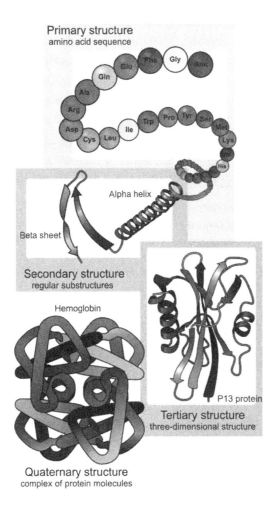

Figure 4.1: Protein structures. (a) Primary structure: amino acids and peptide bonds (by convention, the amino end is where the polypeptide chain begins). (b) Secondary structure: the ribbon represents α-helix, also shown β-pleats. (c) Tertiary structure. (d) Quaternary structure. (Source: http://en.wikipedia.org/ wiki/Protein_structure.html.)

with an imino H of the other strand), leading to β-sheets. The α-helix has 3.6 amino acid residues per turn; every fourth residue is involved in hydrogen bond formation — α-helix length varies in the range 1.5–11 turns, involving 5–40 amino acids. In *parallel* β sheet, one amino acid residue segment CO-‖**NH**-CH(R_i)-**CO**‖-NH of one strand forms hydrogen bonds with *two* amino acid residues CO‖NH-CH(R_j)-**CO**‖-NH-CH

Figure 4.2: Left: Protein loops. Human triosephosphate isomerase. 30.791 kDa. DOI : 10.2210/pdb4pod/pdb. Right: Human erythrocytic ubiquitin, 1UBQ.pdb. Ubiquitin consists of 76 amino acids, ≈ 8.58 kDa.

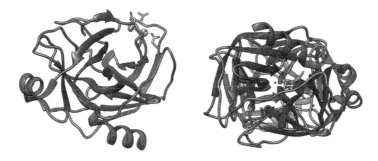

Figure 4.3: Left: Bovine trypsin. 2PTN.pdb. 25.785 kDa. Right: Bovine chymotrypsin. 2CHA.pdb. 25.666 kDa.

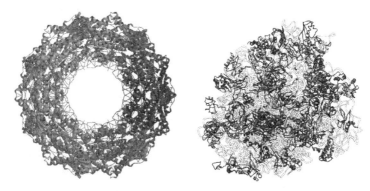

Figure 4.4: Left: *Escherichia coli* GroEL, Homo 14-mer, 1GRL.pdb. (Two nearly 7-fold rotationally symmetrical rings stacked back-to-back with dyad symmetry.) 800–914 kDa. Right: *Thermus thermophilus* ribosome. 4v4J.pdb. ≈ 2.22 MDa.

(R_{j+1})-CO‖NH-CH(R_{j+2})-CO‖-NH of the other strand (for convenient visualization, a residue here is delineated with ‖ lines); for this type of bonding to be possible the *progression* of the two strands needs to be toward *opposite* directions. (Atoms shown in bold are the atoms involved in hydrogen bond formation.) In *antiparallel* β sheet, CO-‖**NH**-CH(R_i)-**CO**‖-NH segment of *one* strand forms hydrogen bonds with *one* amino acid NH‖**CO**-CH(R_j)-**NH**‖-CO of the other strand; for this to be possible the progressions of the *two* strands need to be toward the *same* direction. There can also be *mixed* β-sheets. The α-helix is often depicted as a coiled ribbon; the β-strand — by broad arrows pointing toward the C-terminus. Under certain conditions, it is possible to have hairpin turns of the peptide strand, usually caused by (CO) of one residue hydrogen bonding with (NH) of another residue at 3 peptide bonds' difference; CO‖NH-CH(R_i)-**CO**‖-NHCH(R_{i+1})-CO‖NH-CH(R_{i+2})-CO‖-N**H** -CH(R_{i+3})-CO‖-NH; this is simply termed a *turn, hairpin* turn, β-turn, or *reverse* turn; additionally, relatively complex structures called *loops* (such as Ω-loops) can also cause chain reversals while providing structural rigidity (Fig. 4.1(b)).

Tertiary structure: In hydrophilic environment, amino acid residues that are hydrophobic stay away from water and together collect around the center area; amino acids with hydrophilic side chains stay on the surface area—resulting the protein folding into a compact structure. In tertiary structure, the -S-S- disulfide bonds, linking two strands through two amino acid cysteine residues, contribute toward compacting and formation of a hydrophobic core. In hydrophobic environment, such as in membranes, the reverse effect takes place (Fig. 4.1(c)).

Quaternary structure: Multiple folding, coiling, coming together of multiple chains, all acting together lead to, via sometimes interchain interactions, the final structure which is either fibrous or globular. Quaternary structures are named in terms of the *number* of *subunits* in the complex, such as trimer (3), hexamer (6), octamer (8), dodecamer (12), and 21-mer (21) (Fig. 4.1(d)).

The term *proteoform* has been used to express the degree of complexity of a protein.[194]

4.1.2 Amino acid data

The structures of the 20 naturally occurring amino acids are given in Fig. 4.5 and some basic data in Table 4.1 (also see Table. 4.3). Of the 20 amino acids, *neutral nonpolar* amino acids are glycine, alanine, valine, leucine, isoleucine, methionine, proline, phenylalanine, and tryptophan; *neutral polar* amino acids are serine, threonine, tyrosine, asparagine, cysteine (*cystine* is two

---

[194]Smith, L. M., Kelleher, N. L., and the Consortium for Top Down Proteomics. (2013) Proteoform: a single term describing protein complexity. *Nat. Methods*, **10**, 186–187.

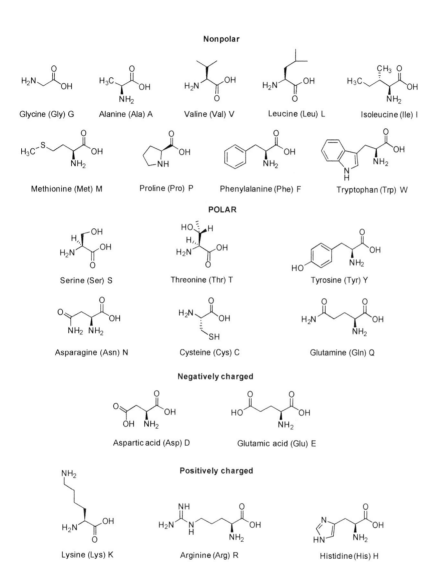

Figure 4.5: Structures of the 20 essential amino acids. They are grouped into nonpolar, polar, positively charged, and negatively charged. Two codes are in use: one with only a single letter (such as G, A, ...), the other with three-letter combinations (such as Gly, Ala, ...). The residue masses of these amino acids, and masses of their corresponding immonium ions are given in Table 4.1.

Table 4.1: Amino Acid Residue Names, Masses, and Immonium Ions.

Residue	1-Letter Code	3-Letter Code	Residue Mass	Immonium Ion (m/z)
Alanine	A	Ala	71.04	44.05
Arginine	R	Arg	156.10	129
Asparagine	N	Asn	114.04	87.09
Aspartic Acid	D	Asp	115.03	88.04
Cysteine	C	Cys	103.01	76
Glutamic Acid	E	Glu	129.04	102.06
Glutamine	Q	Gln	128.06	101.11
Glycine	G	Gly	57.02	30
Histidine	H	His	137.06	110.07
Isoleucine	I	Ile	113.08	86.1
Leucine	L	Leu	113.08	86.1
Lysine	K	Lys	128.09	101.11
Methionine	M	Met	131.04	104.05
Phenylalanine	F	Phe	147.07	120.08
Proline	P	Pro	97.05	70.07
Serine	S	Ser	87.03	60.04
Threonine	T	Thr	101.05	74.06
Tryptophan	W	Trp	186.08	159.09
Tyrosine	Y	Tyr	163.06	136.08
Valine	V	Val	99.07	72.08

Table 4.2: Translation: Amino Acid Codons.

Ala	GCT	Cys	TGT	His	CAT	Met	ATG	Thr	ACT
	GCC		TGC		CAC				ACC
	GCA					Phe	TTT		ACA
	GCG	Glu	GAA	Ile	ATT		TTC		ACG
			GAG		ATC				
Arg	CGT				ATA	Pro	CCT	Trp	TGG
	CGC	Gln	CAA				CCC		
	CGA		CAG	Leu	TTA		CCA	Tyr	TAT
	CGG				TTG		CCG		TAC
	AGA	Gly	GGT		CTT				
	AGG		GGC		CTC	Ser	TCT	Val	GTT
			GGA		CTA		TCC		GTC
Asn	AAT		GGG		CTG		TCA		GTA
	AAC						TCG		GTG
				Lys	AAA		AGT		
Asp	GAT				AAG		AGC	Stop	TAA
	GAC								TAG
									TGA

cysteines bound through a -S-S- disulfide bond), and glutamine; *acidic polar* amino acids are aspartic acid and glutamic acid; and *basic polar* amino acids are lysine, arginine, and histidine. A list of translation amino acid codons is in Table 4.2.

An amino acid *residue mass* (Table 4.1) is the mass of the intact amino acid molecule *minus* the mass of the OH group at its C-terminus and mass of one of the H atoms at its N-terminus.[195] Immonium ion mass calculation is shown in Section 4.1.4.

4.1.3 Nomenclature of ions generated in peptide fragmentation

It is from a knowledge of the ions produced from MS/MS fragmentation of a peptide that its amino acid sequence is obtained; then, from a knowledge of the sequences of all the peptides obtained in the initial proteolysis, the original protein is identified. The *nomenclature* for MS/MS peptide fragments first proposed by Roepstorff and Fohlman,[196] later modified by Biemann,[197,198] is presented in Fig. 4.6. The fragments produced from CID are given in some detail in Fig. 4.9. There are *sequence* ions (which are various direct backbone fragments of the peptide) and *satellite* ions (which are from partial or complete loss of side chains from the sequence ions). The a, b, c, x, y, z ions, for example, are sequence ions, whereas d, v, w are satellite ions. The series of y ions (y_1–y_{14}) and b ions (b_1–b_{14}) obtained in fragmentation of *doubly* charged Glu-fibrinopeptide (EGVNDNEEGFFSAR) are shown in Fig. 4.10 and Fig. 4.11 respectively; the spectrum obtained after water/ammonia loss from the primary residues EDQN is shown in Fig. 4.12.

The nature of the amino acids present in a peptide sequence influences the fragmentation pattern. For instance, the presence of a *basic* amino acid residue (such as Arg, His, Lys) near or at either C- or N-terminus generates in high-energy CID mass spectra fragment ions containing *that* particular terminus; in the same way, their presence at or near the N-terminus generates a- and d-ions, whereas at or near the C-terminus generates y-, z-,

---

[195]To calculate an amino acid residue mass, take the molecular mass of the amino acid and subtract from that 17 (for C-terminus OH) and 1 (for the N-terminus H). Take, for example, Alanine (Table 4.1). Alanine molecular mass is 89.04 Da. Subtracting 18 (17+1) from that gives 71.04. This calculation is done on the basis of monoisotopic masses; monoisotopic residue mass — required for high-resolution mass spectral analysis — calculated on the basis of the most prevalent isotope of an atomic species is given in Table 4.3.

[196]Roepstorff, P. and Fohlman, J. (1984) Proposal for a common nomenclature for sequence ions in mass spectra of peptides. *Biomed. Environ. Mass Spectrom.*, **11**, 601.

[197]Biemann, K. (1988) Contributions of mass spectrometry to peptide and protein structure. *Biomed. Environ. Mass Spectrom.*, **16**, 99–111.

[198]Biemann, K. (1990) Nomenclature for peptide fragment ions (positive ions). *Methods in Enzymology*, **193**, 886–887.

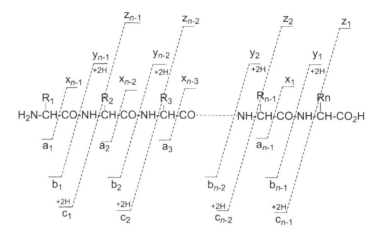

Figure 4.6: Nomenclature for peptide CID fragment ions (first proposed by Roep-storff and Fohlman,[196] later modified by Biemann[197]). Typically, all of the above fragment ion types are observed in high-energy CID tandem mass spectra (although usually not in the same spectrum), but only b, y, and, less frequently, a and z fragments are observed in low-energy CID spectra.[199] Reproduced by permission of John Wiley & Sons, Inc. from *Mass Spectrom. Rev.*, Vol. 14, 49–73 (1995).

and w-ions. (The amino acid sequence in a peptide is reported from its N-terminal.)

Similarly, specific observations can be made for all the amino acids. For example, if there is a sequential Gly-Gly or Gly-Ala presence, then ions generated owing to any cleavage between the residues are *not* observed to be abundant.

When the hydroxy amino acids serine and threonine are present, an important effect — more prominent in low energy CID — is loss of water from the residue, regardless of where the cleavage takes place; this means observation of several pairs of ions differing by -18 Da. In *high-energy* CID spectra, d- and w-fragment ions are observed. For Thr, two such fragments result because of its side chain branched at the β-carbon atom. Also, though *not* common, in high-energy CID spectra, Thr (and to a somewhat lesser extent, Ser) present in the n-th position of the peptide appears to produce — with the peptide fragmenting into predominantly N-terminal fragments — c-type fragment ions (abundant c_{n-1} ions). Any sulfation or phosphorylation of Thr or Ser can be identified from the mass of the modified amino acid obtained. SIM and MS/MS can be used to pinpoint the amino acid location.

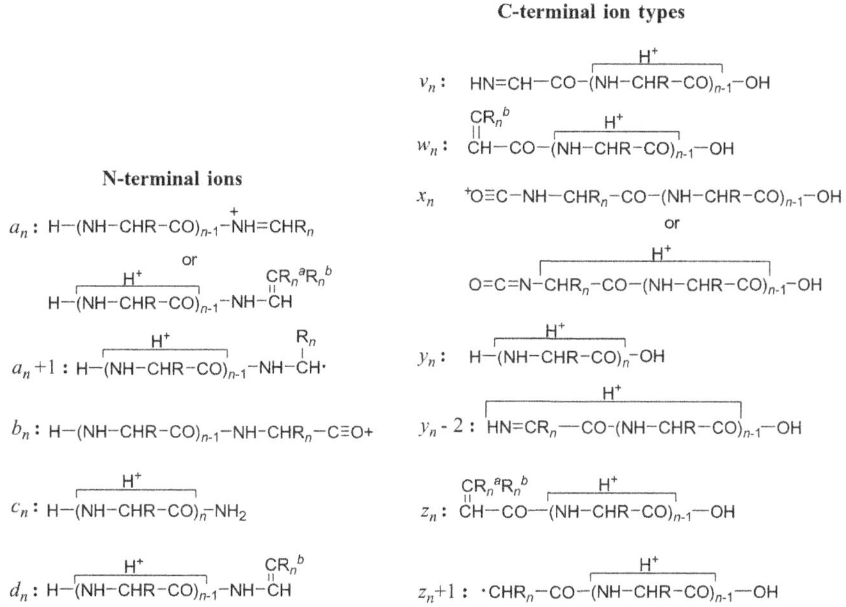

C-terminal ion types

$$v_n: \quad \text{HN=CH—CO-(NH—CHR—CO)}_{n\text{-}1}\text{—OH} \quad \overset{\text{H}^+}{\overbrace{}}$$

$$w_n: \quad \overset{\text{CR}_n{}^b}{\overset{||}{\text{CH}}}\text{—CO—(NH—CHR—CO)}_{n\text{-}1}\text{—OH} \quad \overset{\text{H}^+}{\overbrace{}}$$

$$x_n \quad {}^+\text{O}\equiv\text{C—NH—CHR}_n\text{—CO—(NH—CHR—CO)}_{n\text{-}1}\text{—OH}$$

or

$$\text{O=C=N—CHR}_n\text{—CO—(NH—CHR—CO)}_{n\text{-}1}\text{—OH} \quad \overset{\text{H}^+}{\overbrace{}}$$

N-terminal ions

$$a_n: \quad \text{H—(NH—CHR—CO)}_{n\text{-}1}\text{—}\overset{+}{\text{NH}}\text{=CHR}_n$$

or

$$\text{H—(NH—CHR—CO)}_{n\text{-}1}\text{—NH—}\overset{\text{CR}_n{}^a\text{R}_n{}^b}{\overset{||}{\text{CH}}} \quad \overset{\text{H}^+}{\overbrace{}}$$

$$a_n+1: \quad \text{H—(NH—CHR—CO)}_{n\text{-}1}\text{—NH—}\overset{\text{R}_n}{\overset{|}{\text{CH}}}\cdot \quad \overset{\text{H}^+}{\overbrace{}}$$

$$b_n: \quad \text{H—(NH—CHR—CO)}_{n\text{-}1}\text{—NH—CHR}_n\text{—C}\equiv\text{O+}$$

$$c_n: \quad \text{H—(NH—CHR—CO)}_{n}\text{—NH}_2 \quad \overset{\text{H}^+}{\overbrace{}}$$

$$d_n: \quad \text{H—(NH—CHR—CO)}_{n\text{-}1}\text{—NH—}\overset{\text{CR}_n{}^b}{\overset{||}{\text{CH}}} \quad \overset{\text{H}^+}{\overbrace{}}$$

$$y_n: \quad \text{H—(NH—CHR—CO)}_{n}\text{—OH} \quad \overset{\text{H}^+}{\overbrace{}}$$

$$y_n\text{-}2: \quad \text{HN=CR}_n\text{—CO—(NH—CHR—CO)}_{n\text{-}1}\text{—OH} \quad \overset{\text{H}^+}{\overbrace{}}$$

$$z_n: \quad \overset{\text{CR}_n{}^a\text{R}_n{}^b}{\overset{||}{\text{CH}}}\text{—CO—(NH—CHR—CO)}_{n\text{-}1}\text{—OH} \quad \overset{\text{H}^+}{\overbrace{}}$$

$$z_n+1: \quad \cdot\text{CHR}_n\text{—CO—(NH—CHR—CO)}_{n\text{-}1}\text{—OH} \quad \overset{\text{H}^+}{\overbrace{}}$$

Figure 4.7: Left: N-terminus ions; right: C-terminus ions. (R represents the side chains of the amino acids; R_n^a and R_n^b are the beta substituents of the nth amino acid.) Reproduced by permission of Elsevier Limited from *Methods in Enzymology*, Vol. 193, Appendix 5. 1990.

Proline is associated with very abundant y-type ions. In a CID mass spectrum that contains abundant C-terminal fragment ions, the y-ions that result from cleavage adjacent to Pro are often of high intensity, and are frequently — especially in high-energy CID spectra — accompanied by a $y-2$ ion owing to the loss of hydrogen from proline. Hydroxyproline (which is of relatively rare occurrence) is isobaric with Leu and Ile; to differentiate, in high energy, its side-chain fragmentation can be used. Peptides with Pro at neither the N- nor the C-terminus often yield, in low-energy CID, abundant *internal* fragment ions $y_i b_j$.

It is imperative to derivatize any Cys residue since otherwise Cys, when oxidized and linked intermolecularly or intramolecularly to another Cys via disulfide bond, complicates interpretation of spectra. Iodoacetamide, iodoacetic acid, and vinyl pyridine are common derivatives; they result respectively in 159, 160, and 208 unique amino acid residue masses. (Reduced Cys behaves like Ser.) Pyridylethyl Cys derivative gives an abundant ion at 106 from the ethyl pyridyl ($C_5H_4NCH_2CH_2$) group. The modification arising from peptide cleavage between Cys and its preceding amino acid

Non sequence specific ions

Internal acyl ions (denoted by single-letter codes) (e.g., GA at *m/z* 129 for R = H, CH$_3$):

$$H_2N\text{-}CHR\text{-}CO\text{-}NH\text{-}CHR\text{-}C\equiv O^+$$

Internal immonium ions (denoted by single-letter code followed by the notation - 28):

$$H_2N\text{-}CHR\text{-}CO\text{-}\overset{+}{N}H\!=\!CHR$$

Amino acid immonium ions (denoted by single-letter code) (e.g., F at *m/z* 120 for R = C$_6$H$_5$CH$_2$):

$$\overset{+}{H_2N}\!=\!CHR$$

Loss of amino acid side chains from (M+H)$^+$ is denoted as the single letter code of the amino acid involved, preceded by a minus sign (e.g., -V for [M + H - 43]$^+$).

Figure 4.8: Non sequence specific ions. Reproduced by permission of Elsevier Limited from *Methods in Enzymology*, Vol. 193, Appendix 5, 1990.

results in 2-iminothiozolidine-4-carboxylic acid at the peptide N-terminus and contributes to abundant *a*- and *b*-type ions.

For identification of Val, Leu, and Ile, *high-energy* CID aliphatic side-chain fragmentation of *d* (*b*→*a*→*d*) and *z* (*z*→*v*→*w*) ions is utilized. Val, because of its symmetrical side chain branched *at the β*-carbon, produces *two d*- or *w*-ions 14 Da below the two corresponding *a*- or *z*-ions. Leu, because its side chain is *not* branched at the *β*-carbon, produces only *one d*- or *w*-ion at 42 Da below the corresponding *a*- or *z*-ion. Ile, with a branched side chain *at the β*-carbon, produces *two d*- or *w*-ions at 14 Da and 28 Da below the corresponding *a*- or *z*-ions. This characteristic can be used to differentiate between Leu and Ile. When Leu-Ile differentiation is not possible either in low-energy or high-energy spectra, the designation Xle (X) is used for Leu/Ile.

Between Asn and Asp the mass difference is 1 Da; the same small mass difference is between Asn and Leu/Ile; Asn$_2$ and (Leu, Asp) are isobaric. Therefore, here mass accuracy in CID spectra is very important. Asn is isomeric to Gly$_2$; additional data such as on any *d*- or *w*-ions can be of help here. Knowledge of *high-energy* CID *fragments* can facilitate differentiation between Asp and isomeric *β*-Asp. For negatively charged amino acid residues, such as Asp or Glu, esterification (such as methylation) of the peptide carboxylic acids can be used to confirm the presence of an acid and also for counting the number of Asp or Glu residues.

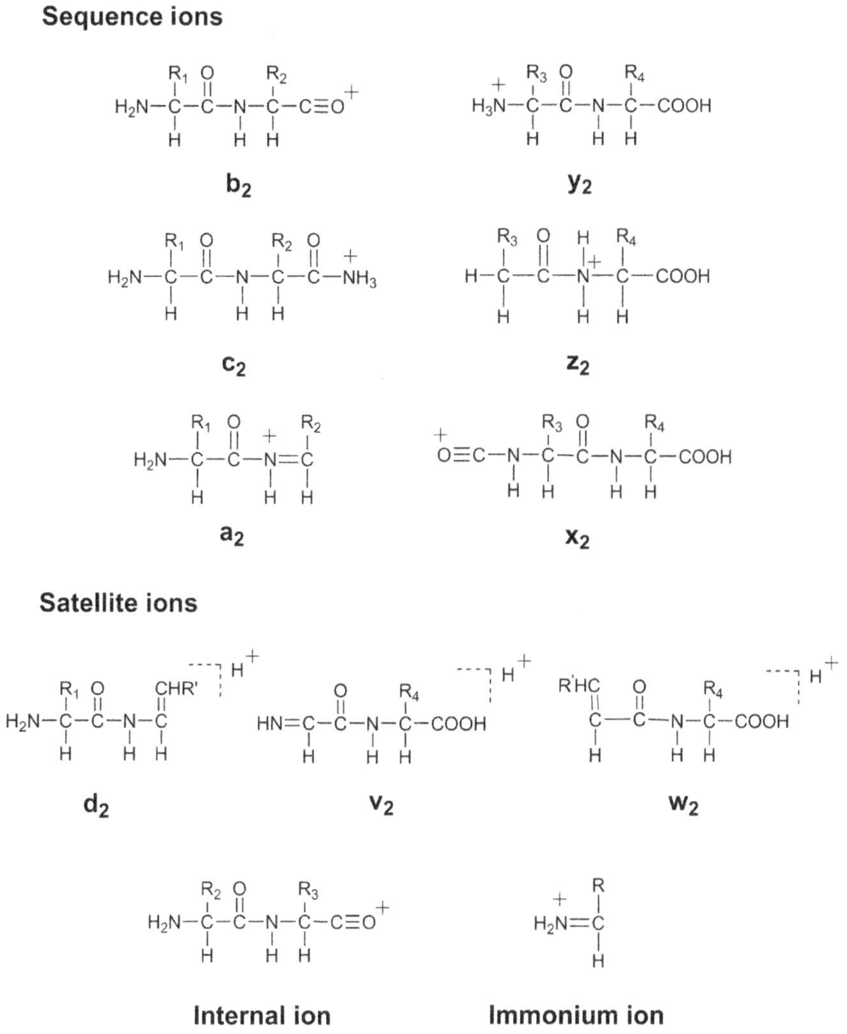

Sequence ions

b_2

y_2

c_2

z_2

a_2

x_2

Satellite ions

d_2

v_2

w_2

Internal ion

Immonium ion

Figure 4.9: Some fragment ions from a tetrapeptide. Ion types: $a \rightarrow$ [N]+[M]−CHO; $b \rightarrow$ [N]+[M]−H; $c \rightarrow$ [N]+[M]+NH$_2$; $x \rightarrow$ [C]+[M]+CO−H; $y \rightarrow$ [C]+[M]+H; $z \rightarrow$ [C]+[M]−NH$_2$; $a^* \rightarrow a$−NH$_3$; $b^* \rightarrow b$−NH$_3$; $y^* \rightarrow y$−NH$_3$; $a° \rightarrow a$−H$_2$O; $b° \rightarrow b$−H$_2$O; $y° \rightarrow y$−H$_2$O; $d \rightarrow a−$ partial side chain; $v \rightarrow y−$ complete side chain; $w \rightarrow z−$partial side chain.

Lys and Gln are isobaric, differing in mass by 0.03639 Da, and, therefore, in the interpretation of tandem mass spectra, the problems are similar as in Leu and Ile. Also, Gln (128.05858) is isomeric and Lys (128.09497) isobaric, with (Gly, Ala 128.05859). Because of their 1 Da mass difference, mass accuracy is very important in correctly assigning fragment ions to Glu or Gln (or Lys). Lys, Gln *can be* differentiated using enzyme specificities if the particular peptide was obtained by enzymatic cleavage of a protein (for example, trypsin and endo Lys-C cleave at Lys but not Gln) or chemical derivatization (Lys can be acetylated but Gln cannot). In *high-energy* CID, Lys, as a result of cyclization of its immonium ion side chain with loss of ammonia, gives a very abundant ion at 84; the 84 ions from Gln and Glu are usually not very abundant. The MS/MS of a peptide with Gln *at the* N-terminus — if this peptide dissociates into predominantly N-terminal fragments — usually contains abundant fragment ions 17 Da below the *a-*, *b-*, and *c-*ions; this characteristic can be used to differentiate Gln from Lys at the peptide N-terminus.

In methionine-containing peptides, in *high-energy* CID, for a_{n+1} + 1-, z_{n+1} + 1-ions one amino acid past the Met residue, additional ions 46 Da below the a_{n+1}-, z_{n+1}-ions are observed resulting from loss of methyl sulfide radical.

Histidine immonium ion at 110 is abundant in low- as well in high-energy CID spectra; therefore, it is a good way to spot histidine in the sequence.

The immonium ions of phenylalanine (at 120, note a possible conflict can arise from oxidized methionine which also has an immonium at 120 — methionine immonium is at 104 which with 16 of oxygen becomes 120) and tyrosine (136) are relatively abundant and can help in detecting their presence in the peptide. In *high-energy* CID spectra from phenylalanine, benzyl ion (91, $C_6H_5CH_2$) and from tyrosine, hydroxybenzyl ion (107, $C_6H_4(OH)CH_2$) are usually present.

Presence of Arg at position 2 of the peptide (and, less prominently at position 3, but not at position 1) causes abundant b_2-17 (and less so b_3-17) ions; this can help determining the order of the first 2–3 amino acids of the sequence when fragment ions between the first and second amino acids in peptides with predominant N-terminal fragments are absent in the spectrum. Although Arg is isobaric with (Val, Gly), the fact that the (Val, Gly) do not direct peptide fragmentation whereas Arg does, identification of Arg is not hindered. Presence of the basic Arg, Lys, His — particularly Arg — near the N- or C-terminus directs fragmentation and causes a number of low mass ions, which are common with immonium ions of some other amino acids; they are useful for inference only when Arg is *absent*.

Some abundant low-mass ions originating from Trp — which is isobaric to (Glu, Gly) and (Asp, Ala) — can be used for its identification; the problem of conflict is not serious, as Trp occurs relatively rarely.

These are a few observations;[199] further details with relevant references are available in the paper of Papayannopoulos.

---

[199]Papayannopoulos, I. A., (1995) The interpretation of collision-induced dissociation tandem mass spectra of peptides. *Mass Spectrom. Rev.*, **14**, 49–73.

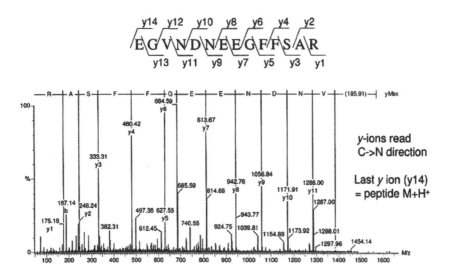

Figure 4.10: Low energy CID (Q-TOF) of doubly charged Glu-fibrinopeptide: *y*-series ions. Courtesy: Kevin Blackburn, Molecular and Structural Biochemistry, North Carolina State University. Reproduced by permission.

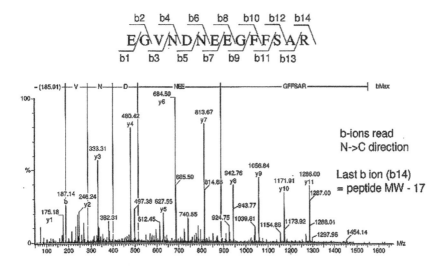

Figure 4.11: Low energy CID (Q-TOF) of doubly charged Glu-fibrinopeptide: *b*-series ions. Courtesy: Kevin Blackburn, Molecular and Structural Biochemistry, North Carolina State University. Reproduced by permission.

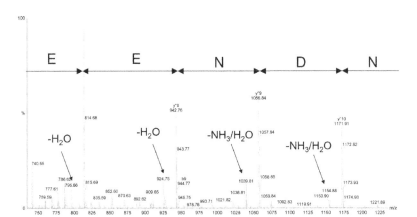

Figure 4.12: Side chain losses of H_2O and NH_3. Primary residues: E, D, Q, N. Courtesy: Kevin Blackburn, Molecular and Structural Biochemistry, North Carolina State University. Reproduced by permission.

4.1.4   Immonium ions

Immonium ions have the structure: $^{+}NH_2=CH\text{-}R$; these ions are produced from a-ions and y-ions and internal cleavage (see Fig. 4.6 and Fig. 4.9); the side-chain structure determines the stability of the ion. Immonium ion mass for glycine: $^{+}NH_2=CH_2$, $16+14=30$; for alanine: $^{+}NH_2=CHCH_3$, $16+13+15=44$.

Presence of the immonium ion of an amino acid in an MS/MS spectrum indicates the presence of that amino acid in the sequence of the peptide. A couple of applications of this will be shown later in Section 4.10.

4.1.5   Molecular weight determination of an intact protein from its ESI spectrum

In this section, we show how the molecular weight of an intact protein can be determined from its ESI spectrum. We use the spectrum of Fig. 4.13 for this purpose.

Molecular weight m of an intact protein can be determined from its ESI spectral peaks by picking *any* two peaks, writing equations for the m/z of the two peaks in terms of m and respective charges of the two peaks, and solving the two equations simultaneously. Here, we use two *successive* peaks $(m/z)_z$ and $(m/z)_{z+1}$ (which differ by $+1$ charge, the *lower m/z*

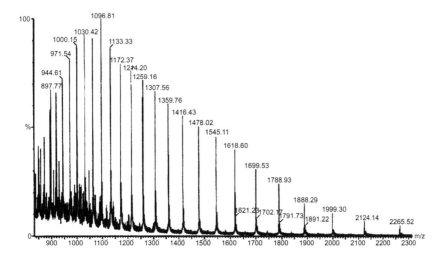

Figure 4.13: ESI spectrum of an intact protein. Courtesy: Kevin Blackburn, Molecular and Structural Biochemistry, North Carolina State University. Reproduced by permission.

peak carries one *more charge* (proton) with respect to the peak that is at its immediate *higher m/z* position).

We construct two equations, eqn 4.1 and eqn 4.2, the first one for the higher m/z peak, $(m/z)_z$, and the second one for the immediately lower m/z peak, $(m/z)_{z+1}$. For the ionic mass in the numerator in eqn 4.1 and eqn 4.2, we write the true mass m of the protein plus the proton masses depending on how many protons are present on that ion. That is,

$$\left(\frac{m}{z}\right)_z = \frac{m + z\,(1.0079)}{z} \qquad (4.1)$$

where 1.0079 is the mass of one proton in Dalton. Similarly, for the ion with $(z + 1)$ charge,

$$\left(\frac{m}{z}\right)_{z+1} = \frac{m + (z + 1)\,(1.0079)}{z + 1} \qquad (4.2)$$

To calculate the molecular weight m, let us take the peaks at 1133.33 Th (thomson, unit of mass-to-charge ratio) and 1096.81 Th from Fig. 4.13 as two successive peaks (the 1096.81 Th ion would be carrying one unit more charge than that of 1133.33 Th). By taking the 1133.33 Th as the higher mass peak (charge z), solving for m using eqn 4.1 we obtain

$$m = 1133.33\,z - 1.0079\,z = 1132.3221\,z \qquad (4.3)$$

We now substitute this for mass m in eqn 4.2, use 1096.81 for $(m/z)_{z+1}$, and obtain, by solving for z,

$$1096.81 = \frac{m}{z} = \frac{m + (z+1)\,(1.0079)}{z+1} = \frac{1132.3221\,z + (z+1)\,(1.0079)}{z+1}$$

$$(4.4)$$

$z = 30$. This means that the charge state of the m/z 1131.33 peak is $+30$. To solve for m of the unknown protein, we use eqn 4.1:

$$\left(\frac{m}{z}\right)_z = \frac{m + z\,(1.0079)}{z} = 1133.33 = \frac{m + 30\,(1.0079)}{30} \quad m = 33,969.663$$

$$(4.5)$$

We consider another example for calculation by taking another pair of peaks 1359.76 Th and 1307.56 Th from the same spectrum. Using eqn 4.1, we obtain, $m = 1358.7521\,z$. Substituting this in eqn 4.2, we obtain $z = 25$. Employing eqn 4.1, we obtain $m = 33,968.803$.

In Section 4.5, a summary of experimental observations on peptide fragmentation patterns will be given. In Section 4.6, we will discuss what is known about the mechanism of fragmentation. Finally, in Section 4.10, procedures for manual *de novo* sequencing from MS/MS data along with a few examples will be presented.

4.2 Alternative analysis methods

4.2.1 SRM, MRM

In selected reaction monitoring (SRM)[200] tandem mass spectrometry, in the first stage of the analysis, an ion of a particular mass is selected and is subjected to fragmentation (CID/CAD). In the next stage, an ion which is a product of the fragmentation (sometimes called *transition ion*) is selected for detection (continuous monitoring).

In multiple reaction monitoring (MRM), the first stage is the same as in SRM. In the second stage, unlike MRM, instead of a single ion, a small number (more than one — *multiple*) of (usually sequence-specific) fragment ions are continuously monitored. The analysis is rapid because a full scan is *not* required; also, it can increase the lower detection limit.

[200]Lange, V., Picotti, P., Domon, B., and Aebersold, R. (2008) Selected reaction monitoring for quantitative proteomics: a tutorial. *Mol. Syst. Biol.*, **4**, 222, October 14, 2008 doi: 10.1038/msb.2008.61.

4.3 Quantitation

4.3.1 Label-free, ICAT, SILAC, iTRAQ, TMT, AQUA

In quantitation, those using mass-difference approaches such as ICAT,[202] SILAC,[204] and ICPL[201] fall in one group — they are handled in one way, and those using isobaric reporter ion methods such as iTRAQ[205] and TMT[210] are handled in a different way. In the category of *relative quantitation* are (i) metabolic labeling, e.g., ^{15}N, SILAC; (ii) post harvest derivatization, e.g., ICAT, iTRAQ; (iii) enzymatic labeling, e.g., trypsin + $H_2^{18}O$. In relative quantitation, *control* and *treated* are mixed, the mix is digested, and MS analysis is carried out. From the ratio of peak intensities, quantitation is achieved. But in AQUA (*Absolute Quantitation*),[214] control and treated are *independently* digested, internal standard is added to both of them, and independent mass analysis is carried out (in both channels, internal standard peaks appear).

Label-free

In the label-free method of quantitation, differential expression level of two or more proteins in biological samples is determined *without* using any stable isotope to bind to peptide/proteins being analyzed.

ICAT — Isotope-coded affinity tags

This is a gel-free method for quantitative proteomics. There are two basic principles which are used in the ICAT (isotope-coded affinity tag)[202] method. The first one is the supposition that from a knowledge of a short sequence of contiguous amino acids (5–25 residues) from a protein, that unique protein can be identified. The second principle is that if two chemically identical (though isotopically different) tags are used, the tagged molecules not only would behave similarly in chromatography, but also would serve as ideal mutual internal standards for accurate quantification. The chemistry involves employing ICAT reagents which have three parts: an *affinity tag* (biotin), which is used to isolate ICAT-labeled peptides, a *linker* in which stable isotopes can be incorporated, and a reactive group with specificity toward thiol groups (cysteine). The linker reagent exists in two forms, light reagent (containing hydrogen) and heavy reagent (containing eight deuterium).[202] The ICAT procedure is shown in Fig. 4.14.

---

[201] Lottspeich, K. and Kellermann, J. (2011) ICPL (isotope coded protein label) labeling strategies for proteome research. *Methods Mol. Biol.*, **753**, 55–64.

[202] Gygi, S. P., Rist, B., Gerber, S. A., Tureček, F., Gelb, M. H., and Aebersold, R. (1999) Quantitative analysis of complex protein mixtures using isotope-coded affinity tags. *Nat. Biotechnol.*, **17**, 994–999.

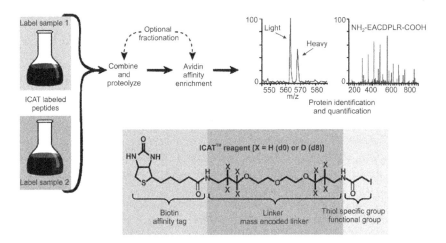

Figure 4.14: Quantitative proteomics with ICAT reagents. Source: www.imsb. ethz.ch/researchgroup/rudolfa/research. Courtesy: Ruedi Aebersold, Institute of Molecular Systems Biology, ETH Zürich. Reproduced by permission.

In the ICAT strategy, two protein mixtures representing two different cell states are treated with the isotopically light and heavy ICAT reagents — the ICAT reagent gets covalently attached to each cysteinyl residue in every protein (in one cell state all cysteines labeled with light ICAT, and in the other cell state they are labeled with heavy ICAT). The two cell states are then combined, optionally fractionated, and proteolyzed to peptides; ICAT-labeled peptides are isolated utilizing the biotin tag. (The biotin tag is there because biotin binds with avidin — a tetrameric biotin-binding protein — on the column and forms a strong complex, that is, a selective adsorption takes place. Later, using a buffer, complete elution of the adsorbed tagged peptides can be effected.) These peptides are separated by microcapillary HPLC. Since the ICAT-labeled peptides are chemically identical, they coelute in chromatography — they only show up differently in a screening mass scan with their intensity ratios expressing the strictly maintained ratios in the composition of the peptides (hence the proteins) in the mixture since MS intensity response is independent of the isotopic composition of the ICAT reagents. In the final MS/MS stage, the mass spectrometer operates in a dual mode — in successive scans it alternates between measuring relative quantities of peptides from the capillary column and determination of sequences of selected peptides. The sequence information is automatically generated from selection of a peptide ion of a certain m/z, CID of the latter, and automatically correlated with sequence databases and identification of the protein involved.

An increased number of protein identifications per experiment, without losing the accuracy of protein quantification, has been achieved using a new version of the ICAT reagent with an acid-cleavable bond, that allows removal of the biotin moiety prior to MS, and which utilizes ^{13}C substitution for ^{12}C in the heavy ICAT reagent rather than ^{2}H (for ^{1}H) as in the original reagent. This employs a single survey (precursor) ion scan and serial CIDs of four different precursor ions observed in the prior survey scan. Use of the new ICAT reagent with ^{13}C as the 'heavy' element rather than ^{2}H eliminates the slight delay in retention time of ICAT-labeled 'light' peptides on a C18-based HPLC separation that occurs with ^{2}H and ^{1}H.[203]

SILAC — Stable isotope labeling by amino acids in cell culture

SILAC was developed in Mathias Mann's laboratory in CEBI by Ong *et al.*[204] for mass spectrometry-based quantitative proteomics by *in vivo* incorporation of specific amino acids into mammalian proteins. In the method, mammalian cell lines are grown in media that lack a standard *essential amino acid* but is provided with a non radioactive, isotopically labeled form of that amino acid. Specifically, they used deuterated leucine (Leu-d3)(incorporation of ^{13}C, ^{15}N also are possibilities). It was found that in terms of cell morphology, doubling time, and ability to differentiate the growth of cells in those media were no different from growth in normal media. After five doublings, complete incorporation of Leu-d3 occurred. After harvesting, the protein populations from experimental and control samples can be mixed directly because the label is encoded directly into the amino acid sequence of every protein, and, since no more synthesis is taking place, no scrambling can take place at the amino acid level and purified proteins or peptides would preserve the exact ratio of the labeled to unlabeled protein. Since *prior* to lysis the unlabeled and labeled samples are combined, after enzymatic digestion *any* method of purification can be used and would not affect final quantitation. Additional purification steps to remove excess labeling reagents are not required. In comparison to cases of chemical incorporation, the amount of labeled protein required in SILAC is far less. Further, subcellular purifications of multiprotein complexes or organellar structures can be carried out as they exist in their native forms, no sample preparation bias because of separate preparation steps introduced. (To be noted here is that whereas primary culture of cells harvested from tissue is possible, SILAC-labeling of tissue samples directly is not

[203]Yi, E. C., Li, X., Cooke, K., Lee, H., Raught, B., Page, A., Aneliunas, V., Hieter, P., Goodlett, D. R., and Aebersold, R. (2005) Increased quantitative proteome coverage with ^{13}C/^{12}C-based, acid-cleavable isotope-coded affinity tag reagent and modified data acquisition scheme. *Proteomics*, **5**, 380–387.

[204]Ong, S. E., Blagoev, B., Kratchmarova, I., Kristensen, D. B., Steen, H., Pandey, A., and Mann, M. (2002) Stable isotope labeling by amino acids in cell culture, SILAC, as a simple and accurate approach to expression proteomics. *Mol. Cell. Proteomics*, **1**, 376-386.

possible.) After this, mass spectrometric identification is carried out. Since every leucine-containing peptide incorporates either all normal leucine or all Leu-d3, the mass spectrometric quantitation is straightforward. Except for the characteristic mass shift of fragments containing the leucine residues, fragmentation spectra of Leu-d0 and Leu-d3 containing peptides are largely identical; the shifts provide additional specificity to the assignment of peptide sequence tags. The SILAC procedure is shown in Fig. 4.15.

In a comparison between ICAT and SILAC, ICAT allows use of protein material from non living sources but requires chemical modification and affinity steps. There are a couple of other major differences between the ICAT and SILAC methods. One of them is that in ICAT, proteins need to be reduced and alkylated *before* mixing; during multiple fractionation steps this can make it difficult to maintain samples in directly comparable states. Another is, whereas ICAT (differentially) labels only somewhat more 20% of the tryptic peptides, SILAC (using leucine) labels more than half of them. In SILAC, fragmentation patterns are the same for both the labeled and the unlabeled peptide; in ICAT they are influenced by the functional group attached to the cysteines. In their initial work, they applied the technique to changes in protein expression during the process of mouse muscle cell C2C12 and NIH 3T3 differentiation; they also tried a variety of other cell lines such as HeLa, CHO-K1, COS-7, PC-12 demonstrating general applicability of the technique.

iTRAQ — Isobaric tag for relative and absolute quantification

To solve sample multiplexing problems iTRAQ (*i*sobaric *t*ags for *r*elative and *a*bsolute *q*uantiifcation)[205] arose. In its workflow, in the first step, after cell lysis to extract proteins (here some standard protein assay is used to estimate the protein concentration of each sample), the samples are reduced, alkylated, and subjected to enzymatic digestion — usually by trypsin — in an amine-free buffer system. After simple adjustment of the digest to 70% organic to control any hydrolysis, the resulting peptides are then labeled with isobaric iTRAQ reagents — *each* digest is labeled with a *different* iTRAQ reagent. The iTRAQ reagents typically have the following components (Fig. 4.16(a)): *a reporter group, a balance group* — these two constituting the isobaric tag, and *a peptide reactive group*. (In the Broad Institute procedure[206] they utilize up to four iTRAQ reagents, each consisting of an N-methyl piperazine reporter group with mass of 114, 115, 116, or 117; a balance group with mass of 31, 30, 29, or 28; and an N-hydroxy succinimide ester group that is reactive with the primary amines of

[205]Wiese, S., Reidegeld, K. A., Meyer, H. E., and Warscheid, B. (2007) Protein labeling by iTRAQ: a new tool for quantitative mass spectrometry in proteome research. *Proteomics*, **7**, 340–350.

[206]http://www.broadinstitute.org/scientific-community/science/platforms/proteomics/itraq.

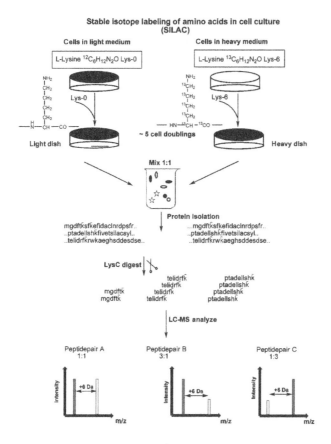

Figure 4.15: SILAC workflow ($\bar{\mathrm{k}}$ light, $\hat{\mathrm{k}}$ heavy counterparts). Courtesy: Marcus Krüger, Biomolecular Mass Spectrometry, Max Planck Institute for Heart and Lung Research, Bad Nauheim. Reproduced by permission.

peptides.) After the samples are labeled, they are combined. In the next step, the peptide mixtures are then identified and quantitatively analyzed by LC/MS/MS. The peptide sequence in the proteins is obtained from trypsin cleavage properties. Relative quantities of the peptides are obtained from the relative intensities of the reporter ions in MS/MS. In the case of simple samples, reagent by-products are removed using one-step elution from a cation exchange column and are then directly analyzed via LC/MS/MS; for complex samples, to reduce overall peptide complexity, cation exchange chromatographic fractionation is used.

The iTRAQ Reagent–8-plex workflow[207] for a duplex type experiment, the 8-plex methodology is as follows: In one part the control sample and

---

[207]www.absciex.com, p. 22.

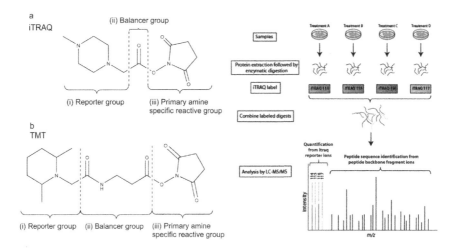

Figure 4.16: Left: iTRAQ reagent. Top (a) Applied Biosystems. Bottom (b) Thermo-Scientific. Right: iTRAQ Workflow, Broad Institute. Reproduced by permission.

in another part the test sample, in a *single* tube (the latter part ensures if any sample loss occurs it will affect all samples equally and inaccuracies in quantitation will be avoided) is reduced, cysteine is blocked, the sample is digested, and is labeled with an iTRAQ Reagent-8-plex. The control and the test sample digests are then combined into one sample mixture (here also ratios would be preserved in case of any sample loss), followed by clean-up and at times high-resolution fractionation, and finally, for protein identification and quantitation, analysis of the mixture by LC/MS/ MS. In the 8-plex method, in a single experiment up to eight samples can be analyzed.

Commercially available isobaric chemical tags: [From Applied Biosystems] Fig. 4.16 left top (a) 4-Plex iTRAQ, with masses of reporter groups of 114–117, balancer group of 28–31, for a constant total mass of 145; note that the 8-plex iTRAQ (not shown) has reporter groups of 113–121 (120 skipped) and balancer groups of 184–192 (185 skipped). [From Thermo-Scientific] Fig. 4.16 left (b) 6-Plex TMT (next section) with masses of reporter groups of 126–131, balancer group of 99–104, for a constant total mass of 230. Carbon, nitrogen, and/or oxygen isotopes are included in the reporter and/or balancer groups to obtain the required mass.

Pappin makes a number of helpful recommendations.[208] In sample preparation and the iTRAQ reaction: for very complex samples, a 2-D separation

[208]Pappin, D. (2011) An iTRAQ Primer — www.ushupo.org/Portals/0/USHUPO_ Tech_Talk_iTraq.pdf (May 5, 2011).

is an absolute must. For the first dimension, the most practical approach is to increase the number of SCX fractions. For the OFFGEL device, it is desirable to opt for 24 fractions; 12–16 salt-step gradients, when classical MudPIT is used. For the second dimension, gradient profiles should be lengthened to 2 hours or more, longer — 20 cm or greater — analytical columns with finer — $< 5\,\mu$ — particle size. It is important to use the right buffers. Whereas the iTRAQ chemistry can stand considerable variations in salts and detergents, primary (Tris) or secondary amine buffers and to some extent free thiols very rapidly quench it. It is important to work at high alcohol concentration. At $\sim 70\,\%$ v/v alcohol, half-life of iTRAQ is 15 minutes, at $\sim$ pH 8 it is around 15 seconds. As regards changing to acetonitrile, at 70% v/v can severely suppress the reaction rate. To improve solubility of hydrophobic peptides or proteins, longer-chain alcohols can work well, even DMSO and DMF, though they are more difficult to remove. Seventy percent aqueous pyridine also can work well; it is a good peptide solvent and also its own buffer. The reaction buffer is kept at 120–150 mM in the reaction mixture. Since the iTRAQ reagents hydrolyze to form N-methyl piperazine acetic acid and N-hydroxysuccinimide, without adequate buffering the medium would get acidic and the reaction with peptide amine will stop as the latter gets fully protonated. Triethylammonium bicarbonate (TEAB) is chosen because it is a good tertiary amine buffer and if required can be easily removed since it is very volatile. The likely higher hydrophobicity (and therefore relative insolubility) of iTRAQ-labeled peptides — because they are larger in size in comparison to the corresponding native peptides — can be dealt with by increasing the ACN content to 5%/0.1% (or greater) formic acid in LC loading solutions. Also steepening LC ACN gradients (by $\sim 10\%$ at the final %B concentration) can help improve chromatographic resolution of labeled peptides. For both SCX and OFFGEL, the total salt concentration at sample loading must be kept below ~ 10 mM to ensure proper binding (SCX) or focusing (OFFGEL). After the iTRAQ reaction, some reagents, principally the hydrolysis product N-methylpiperazine acetic acid present at ~ 40 mM concentration in the standard reaction volume, remain behind — not easily removed — they require removal by C_{18} cartridge or stage-tip cleanup before SCX or OFFGEL loading. In data acquisition: since iTRAQ-labeled peptides are more stable to CID, it is helpful to increase by 10–15% the collision energy relative to native peptides — higher energy would give stronger reporter signals. To minimize precursor overlap, the mass isolation window should be optimized. It is preferable to fragment the $[M+2H]^{2+}$ ions than a higher charge ion, because this leads to higher-intensity iTRAQ reporter ions relative to backbone fragments, unless a higher charge ion is very much more intense. In data processing, better peptides should be used (expectation values > 0.05) for calculating protein level iTRAQ ratios, or even unique peptides, even though this might reduce the number of measurements available for ratio calculations. Intensity weighting should be applied such that in the final calculated ratio the more intense spectra contribute more. Appropriate normalization should be applied — only global normalization with large datasets ($> 1,000$ proteins) — by first summing, in each iTRAQ channel, all the absolute reporter ion intensities. And correction for overlapping isotope contributions, from natural isotope abundance ($+1$, $+2$ Da) as well as from any incomplete enrichment at any carbon and nitrogen position (-1, -2 Da). Fold changes can be ascribed in terms of, say, $> 2\,SD$ up or down, and this can be used to quickly judge the quality of an experiment undertaken. To arrive at significance, all peptides $p > 0.05$, in a given iTRAQ channel, could be taken for all ratio measurements, mean and standard deviations calculated and normalized. The mean should be at or around unity, SD in the range 0.15–0.5 — in place of a detailed statistical analysis.[209] Interference by fragment ions is sequence specific, thus rare; its effects are canceled out from averaging of multiple peptide measurements. Isotope impurity overlaps are of the order of $\sim 3\%$ for the 4-plex reagents and $\sim 1\%$ for the 8-plex reagents; this would be negligible because in real biological samples global standard deviations are of $\sim 25\%$, and the corrections would be only one-tenth of the latter.

TMT — Tandem mass tag

TMT[210] is isobaric mass tag mass spectrometry, similar to iTRAQ; *tandem* refers to the use of MS/MS for the analysis of these tags. Pairs of TMT-tagged peptides are chemically identical — designed to help in comigration in chromatographic separations; they also have same overall mass. TMT quantification of tagged analytes depends on use of CID-based technique; the operation is entirely in MS/MS mode and it ignores untagged material thus greatly improving data quality.

TMT tag, shown in Fig. 4.17, comprises a *sensitization group* and a *mass differentiation group* (such as CH_3 or CD_3 in the figure) which together comprise the TMT fragment that is actually detected. The TMT fragment is linked to a *mass normalization group* (to match the mass differentiation group, correspondingly CD_3 or CH_3) which ensures each tag in a pair of tags shares the same overall mass and atomic composition. Lastly, for the tag to be coupled to any peptide, the tag comprises a *reactive*

[209]Karp, N. A., Huber, W., Sadwoski, P. G., Charles, P. D., Hester, S. V., and Lilley, K. S. (2010) Addressing accuracy and precision issues in iTRAQ quantitation. *Mol. Cell. Proteomics*, **9**, 1885–1897.
[210]Thompson, A., Schäfer, J., Kuhn, K., Kienle, S., Schwarz, J., Schmidt, G., Neumann, T., and Hamon, C. (2003) Tandem mass tags: a novel quantification strategy for comparative analysis of complex protein mixtures by MS/MS. *Anal. Chem.*, **75**, 1895–1904.

functionality R. For R, they used one of a number of peptide sequences. In second generation tags, an additional fragmentation-enhancing group, proline, was incorporated. The TMT fragment (b_2-ion) that is expected to result from the markers—obtained based on theories of protonation-dependent mechanisms of backbone fragmentation—is shown in Fig. 4.17 (c).[210]

Dayon *et al.*[211] carried out proof of principle studies with 6-plex version of the TMT, first with relative quantification of proteins on a model protein mixture, then on human cerebrospinal fluid (CSF) samples. Human postmortem (PM) CSF, taken as a model of massive brain injury, was compared with antemortem (AM) CSF. After immunoaffinity depletion, AM and PM CSF pooled samples were reduced, alkylated, digested by trypsin, and labeled, respectively, with six isobaric variants of TMT. The samples were pooled and fractionated by SCX and, after separation by RPLC, were quantified by MS/MS. In PM CSF samples 78 identified proteins (some—such as GFAP, S100B, PARK7—previously described as brain damage biomarkers) showed clear increase over AM CSF samples; ELISA of these proteins confirmed their elevated concentrations.

Nilsson *et al.*[212] have used TMT tagging (TMTsixplex—Isobaric Mass Tag kit, Part No. 90064) in quantitative phosphoproteomics to follow alterations in key protein expression—they identified 3414 proteins, mass spectrometrically observed alterations in key protein expressions confirmed by rapid (< 1 h) Western blotting technique—in their study of GSC11 glioblastoma stem cells.

Three varieties of TMT are available at present: (i) TMTzero—a nonisotopically substituted core structure; (ii) TMTduplex—an isobaric pair of mass tags with a single isotopic substitution;[210] and (iii) TMTsixplex—an isobaric set of six mass tags with five isotopic substitutions.[211] In Fig. 4.18, the workflow in quantitative proteomics is indicated and also some precision/accuracy benchmarking data in stable isotope labeling based quantitative proteomics.[213]

---

[211]Dayon, L., Hainard, A., Licker, V., Turck, N., Kuhn, K., Hochstrasser, D. F., Burkhard, P. R., and Sanchez, J. C. (2008) Relative quantification of proteins in human cerebrospinal fluids by MS/MS using 6-plex isobaric tags. *Anal. Chem.*, **80**, 2921–2931.

[212]Nilsson, C. L., Dillon, R., Devakumar, A., Shi, S. D., Greig, M., Rogers, J. C., Krastins, B., Rosenblatt, M., Kilmer, G., Major, M., Kaboord, B. J., Sarracino, D., Rezai, T., Prakash, A., Lopez, M., Ji, Y., Priebe, W., Lang, F. F., Colman, H., and Conrad, C. A. (2010) Quantitative phosphoproteomic analysis of the STAT3/IL-6/HIF1alpha signaling network: an initial study in GSC11 glioblastoma stem cells. *J. Proteome Res.*, **9**, 430–443.

[213]Altelaar, A. F. M., Frese, C. K., Preisinger, C., Hennrich, M. L., Schram, A. W., Timmers, T. M., Heck, A. J. R., and Mohammed, S. (2013) Benchmarking stable isotope labeling based quantitative proteomics, *J. Proteomics*, **88**, 14–26.

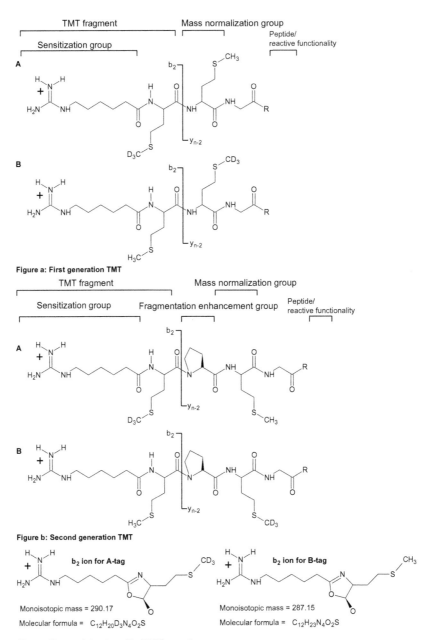

Figure 4.17: The structures of two versions of the TMT markers are shown in parts (a) and (b). The proposed TMT fragment that results from the markers, based on theories on protonation-dependent mechanism, is shown in part (c). Reproduced by permission of the American Chemical Society from *Analytical Chemistry*, Vol. 75, 1895–1904, 2003.

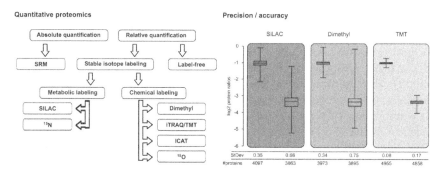

Figure 4.18: Benchmarking stable isotope labeling-based quantitative proteomics. Left: Workflow. Courtesy: Albert J. R. Heck, Bijvoet Center for Biomolecular Research and Utrecht Institute for Pharmaceutical Sciences, Utrecht University, Utrecht. Right: Comparison scheme. Reproduced by permission of Elsevier B. V. from *Journal of Proteomics*, Vol. 88, 14–26 (2013).

AQUA — Absolute quantification

AQUA[214] quantification (up to copy number per cell) is based on SRM/ MRM. An *AQUA peptide* is a synthetic tryptic peptide, which incorporates one stable isotope labeled amino acid and is a few daltons heavier than the peptide of interest. In a quantitation experiment based on AQUA strategy, a known amount of the AQUA peptide is added as an internal standard to the biological protein sample. The sample mixture is then digested and analyzed by HPLC-MS. A comparison of peak ratios in the ion chromatograms generated for the native peptide and the AQUA peptide yields the quantity of the native peptide. In the following are some details of the method mostly from a protocol by Kettenbach *et al.*[215] The major aspects of the protocol are the peptide design, peptide synthesis and validation, method optimization, sample preparation, and analysis.

For the internal standard required, any peptide that is a proteolysis product of the protein to be analyzed can be selected — one or two additional peptides could be used to ensure accurate quantification if unknown modifications of the endogeneous protein are possible. Most important is that the chosen peptide should be unique to the target protein under quantification. Also, it is important that the abundance of *all* forms of the peptide be measured for absolute quantification. Ideally, two or three peptides are chosen per target protein to ensure accurate quantification and exclude

---

[214]Gerber, S. A., Rush, J., Stemman, O., Kirschner, M. W., and Gygi, S. P. (2003) Absolute quantification of proteins and phosphoproteins from cell lysates by tandem mass spectrometry. *Proc. Natl. Acad. Sci., U.S.A.*, **100**, 6940–6945.

[215]Kettenbach, A. N., Rush, J., and Gerber, S. A. (2011) Absolute quantification of protein and post-translational modification abundance with stable isotope-labeled synthetic peptides. *Nature Protocols*, **6**, 175–186.

artifacts that can originate from unknown modifications of the endogenous protein. When the quantitation target is a specific splice variant, isoform, or post translational modification, the site of modification limits the choice of the peptide which only alternative proteases for digestion can marginally modify. Usually, peptides with optimal chromatographic behavior, known ionization, fragmentation properties, and SRM transitions are chosen. Here it is important to take into consideration effects of certain amino acid sequences of the selected protease on the cleavage pattern. Such as, whereas trypsin normally cleaves at the carboxylic side of arginine and lysine, with proline present at the carboxylic side the bond is resistant to trypsin cleavage. Or, trypsin cleavage may be inhibited if the amino acid on the amino-terminal side is phosphorylated. Or, when a series of arginines and lysines are present in a protein, trypsin might cleave after the first arginine, or lysine, or after any of the following those thus creating so-called *ragged ends*. Besides ragged ends, it is desirable to avoid chemically reactive residues like Trp, Met, Cys, and peptides with very few residues. If by discovery experiments all the resulting peptide sequences can be established, several AQUA peptides can be synthesized on that basis; or one AQUA peptide that spans the cleavage site can be synthesized and prior to digestion introduced into the sample, expecting the digestion to be similar to its native counterparts. It is desirable to avoid methionine-containing peptides because methionine may be differentially oxidized during sample preparation; if that is not possible, all methionines should be chemically oxidized to reduce the number of peptide species. For proteins with post translational modifications, the absolute amount of modified peptide can be determined; then using a second AQUA peptide and measuring the amount of unmodified peptide, the site occupancy can additionally be ascertained.

In AQUA peptide synthesis, peptides with peptide sequence corresponding to the ones generated during digestion of the endogeneous protein are chosen and stable isotopes incorporated. Although ^{18}O, ^{13}C, ^{2}H, or ^{15}N all are possibilities, ^{13}C or ^{15}N are by preference used, because they do not lead to chromatographic retention shifts. Usually, into an AQUA peptide, one heavy isotope-labeled leucine, proline, valine, phenylalanine, or tyrosine is incorporated, which leads to a mass shift of 6–8 Da. When tryptic peptides are involved, often the C-terminal arginine or lysine is heavy isotope-labeled, for convenience in monitoring the resulting y-ion series.[216]

For validation, the AQUA peptide synthesized is LC-MS-MS analyzed for verification of its chromatographic behavior and fragmentation spectra.

For method optimization, in SRM-based methods, the MS/MS spectra of the most intense precursor ions of the AQUA peptides are collected for

---

[216]Sigma-Aldrich, Thermo Fisher Scientific or Cell Signaling Technologies are some commercial vendors that conduct peptide synthesis.

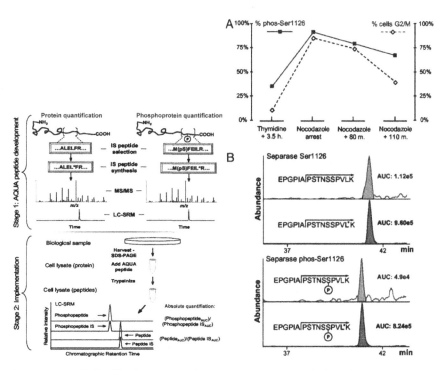

Figure 4.19: Left: Absolute quantification of proteins and phosphoproteins using the AQUA strategy. The strategy has two stages. Stage 1 involves the selection and standard synthesis of a peptide (or phosphopeptide denoted by pS) from the protein of interest. These peptide internal standards are analyzed by MS/MS to examine peptide fragmentation patterns. Stage 2 is the implementation of the new peptide internal standard for precise quantification. Reproduced by permission of the National Academy of Sciences, U.S.A. from *Proc. Natl. Acad. Sci., U.S.A.* Vol. 100, 6940–6945, 2003. Right: Quantitative analysis of the phosphorylation state of Ser-1126 phosphorylation from human separase protein as a function of cell cycle. Some of the data used for the figure are from Stemmann *et al.*[218] Courtesy: Steven Gygi, Department of Cell Biology, Harvard Medical School. Reproduced by permission.

potential later use. Fragment ions at m/z ratios higher than the precursor ion have been found to be often more suitable for monitoring because of reduced noise compared with the lower m/z ions.

In sample preparation, the biological samples are lysed and protease digested — tryptic digests generate peptides with charge states ideal for MS/MS. The AQUA peptide is introduced during digestion; this covers all likely peptide losses in the steps that follow. With complex samples, sometimes, to reduce complexity, the samples are fractionated before digestion.

Also, complete proteolysis is essential, since the abundance measurement is based on the abundance of the resulting peptides. This may require increasing amounts of protease and/or longer time periods.

High attomole detection limits have been possible using AQUA methods. Yeast proteins of low abundance (CBV:01) can be quantitated, using AQUA, from ∼ 50 μg total cell lysate. It has also been reported that triplicate measurements for well-designed AQUA experiments typically produce coeffcients of variation between 8% and 15% relative standard deviation. The strategy of AQUA analysis is shown in Fig. 4.19.

There have been a number of AQUA/SRM/MRM studies, e.g., by Lange *et al.*,[200] Barr *et al.*,[217] Stemmann *et al.*,[218] Gerber *et al.*,[214] Kuhn *et al.*,[219] Kirkpatrick *et al.*,[220] and Kirkpatrick *et al.*[221] Also, Mayya *et al.* examined[222] AQUA applicability to quantitative study of multisite phosphorylation. They studied the conserved inhibitory site of the cyclin-dependent kinases (CDKs) — modulation of inhibitory phosphorylation at Thr[14] and Tyr[15] by complex regulatory mechanisms. They followed quantitative dynamics among the four possible phosphorylated and non-phosphorylated versions of CDKs — T14p-Y15p, T14p-Y15, T14-Y15p, and T14-T15 by using heavy isotope-labeled tryptic peptides spanning the inhibitory site as internal standards and quantified all four versions by LC-SRM. From quantification of the phosphorylation status of the inhibitory site in the cell extracts, they observed that the transition to mitotic phase was dominated by the conversion of T14p-Y15p to the T14-Y15 form, whereas the two monophosphorylated forms were considerably lower in abundance. Further, the amount of all four forms decreased during the progression of apoptosis but with differing kinetics. From quantitative immunoblotting using antibodies to Cdk1 and Cdk2 and to the T14-Y15p, their conclusion was that

[217]Barr, J. R., Maggio, V. L., Patterson, D. G. Jr., Cooper, G. R., Henderson, L. O., Turner, W. E., Smith, S. J., Hannon, W. H., Needham, L. L., and Sampson, E. J. (1996) Isotope dilution–mass spectrometric quantification of specific proteins: model application with apolipoprotein A-I. *Clin. Chem.*, **42**, 1676–1682.

[218]Stemmann, O., Zou, H., Gerber, S. A., Gygi, S. P., and Kirschner, M. W. (2001) Dual inhibition of sister chromatid separation at metaphase. *Cell*, **107**, 715–726.

[219]Kuhn, E., Wu, J., Karl, J., Liao, H., Zolg, W., and Guild, B. (2004) Quantification of C-reactive protein in the serum of patients with rheumatoid arthritis using multiple reaction monitoring mass spectrometry and 13-C labeled peptide standards. *Proteomics*, **4**, 1175–1186.

[220]Kirkpatrick, D. S., Gerber, S. A., and Gygi, S. P. (2005) The absolute quantification strategy: a general procedure for the quantification of proteins and post-translational modifications. *Methods*, **35**, 265–273.

[221]Kirkpatrick, D. S., Hathaway, N. A., Hanna, J., Elsasser, S., Rush, J., Finley, D., King, R. W., and Gygi, S. P. (2006) Quantitative analysis of *in vitro* ubiquitinated cyclin B1 reveals complex chain topology. *Nat. Cell Biol.*, **8**, 700–710.

[222]Mayya, V., Rezual, K., Wu, L., Fong, M. B., and Han, D. K. (2006) Absolute quantification of multisite phosphorylation by selective reaction monitoring mass spectrometry: determination of inhibitory phosphorylation status of cyclin-dependent kinases. *Mol. Cell. Proteomics*, **5**, 1146–1157.

quantification by AQUA was reliable and accurate. Langenfeld *et al.* described[223] the assay of human cytochrome P450 2D6 in a set of 30 genotyped liver samples using AQUA quantitation and found approximately 30 fmol CYP2D6 per μg of microsomal protein, with the values spanning from 0 to nearly 80 fmol/μg. They validated their results by quantitative Western blotting, calibration standards, and activity assays and concluded that by AQUA technique a true assay of CYP2D6 was possible. Xu *et al.*[224] monitored polyubiquitin chains of different linkage by AQUA. They observed that the unconventional linkages are abundant *in vivo* and that all non-K63 linkages may target proteins for degradation. They profiled both the entire yeast proteome and ubiquitinated proteins in wild-type and ubiquitin K11R mutant strains. From that, they identfied K11 linkage-specific substrates, including a ubiquitin-conjugating enzyme involved in endoplasmic reticulum-associated degradation. Papaioannou *et al.*,[225] using AQUA, validated that a large proportion of protein (50 out of 130) are up-regulated by more than 1.3-fold in testes lacking Sertoli cell-Dicer — an RNaseIII endonuclease required for microRNA biogenesis.

4.4 Peptide fragmentation: Experiments

4.4.1 MS/MS methods—CAD (CID), ECD, ETD, EDD

In this section, we mention some common MS/MS methods and give a few examples of ECD, ETD and EDD reactions.

Dissociation via vibrational activation: In low energy CAD/CID, this can be carried out in a collision cell without confinement or in a trap-type environment. In the former, the number of collisions (target is usually nitrogen or argon) is necessarily small, 1–5, because there is *no* confinement; *laboratory* kinetic energy is in the range 20–100 eV. In the trap environment, there can be hundreds of collisions (helium is usually the target gas) with laboratory kinetic energy again in the range 20–100 eV. In the former, the collisions could be induced after a precursor ion is chosen from a first-stage of mass analysis (using Q, 2D-LT), then the final analysis of the fragmentation products carried out by any of Q, TOF, ICR, Orbitrap. In

---

[223]Langenfeld, E., Zanger, U. M., Jung, K., Meyer, H. E., and Marcus, K. (2009) Mass spectrometry-based absolute quantification of microsomal cytochrome P450 2D6 in human liver. *Proteomics*, **9**, 2313–2323.

[224]Xu, P., Duong, D. M., Seyfried, N. T., Cheng, D., Xie, Y., Robert, J., Rush, J., Hochstrasser, M., Finley, D., and Peng, J. (2009) Quantitative proteomics reveals the function of unconventional ubiquitin chains in proteasomal degradation. *Cell*, **137**, 133–145.

[225]Papaioannou, M. D., Lagarrigue, M., Vejnar, C. E., Rolland, A. D., Kühne, F., Aubry. F., Schaad, O., Fort, A., Descombes, P., Neerman-Arbez, M., Guillou, F., Zdobnov, E. M., Pineau, C., and Nef, S. (2011) Loss of Dicer in Sertoli cells has a major impact on the testicular proteome of mice. *Mol. Cell. Proteomics*,**10**, M900587-MCP200.

the latter, either an LT, IT, ICR, or Orbitrap could be used to provide necessary collisional environment in a trap.

CAD of multiply charged polypeptide cations — this is very effective for cationic peptide ions. It is easy to carry out fragmentation; product ions have abundant sequence ions for tryptic peptides, which can be used for peptide and hence protein identification through a database search.

In electron capture dissociation (ECD), a multiply protonated molecule interacts with free electron — fragmentation ensues. An example of electron capture dissociation[226] is given below — the reaction carried out in an ICR trap:

$[M + 3H]^{3+}$ (multiply protonated peptides or proteins) $+ e^-$ (near thermal energy electron, such as those produced by a conventional heated filament) $\rightarrow [C + 2H]^{1+} + [Z + H]^{1+ \cdot}$

The following is another example of ECD of polypeptide cations:

$[M + 3H]^{3+} + e^- \rightarrow [M + 2H]^{2+ \cdot}$; $[M + 3H]^{2+ \cdot} \rightarrow [C + 2H]^{1+}$ (c) + $[Z + H]^{1+}$ (z) — c and z product ions produced; b, y ions are more characteristic of vibrational excitation.

In electron detachment dissociation (EDD),[227] the interacting electron knocks off electrons *from ions:*

$[M + 3H]^{3-}$ (M negative ion) $+ e^-$ (electron kinetic energy ~ 20 eV $\rightarrow [a - 2H]^{1- \cdot} + [X - H]^{1-} + 2e^-$ (more than an electron capture; electron removes electrons from the ion) — reaction performed in FT-ICR trap.

Given below is an example of EDD of polypeptide anions:

$[M - 3H]^{3-} + e^- \rightarrow [M - 3H]^{2- \cdot} + 2e^-$; $[M - 3H]^{2- \cdot} \rightarrow [a - 2H]^{1-}$ (peptide anion) $+ [x - H]^{1- \cdot}$ (peptide anion)

In electron transfer dissociation (ETD),[228] the reagent anion is absorbed by the ion and fragmentation follows:

$[M + 3H]^{3+} + A^-$ (radical anion, prefocused in quadrupole linear ion trap) $\rightarrow [C + 2H]^{1+} + [Z + H]^{1+ \cdot} + A$

Another example of ETD is given below, which involves proton transfer/electron transfer partitioning:

$[M + 3H]^{3+} + A^-$ ... it can proceed in two ways, *proton transfer pathway:* $\rightarrow [M + 2H]^{2+}$, or, *electron transfer pathway:* $\rightarrow [M + 3H]^{2+ \cdot}$. Relative partitioning depends mainly on reagent anion.

There is also negative electron transfer dissociation (NETD) which involves dissociation of negative polypeptide ions by positive ions such as C^+.

[226]Zubarev, R. A., Kelleher, N. L., and McLafferty, F. W. (1998) Electron capture dissociation of multiply charged protein cations, A nonergodic process. *J. Am. Chem. Soc.*, **120**, 3265–3266.

[227]Budnik, B. A., Haselmann, K. F., and Zubarev, R. A. (2001) Electron detachment dissociation of peptide dianions: an electron-hole recombination phenomenon. *Chem. Phys. Lett.*, **342**, 299–302.

[228]Syka, J. E. P., Coon, J. J., Schroeder, M. J., Shabanowitz, J., and Hunt, D. F. (2004) Peptide and protein sequence analysis by electron transfer dissociation mass spectrometry. *Proc. Nat. Acad. Sci., U.S.A.*, **101**, 9528–9533.

These ion/electron and ion/ion reaction-based methods work best with high charge densities — with lower charge densities signal intensity is affected. Some of their strengths are protein ions, non-tryptic peptide cations, highly basic peptide cations (in ECD/ETD), and polypeptide anions yield extensive sequence ions; sequence ion relative abundance variability is less; and post translational modifications are conserved.

4.4.2   Use of H-D exchange

An interesting possibility in probing *protein conformation* in solution is to subject the protein to deuteration and follow the nature and extent of the deuteration by mass spectrometry. There are some inherent advantages as well as disadvantages in this HD-exchange method. The advantage: H exchanges with D in solution easily; the disadvantage: since the exchange process is reversible, equally easily the deuterium that has found a place in the protein can return to the solution. So, it is important that after deuteration the deuterated form is suitably held stable, and this is achieved usually by drastically cutting down the rate of the reverse process, such as by quenching the solution to a much lower temperature. At that lower temperature, proteolysis can be carried out and the resulting peptides then identified by MS or MS/MS.

The digestion must be done at low temperature under acid condition. Trypsin is *not* suitable; pepsin is mostly used. *Online* pepsin digestion also has been used; then, in comparison to in-solution digestion a much higher enzyme:protein ratio can be used. However, in either case, since products of pepsin proteolysis are not predictable, all resulting peptides must be identified by MS/MS. By employing denaturants and reducing agents in the digestion step, the digestion efficiency can be improved. Using improved digestion techniques, even highly resistant proteins can be completely digested within 20–30 s at 0°C, with production of a number of overlapping peptides. Efficiency of hydrogen exchange as a function of pH is shown in Fig. 4.20.[229]

Chromatographic performance at low temperature can be improved by employing ultra-performance liquid chromatography (UPLC). It utilizes sub-2 μm particles — this requires instrumentation that can operate at pressures in the 6,000–15,000 psi range — with high linear solvent velocities. This results in increases in resolution, sensitivity, and also in speed of analysis. For a 10-min separation, the typical peak widths under such condition are in the order of 1–2 s. The recovery of deuterium is more

---

[229]Engen, J. R., Wales, T. E., and Shi, X. (2011) Hydrogen exchange mass spectrometry for conformational analysis of proteins. *Encyclopedia of Analytical Chemistry: Applications, Theory and Instrumentation, Supplementary Volumes S1–S3*, Robert A. Meyers (Editor), 1–17.

satisfactory. Plumb *et al.*[230] have applied this technology to the study of *in vivo* drug metabolism. The peak-width reduction significantly increases analytical sensitivity by three- to five-fold. A narrower peak width brings about a peak capacity increase and significantly reduces spectral overlap. The application of UPLC/MS results in superior separation and has led to detection of additional drug metabolites.

Several experimental parameters need to be controlled stringently, such as the pH, temperature, duration of exchange period, because all these materially affect the degree of exchange. Better pH control through buffers, reproducible pipetting, and transfer time contribute to superior reproducibility in pH-related steps and in turn to overall reproducibility. In the HD exchange experiments, the deuterium labeling time is considerable; deuterium exposure can take several (even 4–8) hours. HD exchange experiments with protein size of 300 kDa are not uncommon these days. Processing of data is by software, and this takes substantial time — order of hours. The time depends on the size of the protein and number of replicates. For a protein of relatively small size (50 kDa), complete exchange and MS can be done in a day; in several days a full set of experiments with several replicates can be completed. There have been several reviews. Baldwin's article[231] traces the beginnings of the HX method, there are several reviews on more recent work,[232,233,234,235,236,237,238,239,240] a "tutorial review" by

[230]Plumb, R., Castro-Perez, J., Granger, J., Beattie, I., Joncour, K. and Wright, A. (2004) Ultra-performance liquid chromatography coupled to quadrupole-orthogonal time-of-flight mass spectrometry. *Rapid. Commun. Mass Spectrom.*, **18**, 2331–2337.

[231]Baldwin, R. L. (2011) Early days of protein hydrogen exchange: 1954–1972. *Proteins*, **79**, 2021–2026.

[232]Woodward, C. (1999) Advances in protein hydrogen exchange by mass spectrometry. *J. Am. Soc. Mass Spectrom.*, **10**, 672–674.

[233]Engen, J. R. and Smith, D. L. (2001) Peer reviewed: Investigating protein structure and dynamics by hydrogen exchange MS. *Anal. Chem.*, **73**, 256A–265A.

[234]Hoofnagle, A. N., Resing, K. A., and Ahn, N. E. (2003) Protein analysis by hydrogen exchange mass spectrometry. *Annu. Rev. Biophys. Biomol. Struct.*, **32**, 1–25.

[235]Englander, S. W. (2006) Hydrogen exchange and mass spectrometry. *J. Am. Soc. Mass Spectrom.*, **17**, 1481–1489.

[236]Wales, T. E. and Engen, J. R. (2006) Hydrogen exchange mass spectrometry for the analysis of protein dynamics. *Mass Spectrom. Rev.*, **25**, 158–170.

[237]Engen, J. R. (2009) Analysis of protein conformation and dynamics by hydrogen/ deuterium exchange MS. *Anal. Chem.*, **81**, 7870–7875.

[238]Brock, A. (2012) Fragmentation hydrogen exchange mass spectrometry: A review of methodology and applications. *Protein Expr. Purif.*, **84**, 19–37.

[239]Iacob, R. E. and Engen, J. R. (2012) Hydrogen exchange mass spectrometry: are we out of the quicksand? *J. Am. Soc. Mass Spectrom.*, **23**, 1003–1010.

[240]Wales, T. E., Eggertson, M. J., and Engen, J. R. (2013) Considerations in the analysis of hydrogen exchange mass spectrometry data. *Methods Mol. Biol.*, Vol. 1007, 263–288. Chapter 11 in Rune Matthiesen (ed.), *Mass Spectrometry Data Analysis in Proteomics*.

Konermann *et al.*,[241] and a special issue with 21 contributions;[242] the review by Iacob and Engen[239] provides a critical assessment of the efficiencies of various aspects of the HX-MS experiments.

The first demonstration of H-D exchange in proteins followed by mass spectrometry was by Katta and Chait. They had earlier studied[243] conformational changes in bovine cytochrome c by mass spectrometry; by adding the H-D exchange method, they examined[244] conformational changes in bovine ubiquitin arising from addition of methyl alcohol to aqueous acidic solutions of the protein — bovine ubiquitin, in its native form, is a small, tightly folded protein having 76 residues, molecular mass 8565.0 Da, no disulfide bonds, and 13 basic sites that is very resistant to denaturation. For those ubiquitin experiments in which it was desired to maintain the protein in a *tightly folded* state, the spray solution was prepared by dissolving bovine ubiquitin in D_2O containing 1% CH_3COOD at concentrations of 10–30 μmol. After dissolution of the protein, mass spectra were obtained as a function of time. For investigations of the *unfolded* states of bovine ubiquitin, an equal volume of CH_3OD was added dropwise to the aqueous ubiquitin solution (1% CH_3COOD in D_2O) immediately after dissolution of the protein — denaturation effected by the addition of CH_3OD exposed a large number of labile hydrogens to exchange. Mass spectra were then obtained as a function of time. It was observed that for a tightly folded protein, the exchange was much less than complete. Katta and Chait[244] reported that the exchange could be followed with < 100 pmol of protein; at present, analysis can be done with as little as 1-5 pmol of protein per injection. Instrument sensitivity is a very minor issue for HX MS.

Zhang and Smith studied[245] horse heart cytochrome c and subjected it to H-D exchange. The protein sample was incubated in D_2O (pD 6.8) as a function of time and temperature to effect isotopic exchange; the reaction mixture was then transferred to slow exchange conditions (pH 2.4, 0°C) and fragmented with pepsin (0°C, 10 min). Quantification of peptide amide deuterium in proteolytic peptides was done from molecular weights through analysis of the digest by HPLC-FABMS (0°C). Hydrogen exchange reactions with half-lives between 0.3 min and 350 h were followed; deuterium content of peptides could be determined with an uncertainty of

---

[241]Konermann, L., Pan, J., and Liu, Y. (2011) Hydrogen exchange mass spectrometry for studying protein structure and dynamics. *Chem. Soc. Rev.*, **40**, 1224–1234.

[242]Engen, J. R. and Jorgensen, T. J. D. eds. (2011) Hydrogen exchange mass spectrometry. *Int. J. Mass Spectrom.*, **302**, 1–174.

[243]Chowdhury, S. K., Katta, V., and Chait, B. T. (1990) Probing conformational changes in proteins by mass spectrometry. *J. Am. Chem. Soc.*, **112**, 9012–9013.

[244]Katta, V. and Chait, B. T. (1991) Conformational changes in proteins probed by hydrogen-exchange electrospray-ionization mass spectrometry. *Rapid Commun. Mass Spectrom.*, **5**, 214–217.

[245]Zhang, Z. and Smith, D. L. (1993) Determination of amid hydrogen exchange by mass spectrometry: a new tool for protein structure elucidation, *Protein Sci.*, **2**, 522–531.

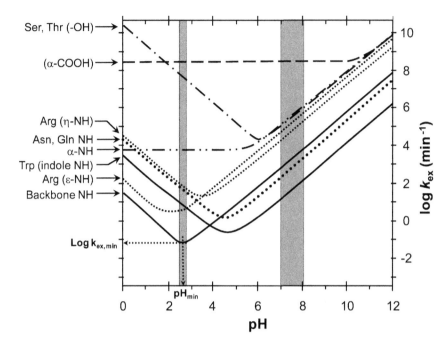

Figure 4.20: Hydrogen exchange as a function of pH. Exchange rate constants (k_{ex}) of several types of labile hydrogens found in proteins and peptides are shown as a function of solution pH. A change of one pH unit modifies the exchange rate constant by ~ 10-fold. The average exchange rate of backbone amide hydrogens is at a minimum ($k_{ex,min}$) when the pH is around 2.5 (pH_{min}), depending on the sequence. The light gray bands denote the quench pH region (left) and the labeling pH region (right). Reproduced by permission of John Wiley & Sons, Ltd. from *Encyclopedia of Analytical Chemistry: Applications, Theory and Instrumentation, Supplementary Volumes S1–S3*, Robert A. Meyers (Editor), pp. 1–17, 2011.

approximately 5%. Though not quite at the single-residue level, sometimes exchange rates could be determined within as short as two residues.

Bai *et al.*,[246] to evaluate the effects of protein side chains on the rate of exchange of peptide group NH with the hydrogens of aqueous solvent, studied the rate of exchange of peptide-group-NH hydrogens with the hydrogens of aqueous solvent employing all 20 naturally occurring amino acids using dipeptide models. Both inductive and steric blocking effects were apparent. The additivity of nearest-neighbour blocking and inductive effects in oligo- and polypeptides were confirmed.

---

[246]Bai, Y., Milne, J. S., Mayne, L., and Englander, S. W. (1993) Primary structure effects on peptide group hydrogen exchange. *Proteins*, **17**, 75–86.

Johnson and Walsh[247] measured the extents of amide deuteration for apo- and holo-myoglobin using liquid chromatography coupled directly to an ESI (LC/MS) source and following that by tandem mass spectrometry to identify rapidly the peptic peptides. ESI provided a more complete coverage of the protein sequence and measurement of deuterium incorporation into intact protein. They found that within 30 s the amides in apo-myoglobin were 47% deuterated, whereas holo-myoglobin was 12% deuterated; removal of the heme did not significantly alter regions represented by peptic peptides encompassing 1–7, 12–29, and 110–134; destabilized regions were within positions 33–106 and 138–153.

In Ehring's[248] approach, which used a probing ligand protein, at first all solvent-accessible amide protons of the (receptor) protein were exchanged with solvent deuterium before binding with the ligand protein. The receptor and ligand proteins were then mixed to form the complex. After binding to the ligand, *those* amide protons that were accessible *even with binding* underwent back exchange to hydrogen while all amide protons involved in the binding remained protected and deuterated. Those regions were then identified by peptic proteolysis, fast microbore LC separation, and ESI/MS. This approach works only for strongly binding molecules. They applied the method to the investigation of structural features of insulin-like growth factor I (IGF-I) and the interaction of insulin-like growth factor I with IGF-I binding protein 1 (IGFBP-1) — the interaction of IGF-I and IGFBP-1 is very strong and long-lived with an association constant of $K_A = 1.7\,[\mathrm{M}^{-1} \times 10^{-9}]$ — and showed that the method can provide information on the location of the hydrophobic core of IGF-I and on two regions that are mainly involved in binding to IGF-I binding protein 1.

Wang *et al.*[249] described how POROS-20AL columns packed with immobilized pepsin — to fragment proteins (leads to an effective pepsin concentration in mmolar range and short digestion times, even when the substrate protein concentration is low) — and a trapping column to concentrate the peptic fragments prior to analysis by HPLC ESIMS can be used online to reduce the digestion time and also to provide an effective means for separation of the pepsin from the isotopically labeled fragments. These columns used online facilitate both rapid digestion of low concentrations of protein and concentration of the peptides. The entire analysis needs substantially less time than the half-life $t_{1/2}$ for isotope exchange at

---

[247] Johnson, R. S. and Walsh, K. A. (1994) Mass spectrometric measurement of protein amide hydrogen exchange rates of apo- and holo-myoglobin. *Protein Sci.*, **3**, 2411–2418.

[248] Ehring, H. (1999) Hydrogen exchange/electrospray ionization mass spectrometry studies of structural features of proteins and protein/protein interactions. *Anal. Biochem.*, **267**, 252–259.

[249] Wang, L., Pan, H., and Smith, D. L. (2002) Hydrogen exchange-mass spectrometry. (Optimization of digestion conditions.) *Mol. Cell. Proteomics*, **1**, 132–138.

most peptide amide linkages ($t_{1/2} \approx$ 20–50 min at pH 2–3, when exchange at peptide amide linkages is slowest).

Ion mobility spectrometry — using Synapt HDMS — has been used by Iacob et al.[250] to probe solution-phase hydrogen/deuterium exchange (HX) of proteins. In this work, the HX profile of the Hck SH3 domain at both the intact protein and the peptic peptide levels was measured. Deuterium loss was found similar with or without use of Synapt IMS and comparable with non mobility Q-Tof measurements. For this small protein and its peptic peptides, deuterium incorporation did not perceptibly change the drift times. However, besides RP-HPLC, the additional orthogonal dimension of IMS allowed deconvolution of overlapping isotopic patterns for co-eluting peptides and in that way improvement in the resolution in mass spectrometry.

Houde et al.[251] showed that by employing UPLC at HX MS quench conditions peptides from the digestion of protein systems in excess of 200 kDa of unique sequence can be efficiently separated. Some additional resolving power could be incorporated — such as unresolved isotropic distributions in overlapping peptides following deuterium incorporation — after the chromatography step, by using ion mobility spectrometry.[250]

Landgraf et al.[252] through single-amide resolution (in contrast to peptide-level resolution) ETD/ECD experiments showed that the problem of deuterium scrambling (as was earlier observed in CID use) can be minimized by controlling instrumental parameters. It was accomplished using a radical-driven ETD fragmentation technique after soft ionization and desolvation. The (bottom-up differential HDX-ETD) technique was applied to determine differential single amide resolution HDX data for the peroxisome proliferator-activated receptor γ (PPARγ) bound with two ligands (full agonist and TZD – thiazolidinedione – rosiglitazone and the partial agonist MRL24) of interest. Single-amide HDX revealed a difference in magnitude of protection between the full agonist rosiglitazone and the partial agonist MRL24. Thus, HDX at single-amide resolution gave structural insight from an important region of the ligand binding pocket of PPARγ, providing site-specific protein-ligand information on complex systems (such as sublocalized deuterium incorporation to individual backbone amides) that was previously not possible to obtain by other HDX methods.

---

[250]Iacob, R. E., Murphy, J. P. III, and Engen, J. R. (2008) Ion mobility adds an additional dimension to mass spectrometric analysis of solution phase hydrogen/deuterium exchange. *Rapid Commun. Mass Spectrom.*, **22**, 2898–2904.

[251]Houde, D. and Demarest, S. J. (2011) Fine details of IGF-1R activation, inhibition, and asymmetry determined by associated hydrogen/deuterium exchange and peptide mass mapping. *Structure*, **19**, 890–900.

[252]Landgraf, R. R., Chalmers, M. J., and Griffin, P. R. (2012) Automated hydrogen/deuterium exchange electron transfer dissociation high resolution mass spectrometry measured at single-amide resolution. *J. Am. Soc. Mass Spectrom.*, **23**, 301–309.

In their review, Iacob and Jengen[239] summarize the status on reproducibility in HX MS experiments citing measurements of Chalmers *et al.*,[253,254] Burkitt *et al.*,[255] and Houde *et al.*[256] In order to quantify the reproducibility and to make the measurements as reproducible as possible, large datasets are essential, and in most cases that means including robotics.[254] Chalmers *et al.*[253] validated the use of HX MS for ligand screening and obtained data over an eight-month period for a large number of deuterated peptides for 127 replicate measurements. In these measurements 96% of all measured percent deuterium values (4039 measurements in all) were within 10% of the mean value, and 3452 values (or 82% of all measurements) were within 5% of the mean. Burkitt *et al.*[255] tested the reproducibility of an automated system over a two-month period; they found the *intra*day repeatability in measured exchange values of deuterated peptides to be between 0.2% and 0.9% deuterium [coefficients of variation (CVs) (CV = standard deviation/expected return (mean)) for *intra*day repeatability ranged from 0.3% to 1.5%]. *Inter*day reproducibility varied between 1.0% and 2.2% deuterium [CVs of 1.9%–3.8%]. Houde *et al.*[256] found the global *error* of each mass measurement of deuterated peptides to be about ± 0.14 Da, *independent* of peptide size, deuteration time, or the magnitude of the mass difference between undeuterated and deuterated. Their calculations — with 98% confidence — showed that differences of ± 0.5 Da between relative measurements were significant differences and were well outside the error range of the experiment. Houde *et al.* too conducted comparability studies, like those of Burkitt *et al.*[255] and Chalmers *et al.*,[253,254] and concluded that *inter*day analyses were significantly more variable than *intra*day analyses. Iacob and Jengen conclude[239] that reproducibility is presently not a limitation to the technique of HX MS.

4.5 Data mining scheme for identifying peptide structural motifs

Pairwise cleavage patterns that have been observed can be utilized to expand our knowledge of peptide gas-phase fragmentation behaviors, and

[253]Chalmers, M. J., Pascal, B. D., Willis, S., Zhang, J., Iturria, S. J., Dodge, J. A., and Griffin, P. R. (2011) Methods for the analysis of high precision differential hydrogen deuterium exchange data. *Int. J. Mass Spectrom.*, **302**, 59–68.

[254]Chalmers, M. J., Busby, S. A., Pascal, B. D., He, Y., Hendrickson, C. L., Marshall, A. G., and Griffin, P. R. (2006) Probing protein ligand interactions by automated hydrogen/ deuterium exchange mass spectrometry. *Anal. Chem.*, **78**, 1005–1014.

[255]Burkitt, W. and O'Connor, G. (2008) Assessment of the repeatability and reproducibility of hydrogen/deuterium exchange mass spectrometry measurements. *Rapid Commun. Mass Spectrom.*, **22**, 3893–3901.

[256]Houde, D., Berkowitz, S. A., and Engen, J. R. (2011) The utility of hydrogen/ deuterium exchange mass spectrometry in biopharmaceutical comparability studies. *J. Pharm. Sci.*, **100**, 2071–2086.

they also have usefulness in algorithm development to predict fragment ion intensities.

There have been work[257,258,259] on designing statistical tools to visualize fragmentation maps through extensive data mining. In one work in particular,[257] data mining was performed on 28,330 unique peptide tandem mass spectra for which sequence assignment was available with high confidence.

Briefly, they divided the spectra into different sets on the basis of structural features and charge states of corresponding peptides, which enabled characterization of chemical interactions involved in promoting specific cleavage patterns in gas-phase peptides.

Taking into account cleavages between all possible Xxx-Zze residue combinations resulting in b and y ions, pairwise fragmentation maps were created. It was observed that for singly charged arginine and lysine residues tryptic peptide dissociation patterns are different — they depend on the differences in basicities between Arg and Lys. For Arg-ending peptides, one dominant protonation form (proton localized) prevails; for Lys-ending peptides, a heterogeneous population of different protonated forms — more facile interconversion forms (proton partially mobile) — exists. Spectra from singly charged peptides that have a localized proton are dominated by cleavages C-terminal to acidic residues; and for those peptides that have a mobile or partially mobile proton, spectra are dominated by cleavages N-terminal to Pro. When a mobile or partially mobile proton is present in peptides but Pro is absent, cleavage at each peptide bond gains prominence.

In multiply protonated peptides, the location of the proton holder determines whether the above patterns can be found in b or y ions, or both. In both b and y ions from peptides that have a mobile proton, as well as in y ions from peptides that have a partially mobile proton, enhanced cleavages C-terminal to branched aliphatic residues (Ile, Val, Leu) are observed; in b ions from peptides that have a partially mobile proton, enhanced cleavages N-terminal to the above mentioned residues are observed.

Quantile maps for a couple of clusters are shown in Fig. 4.21.

[257]Huang, Y., Triscari, J. M., Tseng, G. C., Pasa-Tolic, L., Lipton, M. S., Smith, R. D., and Wysocki, V. H. (2005) Statistical characterization of the charge state and residue dependence of low-energy CID peptide dissociation patterns, *Anal. Chem.*. **77**, 5800–5813.

[258]Huang, Y., Tseng, G. C., Yuan, S., Pasa-Tolic, L., Lipton, M. S., Smith, R. D., and Wysocki, V. H. (2008) A data-mining scheme for identifying peptide structural motifs responsible for different MS/MS fragmentation intensity patterns. *J. Proteome Res.*, **7**, 70–79.

[259]Tseng, G. C. (2010) Quantile map: simultaneous visualization of patterns in many distributions with application to tandem mass spectrmetry. *Comput. Stat. Data Anal.*, **54**, 1124–1137.

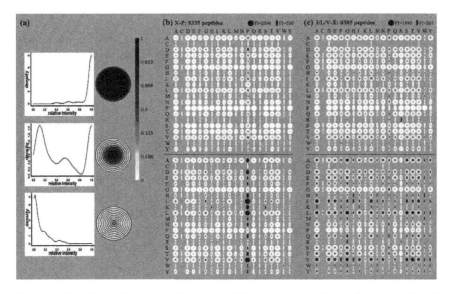

Figure 4.21: Quantile maps for two of five clusters obtained by penalized K-means from 28,330 spectra of unique sequence and charge state. Two quantile maps are plotted for each cluster, one for singly charged *b* ions (top) and one for singly charged *y* ions (bottom). A (alanine), C (cysteine), D (aspartic acid), E (glutamic acid), F (phenylalanine), G (glycine), H (histidine), I (isoleucine), K (lysine), L (leucine), M (methionine), N (asparagine), P (proline), Q (glutamine), R (arginine), S (serine), T (threonine), V (valine), W (tryptophan), Y (tyrosine). Reproduced by permission of the American Chemical Society from *J. Proteome Res.*, Vol. 7, 70–79, 2008.

4.6 Peptide fragmentation: mechanism

4.6.1 Fragmentation pathways: mobile proton model

Understanding of peptide fragmentation pathways that follow the collisional activation step in tandem mass spectrometry has a key use in interpretation of MS/MS spectra. In small peptides it helps in identification of amino acid sequence, and in large peptides and in mixtures of peptides it contributes to the development of proteomics software. However, owing to the complexity of the processes involved, progress in the field has been gradual and much remains ahead. The central question is how the peptide ion which has been energized collisionally distributes its energy and causes certain bonds to fragment preferentially. A couple of key hypotheses have found experimental support with application of hydrogen/deuterium exchange (HDX) kinetics studies, infrared spectroscopy, and ion mobility spectrometry, besides mass spectrometry. Extensive application of theoretical modeling of the fragmentation process also contributed to it. An excellent review of the field in 2005 is by Paizs and Suhai.[260] In this section, we present briefly the current state of understanding of mechanisms of fragmentation of protonated peptide.

A satisfactory model for peptide fragmentation should be able to explain the peak positions and peak intensities in MS/MS spectra on the basis of bond cleavages and any molecular rearrangements that might also be taking place. There have been a number of ongoing combined experimental (CID mass spectra) and theoretical studies (dissociation energies, geometries) mentioned at various points in this chapter; see also the recent work of Obolensky *et al.*[261] Currently the most acceptable model for fragmentation of protonated peptides is the mobile (or roving) proton model. Work from a number of laboratories led to the development of the model: Wysocki,[262] Harrison,[263] Gaskell,[264] Boyd,[265] and others.

[260] Paizs, B. and Suhai, S. (2005) Fragmentation pathways of protonated peptides. *Mass Spectrom. Rev.*, **24**, 508–554.

[261] Obolensky, O. I., Wu, W. W., Shen, R-F., and Yu, Y-K. (2013) Using dissociation energies to predict observability of *b*- and *y*-peaks in mass spectra of short peptides. II. Results for hexapeptides with non-polar side chains. *Rapid Commun. Mass Spectrom.*, **27**, 152–156.

[262] Jones, J. L., Dongré, A. R., Somogyi, Á., and Wysocki, V. H. (1994) Sequence dependence of peptide fragmentation efficiency curves determined by electrospray ionization/ surface-induced dissociation mass spectrometry. *J. Am. Chem. Soc.*, **116**, 8368–8369; Dongré, A. R., Jones, J. L., Somogyi, Á., and Wysocki, V. H. (1996) Influence of peptide composition, gas-phase basicity, and chemical modification on fragmentation efficiency: Evidence for the mobile proton model. *J. Am. Chem Soc.*, **118**, 8365–8374; Tsaprailis, G., Nair, H., Somogyi, Á., Wysocki, V. H., Zhong, W., Futrell, J. H., Summerfield, S. G., and Gaskell, S. J. (1999) Influence of secondary structure on the fragmentation of protonated peptides. *J. Am. Chem. Soc.*, **121**, 5142–5154; Wysocki, V. H., Tsaprailis, G., Smith, L. L., and Breci, L. A. (2000) Mobile and localized protons: A framework for understanding peptide dissociation. *J. Mass Spectrom.*, **35**, 1399–1406.

[263] Tsang, C. W. and Harrison, A. G. (1976) Chemical ionization of amino acids. *J. Am. Chem. Soc.*, **98**, 1301–1308; Harrison, A. G. and Yalcin, T. (1997) Proton mobility in protonated amino acids and peptides. *Int. J. Mass Spectrom. Ion Proc.*, **165**, 339–347.

[264] Burlet, O., Yang, C. Y., and Gaskell, S. J. (1992) Influence of cysteine to cysteic acid oxidation on the collision-activated decomposition of protonated peptides: Evidence for intraionic interactions. *J. Am. Soc. Mass Spectrom.*, **3**, 337–344; Cox, K. A., Gaskell, S. J., Morris, M., and Whiting, A. (1996) Role of the site of protonation in the low-energy decompositions of gas phase peptide ions. *J. Am. Soc. Mass Spectrom.*, **7**, 522–531; Summerfeld, S. G., Whiting, A., and Gaskell, S. J. (1997) Intra-ionic interactions in electrosprayed peptide ions. *Int. J. Mass Spectrom. Ion. Proc.*, **162**, 149–161; Summerfield, S. G., Cox, K. A., and Gaskell, S. J. (1997) The promotion of *d*-type ions during the low-energy collision-induced dissociation of some cysteic acid-containing peptides. *J. Am. Soc. Mass Spectrom.*, **8**, 25–31.

[265] Tang, X. and Boyd, R. K. (1992) An investigation of fragmentation mechanisms of doubly protonated tryptic peptides. *Rapid. Commun. Mass Spectrom.*, **6**, 651–657; Tang, X., Thibault, P., and Boyd, R. K. (1993) Fragmentation reactions of multiply-protonated peptides and implications for sequencing by tandem mass spectrometry with low-energy collision-induced dissociation. *Anal. Chem.*, **65**, 2824–2834.

In electrospray ionization (or MALDI) of peptides (tryptic or produced otherwise), protonated peptides are formed which are directed into the mass spectrometer analyzer, and following the first stage of m/z analysis, they are subjected to collisional activation preparatory to MS/MS. The collisional activation puts energy into the ions which gets distributed inside and bond scissions can then take place, producing fragment ions. In electrospray, the peptide picks up one or more unit of charge, as proton(s). This charge (proton) may be expected to sit in the ion on some basic site either on the peptide backbone or on a side chain if the peptide ion has a side chain of high proton affinity. A preferred position would be on an N atom somewhere. If there is a side chain with a residue of high proton affinity, such as arginine, lysine, or histidine (proton affinities of amino acids are given below[266]), one may expect the proton to sit there rather undisturbed, securely. In such a situation one may not expect any easy movement of the charge — we could then say that the charge is *sequestered*. However, in the absence of such a side chain the proton would locate somewhere on the backbone, on a basic site, probably on some N atom. The same effect is to be expected if the peptide is doubly protonated and one of the protons is sequestered by a side chain. However, if the proton sits on an N atom associated with a peptide bond on the backbone, an effect of that would be to destroy the resonance stabilization of the bond, weaken the bond, and cause it to cleave there. If the peptide is singly protonated and there is no side chain with a strongly basic residue, or the ion has a spare proton beyond any sequestered one, what should it sit on? It could be on any one of the backbone N atoms, with potential to break the peptide bond there. As a starting position, one could just take it to be the N on the N-terminus. The mobile proton model hypothesis says that a non-sequestered proton is free to move over the peptide backbone, and depending on which peptide N atom it goes and sits on, it weakens that peptide bond, increases the potential of cleavage of that peptide bond, giving rise to fragment ions. Concurrently, associated isomerization reactions can also occur.

4.7   Possible pathways in the competition model

For convenience of consideration, various peptide fragmentation pathways (PFPs) can be placed in two major categories (these could be termed "pathways in competition" PIC): the charge-directed PFPs and the charge-remote PFP. In the former are the non sequence PFPs — from which

---

[266]Proton affinities kcal/mol (Hunter, E. P. and Lias, S. G. in Mallard, W. G. and Linstrom, P. J. (Eds.) NIST Standard Reference Database No. 69 (2008)): Arg (251.2), Lys (238), His (236.1), Trp (226.8), Gln (224.1), Met (223.6), Asn (222), Tyr (221), Phe (220.6), Thr (220.5), Pro (220), Ile (219.3), Leu (218.6), Ser (218.6), Glu (218.2), Val (217.6), Asp (217.2), Cys (215.9), Ala (215.5), Gly (211.9).

sequences *cannot* be obtained — such as those involving loss of small molecules like water and ammonia loss, and the sequence PFPs from which peptide sequence could be obtained. The latter are of four types: $a_1 - y_x$, diket.$-y_x$, side chain PFPs, and the $b_x - y_z$ — the last one itself having four types: $b_x \rightarrow b_x - 1$, $b_x \rightarrow a_x$, immonium PFPs, and internal fragmentation PFPs. The charge-remote PFPs are of two types: those which show Asp effect and those which show oxide. met effect.[260] There are essentially two major fragmentation pathways to induce dissociation for formation of sequence ions of protonated peptides. One (the b_x - y_z pathway), in which the N-terminal neighbor amide oxygen, or the other (diketopiperazine pathway), in which the nitrogen of the N-terminal group, attacks the carbon center of the protonated amide bond. We take up the b_x - y_z fragmentation pathway first.[267]

4.7.1 b_x-y_z pathway

The extra proton on the N-terminus N atom moves to the amide N atom and protonates it; there is a nucleophilic attack on the carbon atom of the protonated group by an electron-rich group — in Fig. 4.22 carbonyl oxygen next to R_1; this is followed by formation of a protonated oxazolone derivative. The unstable intermediate rearranges, and when the C-terminal segment (amino acid or peptide) detaches, a *b* ion is left. The dissociation kinetics of the dimer depends on the internal energy distribution of the ion population, the PA of its monomers, etc. and can be approximated by using a linear free-energy relationship (Harrison,[268] Paizs and Suhai[269,270]) $\ln(b_x/y_z) \approx ((PA_{N-term} - PA_{C-term})/RT_{eff})$, where b_x/y_z is the ratio of the abundances of the b_x and y_z ions, PA_{N-term} and PA_{C-term} are the proton affinities of the neutral fragments of the corresponding b_x - y_z pathway (that is, PAs of an oxazolone derivative and a truncated peptide for N- and C-terminal fragments, respectively), and T_{eff} denotes the "effective temperature." Mechanistic considerations involved in the b_x - y_z pathway can explain most of the experimental results obtained from structural and energetic studies on small protonated peptides. Tandem MS experiments have shown that y_z ions are protonated truncated peptides or amino acids. The newly formed y_z ions contain the added proton as well as

---

[267]Harrison, A. G. (2009) To *b* or not to *b*: the ongoing saga of peptide *b*-ions. *Mass Spec. Rev.*, **28**, 640–654.

[268]Harrison, A. G. (1999) Linear free energy correlations in mass spectrometry. *J. Mass Spectrom.*, **34**, 577–589.

[269]Paizs, B. and Suhai, S. (2002) Towards understanding some ion intensity relationships for the tandem mass spectra of protonated peptides. *Rapid Commun. Mass Spectrom.*, **16**, 1699–1702.

[270]Paizs, B. and Suhai, S. (2004) Towards understanding the tandem mass spectra of protonated oligopeptides. 1: Mechanism of amide bond cleavage. *J. Am. Soc. Mass Spectrom.*, **15**, 103–112.

the hydrogen that was originally attached to a nitrogen N-terminal to the cleaved amide bond.

A consideration of general energetic, kinetic, and entropy factors determining the activity of the b_x - y_z pathways shows that under low-energy collision conditions, the reactive configurations of the b_x - y_z pathways (all-*trans*-amide nitrogen protonated species) are energetically accessible. And, once the amide nitrogen protonated species are formed, concerted formation of the oxazolone ring and cleavage of the amide bond can take place on a time-scale characteristic of a rearrangement-type reaction, with the barrier involved of the order of 10–15 kcal/mol. For protonated pentaalanine,[271] the energetic and kinetics of the various b_x - y_z pathways have been theoretically investigated; the results indicate that at low internal energies, the b_4 - y_1 pathway is favored compared to b_3 - y_2 and b_2 - y_3. Metastable ion and low-energy collision-induced dissociation mass spectra support this.[272] At higher energies all the b_4 - y_1, b_3 - y_2, and b_2 - y_3 PFPs are active with contributions from secondary reactions like $b_n \rightarrow b_{n-1}$ and $y_n \rightarrow y_{n-1}$. To estimate the ratio of the b_x and y_z ions on the particular b_x - y_z pathways the above ratio equation, incorporating the necessary proton affinities, was used. It satisfactorily explains the dominance of the b_4 ion over y_1 and the abundance of the b_3 ion over y_2 in the observed mass spectra.

As said earlier, the y_z ions are protonated truncated peptides. Harrison *et al.*[273] were the first ones to propose that most small b_n ions are oxazolones. By IR spectroscopy of gas phase ions, using free electron laser FELIX at Rijnhuizen, on b_4 of YGGFL it has been shown that most of the IR peaks are from oxazolone isomer (0.0 kcal/mol), with only one due to the cyclic isomer (~ 4.0 kcal/mol).[274]

4.7.2 Diketopiperazine pathway

The first step in the diketopiperazine-y_{N-2} pathway is formation of the amide nitrogen protonated species. There is significant difference between cleavages of aa(2) - aa(3) and aa(n) - aa(n+1) (n > 2) peptide bonds. Since diketopiperazines contain two cis-amide bonds, in the case of aa(2) - aa(3)

[271]Paizs, B. and Suhai, S. (2004) Towards understanding some ion intensity relationships for the tandem mass spectra of protonated oligopeptides. 1: Mechanism of amide bond cleavage. *J. Am. Soc. Mass Spectrom.*, **15**, 103–112.

[272]Yalcin, T., Csizmadia, I. G., Peterson, M. B., and Harrison, A. G. (1996) The structure and fragmentation of B$_n$ ($n \geq 3$) ions in peptide spectra. *J. Am. Soc. Mass Spectrom.*, **7**, 233–242.

[273]Yalcin, T., Khouw, C., Csizmadia, I. G., Peterson, M. R., and Harrison, A. G. (1995) Why are *b* ions stable species in peptide spectra? *J. Am. Soc. Mass Spectrom.*, **6**, 1165–1174.

[274]Polfer, N.C., Oomens, J., Suhai, S., and Paizs, B. (2005) Spectroscopic and theoretical evidence for oxazolone ring formation in collision-induced dissociation of peptides. *J. Am. Chem. Soc.*, **127**, 17154–17155.

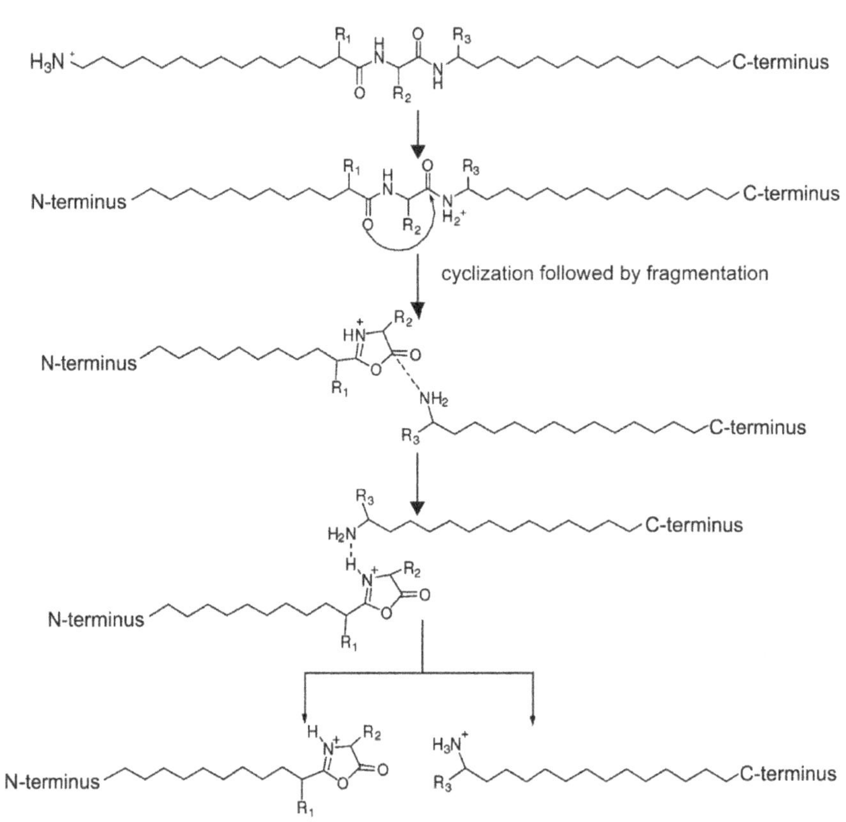

Figure 4.22: The b_x - y_z pathway. Adapted from Paizs and Suhai.[260]

cleavage, first a trans-cis isomerization of the initially *trans* N-terminal amide bond is necessary. In the next step, the mobile proton moves from the N-terminus to the nitrogen of the aa(2) - aa(3) amide bond, and in the next step, the terminal amino group attacks the carbon center of the protonated amide bond. The protonated diketopiperazine derivative and the C-terminal truncated peptide that would leave, form a loose complex, and since the proton affinity of linear peptides is higher than cyclic peptides, a proton transfer to the C-terminal fragment takes place. In the final step, the C-terminal fragment leaves as a y-ion and a neutral diketopiperazine derivative is left behind. (N is the number of amino acid residues in the peptide and n is used to note the amide bond aa(n)–aa(n+1) being cleaved.)

In the diketopiperazine-y_{N-n} $(n > 2)$ pathways, the neutral counterparts can accommodate all of their amide bonds in trans-state, so the additional step of cis-trans isomerization required in the aa(2)-aa(3) case is not

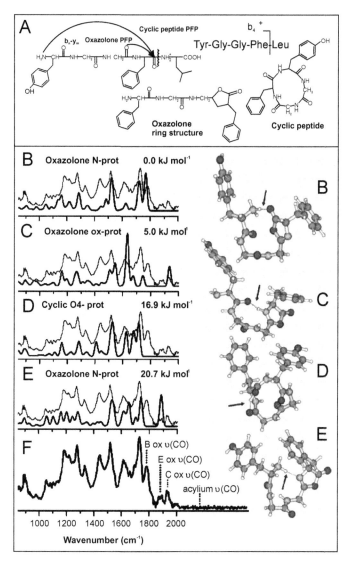

Figure 4.23: (A) Reaction scheme showing peptide fragmentation pathways (PFP) in CID of Leu-enkephalin to form b_4^+ fragment. Calculated spectra for (B) N-terminal protonated oxazolone, (C) oxazolone ring protonated oxazolone, and (D) cyclized peptide protonated on O4, and (E) N-terminal protonated oxazolone. (F) Infrared photodissociation spectrum of Leu-enkephalin b_4^+ fragment depletion. Arrows indicate sites of proton solvation. Reproduced by permission of the American Chemical Society from *Journal of the American Chemical Society*, Vol. 127, 17154–17155, 2005.

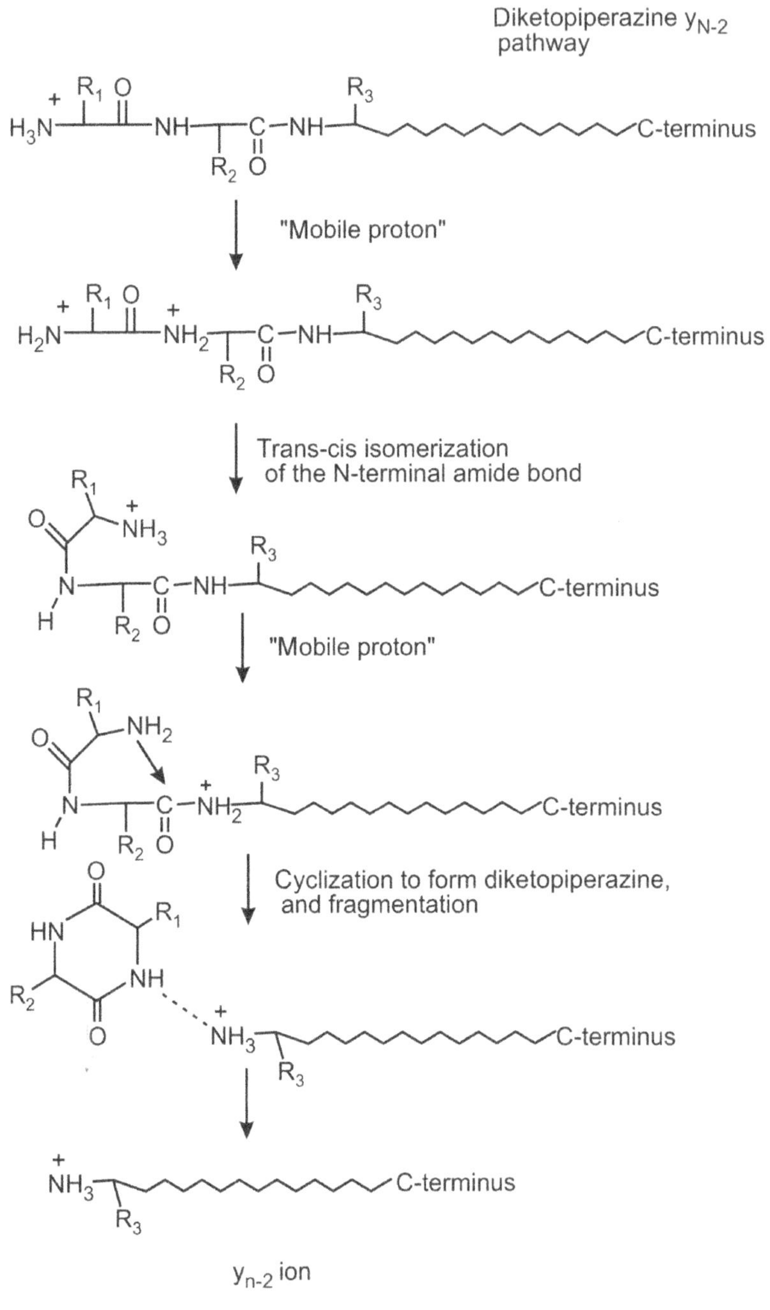

Figure 4.24: The diketopiperazine pathway. Adapted from Paizs and Suhai.[260]

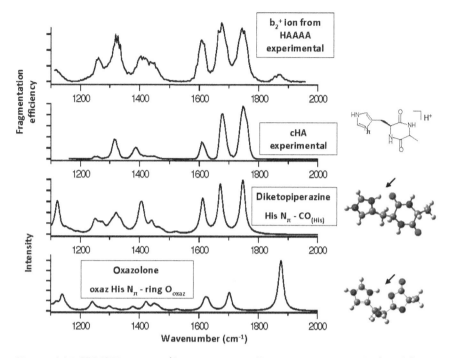

Figure 4.25: IRMPD spectra (fragmentation efficiency vs wavenumber) of (a) the b^{2+} ion produced by CID fragmentation of protonated HA4 and (b) protonated cyclic HA dipeptide. Calculated spectra for selected (c) diketopiperazine and (d) oxazolone structures. Arrows indicate protonation sites. Reproduced by permission of the American Chemical Society from *Journal of the American Chemical Society*, Vol. 131, 17528–17529, 2009.

necessary. Thus, after the mobilization of the added proton, nucleophilic attack of N-terminal amino group on the carbon center of the protonated amide takes place, a cyclic peptide is formed concurrently with the cleavage of the amide bond leading to the formation of the complex of protonated cyclic peptide and the C-terminal fragment, charge transfer to the latter takes place, and the latter leaves as the y_{N-n} $(n>2)$ ion. Location of the cleaved amide bond with respect to the N-terminus determines the size of the cyclic peptide — *cyclo*-tri, *cyclo*-tetra, *cyclo*-penta-peptides are formed corresponding to the y_{N-3}, y_{N-4}, y_{N-5} ion formation.

Supporting evidence on the y_x ions containing the added proton besides a hydrogen that was originally attached to a nitrogen N-terminal (proton transfer reaction between the cyclic peptide and the truncated peptide) has been from MS/MS studies. Neutral fragment reionization studies have shown that the neutral counterpart of the y ion is a diketopiperazine derivative.

There have been suggestions that protonated oxazolones and cyclic peptides like diketopiperazines can interconvert in the ion/molecule complexes formed, but this may not be likely because cis-trans isomerization of the amide bond that would be involved requires significant internal energy and would be expected to be slow.

In diketopiperazine-y_{N-2} pathways, since the necessary *trans-cis* isomerization has to take place before the nucleophilic attack, such paths are kinetically controlled. In diketopiperazine-y_{N-n} pathways, with *average n*, the isomerization step is not necessary; and for n small — because it causes significant ring strain — fragmentation to energetically disfavored products takes place. For n *large*, however, such paths are disfavored because of entropy effects. Because, in the latter case, the amide nitrogen protonated species is effectively solvated by nearby amide oxygens; as a result the terminal amino group is in competition against such charge solvation to reach near the center of the protonated amide bond. Thus though charge solvation of the NH_2^+ moiety by the terminal group is energetically feasible, the number of such species would be relatively small; as a consequence, in those cases where the amide bond is located distant to the N-terminus, entropy factor would disfavor such pathways.

Wysocki *et al.* have discussed evidence of diketopiperazine and oxazolone structures for HA b_2^+ ion (Fig. 4.25).[275] In another work,[276] water-loss products from protonated ArgGly and GlyArg investigated by IRMPD spectroscopy and computational chemistry show that the presence of a basic residue, such as arginine, facilitates the formation of diketopiperazine structures. Further, in the competition between diketopiperazine and oxazolone pathways, the residue order matters. ArgGly exclusively yields diketopiperazine structures, but for GlyArg, a mixture of diketopiperazine and oxazolone structures result.

Effect of third residue on oxazolone or diketopiperazine formation: Morrison *et al.*,[277] by action IRMPD of b_2 ions from tripeptide NAA and pentapeptides QAXIG and NAXIG using CLIO FEL, HDX exchange studies

[275]Perkins, B. R., Chamot-Rooke, J., Yoon, S., Gucinsky, A. C., Somogyi, A., and Wysocki, V. H. (2009) Evidence of diketopiperazine and oxazolone structures for HA b_2^+ ion. *J. Am. Chem. Soc.*, **131**, 17528–17529.

[276]Zou, S., Oomens, J., and Polfer, N. C. (2012) Competition between diketopiperazine and oxazolone formation in water loss products from protonated ArgGly and GlyArg. *Int. J. Mass Spectrom.*, **316−318**, 12−17.

[277]Morrison, L. J., Chamot-Rooke, J., and Wysocki, V. H. (2013) Evidence of third residue involvement in diketopiperazine and oxazolone b_2 ion formation in NAXIG and QAXIG pentapeptides. *Proc. 61st Annual ASMS Conference on Mass Spectrometry and Allied Topics*, Minneapolis, June 9–13. Also see, Morrison, L. J., Chamot-Rooke, J., and Wysocki, V. H. (2014) IR action spectroscopy shows competitive oxazolone and diketopiperazine formation in peptides depends on peptide length and identity of terminal residue in the departing fragment. *Analyst*, **139**, 2137–2143.

and computational modeling (giving theoretical IR spectra — by Monte Carlo conformational search, conformations optimized by Gaussian 09 package), examined the effect of third residue on oxazolone or diketopiperazine formation. This is the first instance of the diketopiperazine structure in systems not containing a proline or basic residue; this also is the first instance showing oxazolone vs diketopiperazine-dependent third residue/precursor hydrogen bridging conformation, and peptide length; (i) bridging plays an important role in the pathways, (ii) length of the chain significant, and (iii) sterics/hydrophobicities important.

Structures of b_2 ions from NAA, QAXIG, and NAXIG peptides were investigated using action IRMPD spectroscopy. In NALIG spectrum relevant diagnostic IR bands present all indicate that the b_2 population is a mixture of the b_2 structures of NAD/EIG and NAA; the experimental action IRMPD spectrum of b_2 of NAA and comparison with theoretical spectrum strongly support an oxazolone structure. Although differences in band intensities between the theoretical diketopiperazine IR spectrum and the experimental NAD/EIG are observed, the spectral positions of these IR vibrations suggest a diketopiperazine structure. In QAXIGs however, QAEIG forms exclusively diketopiperazine and QADIG forms mixtures of oxazolone and diketopiperazine.

ER-HDX was performed for a better appraisal of the relative population abundances of oxazolone and diketopiperazine b_2 ions. The b_2 population from NAEIG and NADIG exhibits an HDX exchange distribution that is consistent with the slower exchanging diketopiperazine structure. The HDX of the b_2 ion from NAA features, in contrast, corresponds to an oxazolone. The b_2 from NALIG shows a bimodal exchange population, indicating both the oxazolone and diketopiperazine structures. Only the NALIG b_2 population is observed to vary as a function of collision energy. Increase of the relative abundance of b_2 oxazolone with increasing collision energy and the thermodynamically more stable diketopiperazine dominating at low collision energies imply lower diketopiperazine ring closure barrier. However, at higher collision energies, more abundant oxazolone formation suggests lower transition point for oxazolone formation.

ER-HDX studies of the b_2 ions from the peptides NAAIG, NA(Abu)IG (Abu is aminobutyric acid), and NA(tbG)IG (tbG is ter-butylglycine) yielded information on how the peptide length and the steric bulk of the side chain of the third amino acid influences oxazolone and deketopiperazine formation. The diketopiperazine b_2 structure — present in all analogues — dominated at all energies in the NAAIG and NA(Abu) systems. The b_2 population substantially shifted by the addition of a branching side chain in NAVIG and Na(tbG)IG; as a result, particularly at high energies, the oxazolone was favored. This indicates steric or hydrophobic effect's significant role in contributing to the relative heights of the diketopiperazine and oxazolone ring closure barriers.

Except QADIG, the QAXIG analogues showed the same trends as their NAXIG counterparts, however, all with a relatively higher abundance of oxazolone. This points to a direct role of the amide side chain in the first position in diketopiperazine and oxazolone formation. The asparagine side chain can form a five-membered ring in bridging to the N-terminus, glutamine a six-membered ring, which may have a role regarding the extent to which the side chain can stabilize certain structures and limit nucleophilic attack from the N-terminus.

4.7.3 Histidine effect

When there is a histidine residue on the backbone of the peptide undergoing fragmentation, one of the nitrogen atoms on the imidazole ring of histidine which is on its side chain (say, the δ_1 one) can get preferentially protonated — because of high PA of histidine, most probably in the peptide's lowest energy structure. Then a mobilization of the proton can take place; this mobilized proton in turn could first transfer to the nitrogen of the C-terminal neighbor amide bond. In the next step, a nucleophilic attack on the carbon of the protonated amide bond (C-terminal side of the histidine) by the imidazole nitrogen can result in cyclization. The cyclic part can separate and form a b_x ion (a bicyclic ion). This ion does not have the classical oxazolone structure; *ab initio* calculations indicate that the non classical b_x ion thus formed is stable. Alternatively, there could be a proton transfer from the ring (say from the ϵ_2 position) to the amide nitrogen (of the C-terminus) resulting in the latter segment separating as a y_z ion. Thus, both b_x and y_z ions can be formed. This is the *histidine effect*. This, instead of backbone nucleophiles like the amide oxygens or the nitrogen of the terminal amino group, entails active involvement of the histidine side chain. The lifetime of the complex holding the N-terminal and C-terminal segments is long enough under low-energy conditions to allow alternative proton transfers between the monomers and formation of both b_x and y_x ions.[278,279,280,281]

[278]Wysocki, V. H., Tsaprailis, G., Smith, L. L., and Breci, L. A. (2000) Mobile and localized protons: a framework for understanding peptide dissociation. *J. Mass Spectrom.*, **35**, 1399–1406.

[279]Farrugia, J. M., Taverner, T., and O'Hair, R. A. J. (2001) Side-chain involvement in the fragmentation reactions of the protonated methyl esters of histidine and its peptides. *Int. J. Mass Spectrom.*, **209**, 99–112.

[280]Tureček, F., Chung, T. W., Moss, C. L., Wyer, J. A., Ehlerding, A., Holm, A. I. S., Zettergren, H., Nielsen, S. B., Hvelplund, P., Chamot-Rooke, J., Bythell, B., and Paizs, B. (2010) The histidine effect. Electron transfer causes different dissociations and rearrangements of histidine peptide cation-radicals. *J. Am. Chem. Soc.*, **132**, 10728–10740.

[281]Chung, T. W. and Tureček, F. (2011) Amplified histidine effect in electron-transfer dissociation of histidine-rich peptides from histatin 5. *Int. J. Mass Spectrom.*, **306**, 99–107.

The histidine effect has been extensively examined by MS^n experiments. Amino acid residues adjacent to His have been varied and their effects studied. For example, the MS/MS spectrum of doubly protonated H-Arg-Val-Tyr-Ile-His-Pro-Phe-OH is found to be dominated by the b_5^+ and y_2^+ ion pair; this suggests that the proton transfer step is possible. Alkylation of the His side-chain shows that this blocks the proton transfer mechanism and leads to dominance of b_x^{2+} ions in the MS/MS spectra. The ions b_5^+ and b_6^+ are found to dominate when Pro is substituted by Ala, as in H-Pro-Phe-OH and H-Ala-Phe-OH; this is because of the difference between the PAs of the two cases. Also it is observed that for many peptides, if the number of added protons is larger than the number of Arg residues present, cleavage at the C-terminal side of histidine is preferred.

4.7.4 Proline effect

In a study of tandem mass spectral effects on replacement of alanine in pentaalanine by proline, Schwartz and Bursey[282] observed that when residue 3 or 4 is proline, there is substantial enhancement of abundance of the y_3 or y_4 ion, respectively. This enhanced abundance of y fragment ions from protonated peptides containing proline — arising from cleavage of the amide bond N-terminal to the proline residue — is known as the *proline effect*. From a thermochemical viewpoint, a y-type ion with N-terminal proline is about 32 kJ mol^{-1} more stable than N-terminal alanine; this constitutes the principal difference. When this proline effect is accompanied with the effect of neutral diketopiperazine loss, the y-type ion is not observed as predicted from the kinetic approach on the basis of the additivity of substituent effects. Their explanation was based on high basicity of proline. They offered no interpretation on intensities of b-type ions.

Vaisar and Urban later examined[283] the proline effect using fragmentations of a series of protonated peptides with the general structure H-Val-Ala-Xaa-Leu-Gly-OH (Xaa = N-alkylated amino acid), where Xaa is sarcosine (N-methylglycine), N-methylalanine, proline, and pipecolinic acid (Pip), besides alanine, which was included for reference. They observed that protonated HValAlaProLeuGly-OH produces an abundant amount of y_3 ion, whereas both protonated H-ValAlaNMeAlaLeuGly-OH and H-ValAlaPipLeuGly-OH — even though Pip differs from Pro only by a CH$_2$ group — fragment to form mainly b_3 ions. According to Vaiser and Urban several factors are responsible for the effect with the unfavorable structure of the b_3 ion playing the major role; the role of high proton affinity

[282]Schwartz, B. L. and Bursey, M. M. (1992) Some proline substituent effects in the tandem mass spectrum of protonated pentaalanine. *Biol. Mass Spectrom.*, **21**, 92–96.

[283]Vaisar, T. and Urban, J. (1996) Probing the proline effect in CID of protonated peptides. *J. Mass Spectrom.*, **31**, 1185–1187.

of proline is secondary; and, because of the highly strained [3.3.0] bicyclic moiety formed, formation of the b ion on the C-terminal side of Pro is not favored. The b_3 ion from corresponding H-ValAlaPipLeuGly-OH has a less strained [4.3.0] bicyclic moiety and the fragmentation is determined by N-methylation effect — that is, methylation of the amide nitrogen.

Paizs and Suhai[260] have made several observations about the proline effect. According to them, the a_1 - y_z pathway rules can explain the cleavage of such amide bonds and the proline residue has no specific effect on the cleavage of the N-terminal amide bond. Preferential formation of y_1 ion in low-energy CID experiment on protonated H-ValPro-OH[284] is according to the basic characteristics of the a_1 - y_1 pathway (they point out that PA of Pro is higher than that of the imine derived from Val). When proline is located at the C-terminal of tripeptides, a certain specific effect is seen. In protonated H-LeuGlyPro-OH, a rather specific fragmentation pattern is observed[285] — the majority of the y_1 ions are formed on the diketopiperazine pathway, whereas in the case of other tripeptides, like H-GlyGlyGly-OH, H-GlyGlyLeu-OH, etc., y_1 ions are produced on the b_2 - y_1 pathway. Nold *et al.*[285] observe that the behavior of H-ValAlaNMeAla LeuGly-OH and H-ValAlaPipLeuGly-OH, which were studied by Vaisar and Urban, indicates that formation of the b ions of the C-terminal neighbor amide bond is promoted by methylation of the amide nitrogen. On the apparently unfavorable energetics of the b ion having the [3.3.0] bicyclic moiety mentioned by Vaisar and Urban, they draw conclusions from results of their own computational studies on the structure and energetics of various fragment ions of H-GlyProGly-OH and also cite the metastable ion spectrum of protonated HGlyProGly-OH[286] which shows a dominant b_2 ion containing the [3.3.0] bicyclic moiety. Availability of other low-energy fragmentation channels leading to y_1 ions from protonated H-GlyProGly-OH is pointed out. They also point out a significant difference between the histidine/aspartic acid effects and the proline effect by commenting that the former two, unlike the latter, can be prevented to occur via chemical modifications such as alkylation/esterification respectively. Overall, citing some experiments of Harrison *et al.* and some from their own investigations on a large number of peptides containing proline, their view is that dominance of a few ions takes place by the proline residues exerting their specific activity via affecting otherwise non specific fragmentation pathways.

---

[284]Harrison, A. G., Csizmadia, I. G., Tang, T. H., and Tu, Y. P. (2000) Reaction competition in the fragmentation of protonated dipeptides. *J. Mass Spectrom.*, **35**, 683–688.

[285]Nold, M. J., Wesdemiotis, C., Yalcin, T., and Harrison, A. G. (1997) Amide bond dissociation in protonated peptides. Structures of the *N*-terminal ionic and neutral fragments. *Int. J. Mass Spectrom. Ion Proc.*, **164**, 137–153.

[286]Ambihapathy, K., Yalcin, T., Leung, H. W., and Harrison, A. G. (1997) Pathways to immonium ions in the fragmentation of protonated peptides. *J. Mass Spectrom.*, **32**, 209–215.

4.7.5   Sequence scrambling in fragmentation

One likely complication in the fragmentation process is scrambling of the original sequence in the peptide. This can happen if the unstable intermediate formed has a cyclic part which opens up in a later step to yield a linear ion. Depending upon how the cyclic part opens (if there is a mobile proton on the ring, it would cause ring opening at a number of alternate points), it can retain the original sequence or yield an altered sequence. In the latter case, it would give misleading cue about the sequence in the original peptide.

Using a combined experimental and theoretical approach, Bleiholder *et al.*[287] probed the scrambling and rearrangement reactions that take place in CID of b and a ions. They studied the dissociation patterns from low-energy CID of b_5 fragments of the peptides YAGFL-NH$_2$, AGFLY-NH$_2$, GFLYA-NH$_2$, FLYAG-NH$_2$, and LYAGF-NH$_2$ and found them to be nearly the same. Besides, under similar experimental conditions, from CID of protonated *cyclo*-(YAGFL), the same fragments are observed with nearly identical ion abundances. This indicates that rapid cyclization of the primarily linear b_5 ions takes place; also, it is the fragmentation behavior of the cyclic isomer that determines the CID spectrum. The cyclic isomer can open up at various amide bonds, and a consideration of a multitude of fragmenting linear structures can contribute toward understanding of the fragmentation behavior. This cyclization-reopening mechanism is fully supported by their computational results; they show that protonated *cyclo*-(YAGFL) is energetically favored over the linear b_5 isomers, and in comparison to conventional bond-breaking reactions, the cyclization-reopening transition structures are energetically less demanding, and it is this which permits fast interconversion among the cyclic and linear isomers. As a consequence of the above reaction possibilities, in principle complete loss of sequence information upon CID can take place, as has been observed for the b_5 ion of FLYAG-NH$_2$. Fragment ion distributions produced from CID of the a_5 ions of the above peptides can be explained by assuming b-type scrambling of their parent population and $a \rightarrow a^*$-type rearrangement pathways.[288] While a ions easily undergo cyclization, predominantly the original linear structure is regenerated when the resulting macrocycle reopens. Computational data indicate that post-cleavage proton-bound dimer intermediates are involved in the $a \rightarrow a^*$-type rearrangement pathways of the linear a isomers in which the fragments reassociate; in the process, the originally C-terminal fragment is transferred to the N-terminus.

[287]Bleiholder, C., Osburn, S., Williams, T. D., Suhai, S., Stipdonk, M. V., Harrison, A. G., and Paizs, B. (2008) Sequence-scrambling fragmentation pathways of protonated peptides. *J. Am. Chem. Soc.*, **130**, 17774–17789.

[288]Vachet, R. W., Bishop, B. M., Erickson, B. W., and Glish, G. L. (1997) Novel peptide dissociation: gas phase intramolecular rearrangement of internal amino acid residues. *J. Am. Chem. Soc.*, **119**, 5481–5488.

Opinion is divided about the extent to which sequence scrambling takes place and is important. According to Zubarev *et al.*,[289] in shotgun proteomics, the extent of sequence scrambling is negligible. Siu *et al.*[290] report that although sequence scrambled *b* ions appeared in about 35% of 43 tryptic peptides under MS/MS conditions, their abundances were low, 8–16%. There were a few tryptic peptides that resulted in abundant scrambled *b* ions, but despite that, protein identification using bottom-up software was not affected. In another work,[291] it has been observed that the propensity for sequence permutation increases with the length of the tryptic peptide and, for larger peptides, increases also with a higher charge state.

4.7.6 Gas-phase fragment ion isomer analysis

Li *et al.*[292] suggested and examined a novel strategy for both qualitative and quantitative analysis of various gas-phase *b* ion isomers.

They combined ETD, IMS with formaldehyde (FH) labeling for ion isomer analysis (formaldehyde labeling results in dimethylation of lysine-amines and free-amino groups at the N-termini of proteins — also known as TAILS: *T*erminal *A*mine *I*sotope *L*abeling of *S*ubstrates — and is carried out in a single step using amine-reactive isotopic reagents, employing either $^{12}CH_2$-formaldehyde (light) or $^{13}CD_2$-formaldehyde (heavy) with sodium cyanoborohydride as catalyst).[293] Peptide molecular ions were selected and fragmented by CID with optimized collision energy in the Traveling-Wave trap cell of Synapt G2 HDMS; the resulting fragment ions were then injected into drift tube for drift times measurement.

By applying the known observation[294] of formation of a unique set of [*z-c*] and [*z-a*] ions from electron capture dissociation of cyclic peptides (the

[289] Goloborodko, A. A., Gorshkov, M. V., Good, D. M., and Zubarev, R. A. (2011) Sequence scrambling in shotgun proteomics is negligible. *J. Am. Soc. Mass Spectrom.*, **22**, 1121–1124.

[290] Saminathan, I. S., Wang, X. S., Guo, Y., Krakovska, O., Voisin, S., Hopkinson, A. C., and Siu, K. W. M. (2010) The extent and effects of peptide sequence scrambling via formation of macrocyclic *b* ions in model proteins. *J. Am. Soc. Mass Spectrom.*, **21**, 2085–2094.

[291] Yu, L., Tan, Y., Tsai, Y., Goodlett, D. R., and Polfer, N. C. (2011) On the relevance of peptide sequence permutations in shotgun proteomics studies. *J. Proteome Res.*, **10**, 2409–2416.

[292] Jia, C., Wu, Z., Lietz, C. B., Liang, Z., Cui, Q., and Li, L. (2013) Gas-phase fragment ion isomer analysis reveals the mechanism of peptide sequence scrambling. *Proc. 61st Annual ASMS Conference on Mass Spectrometry and Allied Topics*, Minneapolis, June 9–13; *Anal. Chem.*, **86**, 2917–2924.

[293] Kleifeld, O., Doucet, A., Prudova, A., auf dem Keller, U., Gioia, M., Kizhakkedathu, J. N., and Overall, C. M. (2011) Identifying and quantifying proteolytic events and the natural N terminome by terminal amine isotope labeling of substrates. *Nat. Protoc.*, **6**, 1578–1611.

[294] Leymarie, N., Costello, C. E., and O'Connor, P. B. (2003) Electron capture dissociation initiates a free radical reaction cascade, *J. Am. Chem. Soc.*, **125**, 8949–8958.

ECD process results in: (i) neutralization of a hydrogen bonded amine, (ii) transfer of the H· to the backbone carbonyl, (iii) cleaving the backbone N-C_α bond and forming a radical on the α-carbon — free radical rearrangement — the process can be used to separate/distinguish cyclic b ion from linear b ion), these diagnostic ions were used to probe the gas-phase isomer composition and structures of cyclic b-ions having shared structures with cyclic peptides. This strategy enabled simultaneous identification of *original linear b*-ion (^{ol}b) (the linear ion formed from the protonated peptide M through lone pair of O of the second amide — counting from the C-terminus — attacking carbonyl carbon of the first amide thereby forming oxazolone ring while losing the C-terminal end, that is, M → ^{ol}b; these lead to *direct* sequence ions), *cyclic b*-ion (^{c}b)(the cyclic b ion formed by N terminal lone pair electrons of the original linear b ion attacking nucleophilically the carbonyl carbon of the oxazolone ring followed by ring closure, ^{ol}b → ^{c}b, or directly the N terminal lone pair electrons of the protonated peptide M attacking the carbonyl carbon at the C-terminus, M → ^{c}b), and *rearranged linear b*-ions (^{rl}b)(^{c}b → ^{rl}b, a series of rearranged linear ions formed depending on the various alternative points the cyclic b ion opens up — with oxazolone ring at one end; these lead to the *nondirect* sequence ions). To this experimental step Li *et al.* added IMS, via a *pseudo* MS^3-IMS-MS^4 scheme, that revealed relative quantitative information on dynamic conversion of various isomers. (The ESI-ISD corresponds to MS^2, ion selection in a quadrupole and CID in Trap Cell constitutes MS^3, IMS drift time measurement comes next, which is followed by CID in a transfer cell that is called MS^4.) Employment of formaldehyde labeling coupled with ETD fragmentation for diagnostic ion detection and IMS for drift time measurement revealed mechanism of blocking cyclic-b ion formation and elimination of sequence scrambling event.

Li *et al.* demonstrated efficacy of this strategy by analyzing three representative b ions from three peptides with diverse isomer components: neurokinin b_9 ion, substance P b_{10} ion, and crustacean hyperglycemic hormone precursor related peptide (CPRP) b_{33} ion. Drift time measurements of CID-produced b_9^{2+} ion of neurokinin using IMS under various collision energy (0–22 eV) showed four peaks at 4.02, 4.29, 4.43, and 4.78 ms corresponding respectively to four b-ion isomers $^{c}b_9^{2+}$, $^{rl}b_{A9}^{2+}$, $^{ol}b_9^{2+}$, and $^{rl}b_{B9}^{2+}$ — in good agreement with predictions from molecular dynamics simulation of respective cross sections Ω 229, 237, 244, 256 Å^2; dynamic conversion among these structures was also probed. Analysis of b_{10}^{2+} ion isomers of substance P RPKPQQFFGL^{2+} is shown in Fig. 4.26. In (E), ETD spectrum on 600.2^{2+} ion and in (G), ETD spectrum on formaldehyde-labeled 628.2^{2+} ion are shown along with IMS peaks at 4.90 ms $^{rl}b_{10}^{2+}$ (F) and at 5.74 ms FH-b_{10}^{2+} (H). E shows [z-b] ions, indicating rearranged lin-

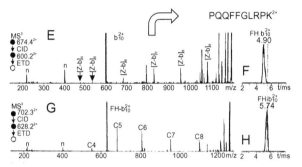

Figure 4.26: Analyses of b_{10}^2+ ion isomers of substance P RPKPQQFFGL^{2+}. Reproduced by permission of the American Chemical Society from *Analytical Chemistry*, Vol. 86, 2917–2924, 2014.

ear ion $^{rl}b_{10}^{2+}$ whereas in G owing to formaldehyde labeling/blocking [z-b] ions are not formed. The large difference between 4.90 ms and 5.74 ms indicates that we are looking at rearranged linear ion in F. Results of fragment ion isomers are (a) neurokinin b_9^{2+} HKTDSFVGL^{2+} products: (i) $^cb_9^{2+}$ cyclo-(HKTDSFVGL)$^{2+}$ 80%, (ii) $^{ol}b_9^{2+}$ (HKTDSFVGL)$^{2+}$ 16%, (iii) $^{rl}b_{A9}^{2+}$ TDSFVGLHK^{2+} 0%, (iv) $^{rl}b_{B9}^{2+}$ SFVGLHKTD^{2+} 4%; (b) substance P b_{10}^{2+} RPKPQQFFGL^{2+} $^{rl}b_{10}^{2+}$ PQQFFGL RPK^{2+} 100%; (c) CPRP b_{33}^{4+} RSAEGLGRMGRLLASLKSDTVTPLRGFEGETGH^{4+} $^{ol}b_{33}^{4+}$ RSAEGLGRMGRLLASLKS DTVTPLRGFEGETGH^{4+} 100%. Their studies reveal the mechanism of peptide sequence scrambling via b ion cyclization. Their strategy can be expected to be applicable in both top-down and bottom-up proteomics for high-throughput screening of ion isomers related to peptide sequence scrambling.

4.7.7 Role of gas phase conformations in ETD fragmentation

To probe the role of gas phase conformations in ETD fragmentation, Pepin *et al.*[295],[296] examined a set of eight peptides: LRGPADK, LKGPADR, RL-

---

[295] Pepin, R., Marek, A., Peng, B., Afonso, C., Lavanant, H., Laszlo, K. J., Bush, M. F., and Tureček, F. (2013) Electron transfer dissociation of rationally chosen heptapeptide cations. *Proc. 61st Annual ASMS Conference on Mass Spectrometry and Allied Topics*, Minneapolis, June 9–13.

[296] For a review of the chemistry of peptide radicals and cation radicals in the gas phase, see: Tureček, F. (2013) Peptide radicals and cation radicals in the gas phase. *Chem. Rev.*, **113**, 6691–6733.

GPADK, and KLGPADR, and the other four were with L in place of P in the fourth position — all chosen for having compact globular forms in the gas phase. The purpose of sequential acid and basic sites in the peptides was to utilize their attraction toward each other, multiple basic sites were to enhance chances of formation of triply charged states, and P in the middle was to force a β-turn. The operating theory has been that the 2+ charge state is a zwitterionic species with the negative ionic site attached to one of the two positive ionic sites leads to a much more compact and globular conformation than in the 3+ species.

Using IMS they measured for the peptides a cross-section of $191.70\,\text{Å}^2$. Their theoretical calculations on low-energy structures for doubly charged ion in gas phase (2+ ETD precursor ions are compact globular shapes in the gas phase) predicted cross sections which were close to this experimentally obtained value. Computational results indicated tight hydrogen bonding between the aspartic acid or C-terminal carboxylic group and the protonated lysine or arginine side chains with a separation of only $1.48\,\text{Å}$ between them.

In the ETD of 2+ ions, z-ions showed mass-shifted satellites. Upon fragmentation, the resultant fragment complex is more closely bound, giving additional time to the z fragment's radical to interact with the labile α-hydrogens in the c fragment and potentially capture a hydrogen. Inter-fragment H-atom transfers appear to increase further when the carboxyl sites are blocked by methyl esters. In the LKGPADR ETD of 2+, $(M+2H)^{+\cdot}$, on deconvolution, indicates the presence of metastables. In LKGLADR a greater fraction of metastable contribution was recognized (L-species are ejected earlier from the trap). The conformational restriction, besides expected backbone fragmentation, resulted in some hydrogen atom transfers between fragments. This effect was far more prevalent in the fragmentation of the 2+ peptides than in the 3+ peptides. Pepin *et al.* observed higher metastable contribution in ester forms than their corresponding native forms; this they attributed to a proline ring opening event which better stabilizes the radical.

Their conclusion was that MS3 ETD fragmentation of the peptides with leucine in the fourth position is charge-driven — with the exception of LGLADR which seems to have a complex path for dissociation — whereas in the peptides with proline at the fourth position the dissociation is radical-driven; presence of proline stabilizes radical and promotes radical fragmentation.

Absence of any difference in fragmentation sequence information from the ester and the native forms indicated that the two are of similar shape. That presence of proline produces less metastable ions indicates that it stabilizes radical; stabilized radical leads to radical fragmentations in MS3 of ETD charge-reduced species. It is believed that owing to the decreased polarity in the ester bond over the acid, this allows the peptide to be more flexible in its conformations and therefore can be even more globular and the fragment complex to be more close in space, further facilitating the hydrogen transfer reaction.

4.7.8 Fragmentation patterns as a measure of antioxidant capacity

Hamdy and Julian[297] have suggested and examined a method for recognizing antioxidant properties of peptides by monitoring their ability to sequester radicals; the method requires tracking the ratio of mobile proton-caused fragments and radical-directed fragments. They defined a figure of merit *Radical Sequestering Score* (RSS) — the ratio of the sum of mobile proton fragments to the sum of radical-directed fragments — and used it to assess the antioxidant capability of peptides, associating larger RSS value (that is, fewer radical directed fragments, implying limited radical migration) with higher antioxidant potential. Radical quenching propensity was examined by photodissociating an iodinated peptide and examining *whether* the radical can be sequestered, which leads to availability of mobile proton and thus results in *b* and *y* ions, *or*, direct radical-directed dissociation (RDD), leading to *a*, *c*, *x*, *z*, *w* ions and side-chain losses. Experimentally, radicals were generated by three alternative ways: by trapping the ion of interest tagged with IBA(2 iodozo benzoic acid)-18C6(18 crown C ether), or iodobenzoic acid, or iodinated tyrosine residue, and then photodissociating the C-I bond by 266 nm radiation from a Nd-YAG laser. Hamdy and Julian examined amyloid-β which is an antioxidant peptide; there, overwhelming number of *b* ions are formed.

With some confirmed antioxidant peptides that they tested, values of RSS (no mobile proton — charges sequestered), RSS (having mobile proton), obtained were respectively as follows: PHCK(IB)RM 0.11, 5.86; LQP-GQGQQG N/A, 7.30; HGPLGPL N/A, +1 0.07, +2 9.89; Aβ 1-42 0.02, 3.21; PSKYEPFV 0.07, 5.62; NGPLQAGQPGER 0.04, 4.15. With *non*-antioxidant peptides, values of RSS (no mobile proton — charges sequestered), RSS (having mobile proton), obtained were respectively as follows: HDSGYEVHHQK 0.71, 0.47; IQTGLDATHAER 0.40, 0.14; YEVHHQK-LVFF 0.41 0.19; DRVYIP 0.02, 0.02; FVNQHLC*(oxidized)GSHLVEA-LYLVC*(oxidized)GERGFFYTPKA (insulin β chain) 0.13, 0.21; HAEGTF TSDVSSYLEGQAAKEFIAWLVKGRG (Glucagon-like peptide) 0.14, 0.15; NSKMAHSSSCFGQK(IB)IDRIGAV SRLGCDGLRLF (brain natriuretic

[297]Hamdy, O. M. and Julian, R. R. (2013) Peptide fragmentation patterns as a measure of antioxidant capacity. *Proc. 61st Annual ASMS Conference on Mass Spectrometry and Allied Topics*, Minneapolis, June 9–13. See also, Hamdy O. M., Lam, S., and Julian, R. R. (2014) Identification of inherently antioxidant regions in proteins with radical-directed dissociation mass spectrometry. *Anal. Chem.*, **86**, 3653–3658.

peptide) 0.06; 0.29. The antioxidant property predicted from this mass spectrometry checked well with methods used in solution. In one case, however, for the suspected antioxidant peptide HFGDPFH, RSS (no MP) N/A, RSS (MP) 0.08, the observation did not seem to have a satisfactory explanation. By taking C- and M-containing peptides and scrambling the sequences, Hamdy and Julian showed that the common belief that the presence of cysteine or methionine ensures antioxidant properties is not always valid; the overall sequence, their order, structure, together seemed to be the determining factors.

Hamdy and Julian also studied human serum albumin (HSA) (20-60% of antioxidant capacity of blood plasma is from proteins, HSA is the most abundant protein in human serum; it constitues ~50% of plasma proteins). In tryptic digest of HSA they identified $\sim$ 44 peptides and studied them. They identified three antioxidant peptides with RSS (MP) values $\gg 1$: RSS (no MP), RSS (MP): DVFLGMFLYEYAR 0.09, 3.79; VPQVSTPTLVEVSR -, 4.21; DDNPNLPR 0.05, 2.1. Their efforts to confirm these predictions from mass spectrometric measurements, by lipid peroxidation assay in solution phase, had however partial success. According to them, DDNPNLPR is a special case. The crystal structure of HSA from PDB shows DDNPNLPR close to a lipid binding site, and since HSA transports lipids (and lipid peroxidation causes disease development), the antioxidant property of HSA with the DDNPNLPR present near a lipid binding site is understandable.

4.8 Studying post translational modifications

In Section 4.10.2 in a manual *de novo* sequencing example, we shall see how a PTM can get recognized. In Section 5.3 in conotoxin analysis, we shall see how a certain PTM was detected. Here we present sections of a summary[298] of some mass spectrometric studies for *large-scale* analysis of protein modifications.

In general, in proteome measurements for characterization of each protein, *several peptides* are generally available. This makes the measurement much more robust. In contrast, in PTM measurements each peptide carrying a modification site of interest stands *on its own*. When the modified peptides are of lower abundance, their identification is more difficult in comparison to nonmodified peptides. For single-amino-acid resolution localization of the modification, the MS/MS spectra also then must contain sufficient information.

Whereas the sample preparation method for PTM analysis is generally the same as for shotgun analysis, it is important to inactivate enzymes that may add or remove the PTM of interest. Trypsin, because of its high

[298]Olsen, J. V. and Mann, M. (2013) Status of large-scale analysis of post-translational modifications by mass spectrometry. *Mol. Cell. Proteomics*, 12:10:1074/mcp.O113. 0341810, 3444–3452.

cleavage specificity and also because the resulting peptides undergo MS/MS well, is the protease of choice. Some PTMs modify significantly the solubility properties of the peptides (such as making them too hydrophilic), potentially affecting the ability of conventional reversed-phase C18 material; under such situations besides trypsin, proteases such as chymotrypsin, Lys-N, or endoproteinase Glu-C with different cleavage specificities can be helpful.

Notwithstanding the general increase in sensitivity in proteomics workflow, *enrichment* of PTMs of interest with respect to the unmodified proteins/peptides is required. Whereas for most PTMs, efficient enrichment strategies still do not exist; protocols with very high specificity for phosphorylated ones are now available that lead to phosphopeptide proportion of more than 90%. Metal-affinity chromatography using titanium dioxide and/or anti-phosphotyrosine antibodies are often used. Recently introduced diglycine-specific antibodies to detect the remnant modification after tryptic digestion of ubiquitinated proteins have made possible specific enrichment analysis of tens of thousands of ubiquitinated sites in a generic workflow.[299,300]

High resolution at both MS and MS/MS levels is particularly effective for PTM. Whereas usual procedures for unambiguous identification apply, since *localization* of the modified amino acids is involved, a separate *localization score* is required. Typically, the mean localization probability in large datasets is greater than 99%. Still, annotated spectra of PTM-bearing peptides are expected to be reported. For site localization, sufficient peptide fragmentation should be available; HCD, ETD (for very labile modifications such as O-GlcNAcs), multistage activation in ion trap instruments,[301] beam-type fragmentation on time-of-flight are necessary. Detailed determination of the structures of very heterogeneous PTMs as in N-linked glycosylation can be considerably more difficult.

Whereas it is possible to determine *complete* model proteomes, for PTMs although almost proteins can be post translationally modified, there has been so far *no* complete PTM catalogs. One intrinsic reason is, it is difficult to define a complete PTM proteome. Propensity of a protein toward modifications does not assure biological function; at the same time,

[299] Udeshi, N. D., Svinkina, T., Mertins, P., Kuhn, E., Mani, D. R., Qiao, J. W., and Carr, S. A. (2013) Refined preparation and use of anti-diglycine remnant (K-epsilon-GG) antibody enables routine quantification of 10,000s of ubiquitination sites in single proteomics experiments. *Mol. Cell. Proteomics*, **12**, 825–831.

[300] Wagner, S. A., Beli, P., Weinert, B. T., Scholz, C., Kelstrup, C. D., Young, C., Nielsen, M. L., Olsen, J. V., Brakebusch, C., and Choudhary, C. (2012) Proteomic analyses reveal divergent ubiquitylation site patterns in murine tissues. *Mol. Cell. Proteomics*, **11**, 1578–1585.

[301] Schroeder, M. J., Shabanowitz, J., Schwartz, J. C., Hunt, D. F., and Coon, J. J. (2004) A neutral loss activation method for improved phosphopeptide sequence analysis by quadrupole ion trap mass spectrometry. *Anal. Chem.*, **76**, 3590–3598.

biologically relevant substrates may undergo modifications under a very restricted set of circumstances beyond those used in the global PTM analysis. Among several protein classes where PTMs have been studied at depth are histones, but even there, there are difficulties of generating peptides of suitable lengths for MS. But there have been advances; presently, in single LC-MS/MS runs, under multiple conditions and replicates very large number (thousands) of sites can be studied.

Quantification of PTMs, since it is based on single peptides, often of very low abundance, is quite challenging. SILAC use is common — the "gold standard" — because of the absence of many workflow-induced sources of error.[204,302] Chemical labeling strategies, which can be applied to any sample type, have been used.[303,304] Currently label-free quantification owing to improvements in data quality and in algorithms is very attractive — here direct comparison of MS signals between any number of samples can be made. Besides relative quantification of the modification site of interest, it is very useful to determine the fraction of proteins that are modified (the *sites occupancy* or *stoichiometry*). In PTM identification, specifying the modified amino acid is normally required in search software; unbiased PTM analysis methods are now there which call for in turn *in silico* addition of the mass difference between the modified and unmodified peptide to each amino acid.

Most of the large-scale PTM mapping has involved the most prominent modifications, especially phosphorylation, ubiquitylation, glycosylation, and acetylation. However, there are more than 200 *in vivo* peptide modifications known, and there are even more modifications that can be induced chemically, as in during sample preparation.

Bioinformatic analysis area is now well developed in both expression proteomics and large-scale PTM analysis; comprehensive and statistically rigorous software tools are available; analysis typically includes comparison to one of the PTM databases. Analysis steps include motif analysis, Gene Ontology enrichment, pathway analysis, and analysis of protein-protein interactions. In the MaxQuant environment, the Perseus software allows extensive statistical and functional analysis of proteome and PTM data. The Net-WorKIN software combines protein-protein information from STRING with information on linear kinase motifs to provide likely kinase-substrate relationships in large-scale data. For references to the software mentioned here see the paper of Olsen *et al.*[298] (Indiscriminate incorporation of all data has been largely stopped by Uniprot; Phosphosite includes.)

[302]Blagoev, B., Kratchmarova, I., Ong, S. E., Nielsen, M., Foster, L. J., and Mann, M. (2003) A proteomics strategy to elucidate functional protein-protein interactions applied to EGF signaling. *Nat. Biotechnol.*, **21**, 315–318.

[303]Mallick, P. and Kuster, B. (2010) Proteomics: a pragmatic perspective. *Nat. Biotechnol.*, **28**, 695–709.

[304]Bantscheff, M. and Kuster, B. (2012) Quantitative mass spectrometry in proteomics. *Anal. Bioanal. Chem.*, **404**, 937–938.

In terms of time requirement, deep phosphoanalysis can now be completed in a single day of measurement. Already, using single-shot approaches [305,306,307] analysis of more than 10,000 sites can be carried out in a few hours,[308] thus an entire project can finish in a few days. Single-shot approaches have additional advantage in that they tend to use less input material and the absence of fractionation steps makes them relatively more robust. The identification and quantification time is *short* for targeted approaches, since they aim a small number of peptides of interest.[309,310] Key PTMS sites are examined under many conditions in the targeted approach.[311,312,313] But in targeted mode, there is *no* scope for discovery of new and unexpected sites. For clinical application, however, much development of the PTM workflow and instrumentation will still be required.

The approach of the signaling community usually is concentration on a very few but biologically highly important PTMs. MS-based proteomics identifies tens of thousands of sites whose biological roles need to be resolved. This means that if properly directed, these large-scale datasets can

[305]Thakur, S. S., Geiger, T., Chatterjee, B., Bandilla, P., Frohlich, F., Cox, J., and Mann, M. (2011) Deep and highly sensitive proteome coverage by LC-MS/MS without prefractionation. *Mol. Cell. Proteomics*, **10**, M110.003699.

[306]Kocher, T., Pichler, P., Swart, R., and Mechtler, K. (2012) Analysis of protein mixtures from whole-cell extracts by single-run nanoLC-MS/MS using ultralong gradients. *Nat. Protoc.*, **7**, 882–890.

[307]Pirmoradian, M., Budamgunta, H., Chingin, K., Zhang, B., Astorga-Wells, J., and Zubarev, R. A. (2013) Rapid and deep human proteome analysis by single-dimension shotgun proteomics. *Mol. Cell. Proteomics*, **12**, 3330–3338.

[308]Lundby, A., Secher, A., Lage, K., Nordsborg, N. B., Dmytriyev, A., Lundby, C., and Olsen, J. V. (2012) Quantitative maps of protein phosphorylation sites across 14 different rat organs and tissues. *Nat. Commun.*, **3**, 876.

[309]Wolf-Yadlin, A., Hautaniemi, S., Lauffenburger, D. A., and White, F. M. (2007) Multiple reaction monitoring for robust quantitative proteomic analysis of cellular signaling networks. *Proc. Natl. Acad. Sci. U.S.A.*, **104**, 5860–5865.

[310]Lange, V., Picotti, P., Domon, B., and Aebersold R. (2008) Selected reaction monitoring for quantitative proteomics: a tutorial. *Mol. Syst. Biol.*, **4**, 222.

[311]Oliveira, A. P., Ludwig, C., Picotti, P., Kogadeeva, M., Aebersold, R., and Sauer U. (2012) Regulation of yeast central metabolism by enzyme phosphorylation. *Mol. Syst. Biol.*, **8**, 623.

[312]Bisson, N., James, D. A., Ivosev, G., Tate, S. A., Bonner, R., Taylor, L., and Pawson, T. (2011) Selected reaction monitoring mass spectrometry reveals the dynamics of signaling through the GRB2 adaptor. *Nat. Biotechnol.*, **29**, 653–658.

[313]Zheng, Y., Zhang, C., Croucher, D. R., Soliman, M. A., St-Denis, N., Pasculescu, A., Taylor, L., Tate, S. A., Hardy, W. R., Colwill, K., Dai, A. Y., Bagshaw, R., Dennis, J. W., Gingras, A. C., Daly, R. J., and Pawson, T. (2013) Temporal regulation of EGF signalling networks by the scaffold protein Shc1. *Nature*, **499**, 166–171.

be used as pointers to key regulatory sites[314,315,316] thus acting as an initial filter in a cell-signaling project. The *cooperative nature* of PTM sites, which classical approaches do not quite include, is receiving more recognition. The MS-based proteomics with its high throughput and accuracy presents the possibility of global-scale quantification of a PTM with scope for its systematic perturbation.

Whereas it is now possible to analyze routinely in single experiments tens of thousands of phosphorylation sites, assignment of specific kinase(s) that directly catalyze the modification of a given phosphorylation site of interest remains difficult. Experimentally this can be addressed via knockdown of kinases involved in the cellular response under investigation or partially bioinformatically via linear kinase motif analyses.[317,318,319]

Availability of rigorous and streamlined bioinformatic analysis of PTM data at present constitutes an important foundation of systems biology. Appropriate integration modes of large-scale PTM quantification data in PTM databases to make it to the community however are still not resolved. Thus, availability of such data will be more used by signaling biologists simply as an initial screen prior to detailed study of key sites. Appropriate functional interpretation and follow-up are required.

For large-scale PTM analysis to make an impact in the clinic, what more is required? A combination of the advantages of targeted approaches with those of shotgun approach that would result in further development of single-shot approach. Additionally, faster, more sensitive, and more reproducible PTM screens, along with greater dynamic range are needed.

---

[314]Hsu, P. P., Kang, S. A., Rameseder, J., Zhang, Y., Ottina, K. A., Lim, D., Peterson, T. R., Choi, Y., Gray, N. S., Yaffe, M. B., Marto, J. A., and Sabatini, D. M. (2011) The mTOR-regulated phosphoproteome reveals a mechanism of mTORC1-mediated inhibition of growth factor signaling. *Science*, **332**, 1317–1322.

[315]Yu, Y., Yoon, S. O., Poulogiannis, G., Yang, Q., Ma, X. M., Villen, J., Kubica, N., Hoffman, G. R., Cantley, L. C., Gygi, S. P., and Blenis, J. (2011) Phosphoproteomic analysis identifies Grb10 as an mTORC1 substrate that negatively regulates insulin signaling. *Science*, **332**, 1322–1326.

[316]Francavilla, C., Rigbolt, K. T., Emdal, K. B., Carraro, G., Vernet, E., Bekker-Jensen, D. B., Streicher, W., Wikstrom, M., Sundstrom, M., Bellusci, S., Cavallaro, U., Blagoev, B., and Olsen J. V. (2013) Functional proteomics defines the molecular switch underlying FGF receptor trafficking and cellular outputs. *Mol. Cell*, **51**, 707–722.

[317]Miller, M. L., Jensen, L. J., Diella, F., Jorgensen, C., Tinti, M., Li, L., Hsiung, M., Parker, S. A., Bordeaux, J., Sicheritz-Ponten, T., Olhovsky, M., Pasculescu, A., Alexander, J., Knapp, S., Blom, N., Bork, P., Li, S., Cesareni, G., Pawson, T., Turk, B. E., Yaffe, M. B., Brunak, S., and Linding, R. (2008) Linear motif atlas for phosphorylation-dependent signaling. *Sci. Signal.*, **1**, ra2.

[318]Gnad, F., Ren, S., Cox, J., Olsen, J. V., Macek, B., Oroshi, M., and Mann, M. (2007) PHOSIDA (phosphorylation site database): management, structural and evolutionary investigation, and prediction of phosphosites. *Genome Biol.*, **8**, R250.

[319]Linding, R., Jensen, L. J., Pasculescu, A., Olhovsky, M., Colwill, K., Bork, P., Yaffe, M. B., and Pawson, T. (2008) NetworKIN: a resource for exploring cellular phosphorylation networks. *Nucleic Acids Res.*, **36**, D695–D699.

4.9 Chemical cross-linking/mass spectrometry

The chemical *cross-linking* method for obtaining structural information on proteins is based on the formation of a *covalent bond* between *two spatially proximate residues*. This bond formation can be within a single or between two polypeptide chains.

There have been several reviews on the growth of the field: Back *et al.*,[320] Sinz,[321] Lee,[322] and Leitner *et al.*[323] Details and references on what follows here can be found in the last reference.

The chemical cross-linking method was introduced by Young *et al.*[324] They used chemical cross-linking of *lysine* residues in bovine basic fibroblast growth factor FGF-2 (heparin-binding growth factor) to provide distance constraints for the computational derivation of the fold of this small (17-kDa) protein. FGF-2 was cross-linked with bis(sulfosuccinimidyl) suberate, the product was purified by size exclusion chromatography, and digested with trypsin. Cross-linked peptides were separated by HPLC and analyzed by ESI-TOF (on line) and by MALDI-TOF (off line) mass spectrometry. On the basis of their precursor masses putative cross-links were then assigned. Fifteen cross-links could be identified that did *not* bridge directly adjacent lysines which provided information on the three-dimensional structure of the protein. By excluding those calculated models that did not fit the distance constraints, the data obtained were used to assign FGF-2 to the β–trefoil family.

Nomenclature of common products of chemical cross-linking reactions and structures of most commonly used amine-reactive cross-linking reagents are shown in Fig. 4.27 which depicts mono-link (also referred to as dead-end link, type 0 cross-link), loop-link (also referred to as intrapeptide link, type-1 cross-link), cross-link (also referred to as inter/intraprotein link, interpeptide link, type-2 cross-link).

---

[320]Back, J. W., de Jong, L., Muijsers, A. O., and de Koster, C. G. (2003) Chemical cross-linking and mass spectrometry for protein structural modeling. *J. Mol. Biol.*, **331**, 303–313.

[321]Sinz, A. (2006) Chemical cross-linking and mass spectrometry to map three-dimensional protein structures and protein-protein interactions. *Mass Spectrom. Rev.*, **25**, 663–682.

[322]Lee, Y. J. (2008) Mass spectrometric analysis of cross-linking sites for the structure of proteins and protein complexes. *Mol. Biosyst.*, **4**, 816–823.

[323]Leitner, A., Walzthoeni, T., Kahraman, A., Herzog, F., Rinner, O., Beck, M., and Aebersold. R. (2010) Probing native protein structure by chemical cross-linking, mass spectrometry and bioinformatics. *Mol. Cell. Proteomics*, **9**, 1634–1649.

[324]Young, M. M., Tang, N., Hempel, J. C., Oshiro, C. M., Taylor, E. W., Kuntz, I. D., Gibson, B. W., and Dollinger, G. (2000) High throuput protein fold identification by using experimental constraints derived from intramolecular cross-links and mass spectrometry. *Proc. Natl. Acad. Sci., U.S.A.* **97**, 5802–5806.

The requirements of an effective cross-linking reagent are stability, reactivity, and sufficient solubility under the relevant biological conditions that favor protein (complex) stability. To enable identification it should not fragment under conditions that induce peptide bond cleavage. Also, in order to generate meaningful spatial information, its extent should not exceed a certain cross-linking distance.

The experimental workflow[325] for the chemical cross-linking of *acidic* side-chains and the identification of cross-linked peptides by MS consists of (i) cross-linking: protein (complex) + ADH or PDH + DMTMM for 30–60 min at pH 7–7.5, (ii) sample processing: quenching, evaporation, Cys reduction/alkylation, digestion (Lys-C, trypsin), (iii) peptide fractionation: size exclusion chromatography, (iv) LC-MS/MS analysis, (v) data analysis: independent database search for AXL and ZLXL (xQuest); estimation (xProphet).

Whereas the aim of the cross-linking reaction is a covalent bond formation between two spatially proximate residues within a single or between polypeptide chains, depending on the sample, it is possible that only one end of the bifunctional cross-linker reacts with the protein because the other end does not come into contact with another cross-linkable residue or it can get deactivated by hydrolysis. A large number of chemical cross-linking reagents have been developed. The conventional ones are amine- or thiol-reactive and homo- and heterobifunctional; the others are by incorporation of additional functional groups — cleavable sites and affinity tags.

In conventional cross-linking reagents, two reactive sites are connected through a spacer or linker region, typically an alkyl chain. Most commonly, the reactive groups of cross-linkers target the primary amino group of lysine (and the protein N termini). To effect this, N-hydroxysuccinimidyl or sulfosuccinimidyl esters are almost exclusively used. Although these esters have high reaction rates, they are also susceptible to rapid hydrolysis in aqueous solutions with half-lives at a time scale of tens of minutes under typical reaction conditions (pH > 7, 25–37 °C). This can be a problem for low protein concentrations. Common succinimide-type linkers are disuccinimidyl suberate (DSS; one six-carbon linker) and disuccinimidyl glutarate (DSG; three-carbon linker) as well as their sulfo analogs bis(sulfosuccinimidyl) suberate (BS3) and bis(sulfosuccinimidyl) glutarate (BS2G), which are more soluble in purely aqueous solutions. DSS and DSG, in contrast, require prior dissolution in small volumes of polar organic solvents such as N,N-dimethylformamide or DMSO before addition to the sample. Among lysine's advantages are relatively high reaction specificity and high prevalence of Lys residues (about 6%).

---

[325]Leitner, A., Joachimiak, L. A., Unverdorben, P., Waltzhoeni, T., Frydman, J., Förster, F., and Aebersold, R. (2014) Chemical cross-linking/mass spectrometry targeting acidic residues in proteins and protein complexes. *Proc. Nat. Acad. Sci. (U.S.A.)*, **111**, 9455–9460.

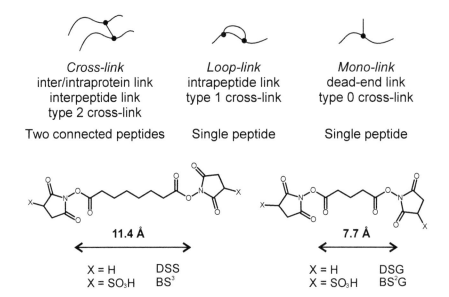

Cross-link
inter/intraprotein link
interpeptide link
type 2 cross-link

Loop-link
intrapeptide link
type 1 cross-link

Mono-link
dead-end link
type 0 cross-link

Two connected peptides   Single peptide   Single peptide

11.4 Å   7.7 Å

X = H   DSS
X = SO₃H   BS³

X = H   DSG
X = SO₃H   BS²G

Figure 4.27: Top: Nomenclature of common products of chemical cross-linking reactions. Bottom: Structures of most commonly used amine-reactive cross-linking reagents. DSS, BS[3], DSG, and bis(sulfosuccinimidyl) glutarate (BS[2]G). Reproduced by permission of the American Society for Biochemistry and Molecular Biology, Inc. from *Molecular & Cellular Proteomics*, **9**, 1634–1649, 2010.

Specific cross-linking reactions can be carried out targeting cysteine residues, e.g., by maleimides, but the low abundance of Cys ($< 2\%$) is a limitation. Arginine-specific cross-linking or acidic cross-linking are not frequently used because reaction products are unstable or inhomogeneous. Besides homobifunctional cross-linkers, several heterobifunctional linkers have been used. In these two different reactive groups, Lys- and Cys-reactive may be used, or chemical and photoinduced cross-linking methods may be used.

Formaldehyde is a notable exception to the general linker designs; this contains only a single aldehyde group but is able to connect two amino acid side chains via a two-step reaction. A less specific reagent, it targets primarily lysine and tryptophan.[326,327] It forms an amide bond between Lys and Asp/Glu residues and requires very close spatial proximity — a so-called "zero-length" cross-link is formed.

---

[326]Sutherland, B. W., Toews, J., and Kast, J. (2008) Utility of formaldehyde cross-linking and mass spectrometry in the study of protein-protein interactions. *J. Mass Spectrom.*, **43**, 699–715.

[327]Toews, J., Rogalski, J. C., Clark, T. J., and Kast, J. (2008) Mass spectrometric identification of formaldehyde-induced peptide modifications under in vivo protein cross-linking conditions. *Anal. Chim. Acta*, **618**, 168–183.

Several types of functionalized cross-linking reagents have been proposed to facilitate the analysis by mass spectrometry. In stable isotope-labeled crosslinkers,[328] a mixture of a cross-linker containing only natural ('light') isotopes and a 'heavy' (usually deuterated) form of the reagent is used; then the reaction products carry a unique isotopic signature.[329] By digestion in $H_2^{18}O$, an isotopic signature can also be introduced into cross-linked peptides.[330,331] Affinity-tagged cross-linking reagents (using such as biotin as the affinity group) could be used followed by isolation of modified peptides by avidin affinity chromatography.[332,333] An azide-containing cross-linking reagent has been used[334] to capture cross-links on a cyclooctyne resin that involves a "click-chemistry"-type reaction. A lysine-reactive linker with a protected thiol group has been used[335] for enrichment after the cross-linking step. Linkers containing labile bonds that are easily cleaved during collision-induced dissociation have been used.[336] This concept, termed protein interaction reporter (PIR), results, upon fragmentation, in the generation of a diagnostic ion from the cross-linker. PIR cross-linkers have been applied[337] *in vivo* cross-linking in Shewanella oneidensis;

[328]Müller, D. R., Schindler, P., Towbin, H., Wirth, U., Voshol, H., Hoving, S., and Steinmetz, M. O. (2001) Isotope tagged cross linking reagents. A new tool in mass spectrometric protein interaction analysis. *Anal. Chem.*, **73**, 1927–1934

[329]Isotope-labeled cross-linking reagents are now commercially available, from vendors such as Creative Molecules and the Pierce division of Thermo Scientific, and more complex reagents have also been prepared in labeled form.

[330]Back, J. W., Notenboom, V., de Koning, L. J., Muijsers, A. O., Sixma, T. K., de Koster, C. G., and de Jong, L. Z. (2002) Identification of cross-linked peptides for protein interaction studies using mass spectrometry and O-18 labeling. *Anal. Chem.*, **74**, 4417–4422.

[331]Huang, B. X. and Kim, H. Y. (2009) Probing Akt-inhibitor interaction by chemical cross-linking and mass spectrometry. *J. Am. Soc. Mass Spectrom.*, **20**, 1504–1513.

[332]Trester-Zedlitz, M., Kamada, K., Burley, S. K., Fenyö, D., Chait, B. T., and Muir, T. W. (2003) A modular cross-linking approach for exploring protein interactions. *J. Am. Chem. Soc.*, **125**, 2416–2425.

[333]Kang, S., Mou, L., Lanman, J., Velu, S., Brouillette, W. J., and Prevelige, P. E., Jr. (2009) Synthesis of biotin-tagged chemical cross-linkers and their applications for mass spectrometry. *Rapid Commun. Mass Spectrom.*, **23**, 1719–1726.

[334]Nessen, M. A., Kramer, G., Back, J., Baskin, J. M., Smeenk, L. E., de Koning, L. J., van Maarseveen, J. H., de Jong, L., Bertozzi, C. R., Hiemstra, H., and de Koster, C. G. (2009) Selective enrichment of azide-containing peptides from complex mixtures. *J. Proteome Res.*, **8**, 3702–3711.

[335]Yan, F., Che, F. Y., Rykunov, D., Nieves, E., Fiser, A., Weiss, L. M., and Hogue Angeletti, R. (2009) Nonprotein based enrichment method to analyze peptide cross-linking in protein complexes. *Anal. Chem.*, **81**, 7149–7159.

[336]Tang, X., Munske, G. R., Siems, W. F., and Bruce, J. E. (2005) Mass spectrometry identifiable cross-linking strategy for studying protein-protein interactions. *Anal. Chem.*, **77**, 311–318.

[337]Zhang, H., Tang, X., Munske, G. R., Tolic, N., Anderson, G. A., and Bruce, J. E. (2009) Identification of protein-protein interactions and topologies in living cells with chemical cross-linking and mass spectrometry. *Mol. Cell. Proteomics*, **8**, 409–420.

more than 20 cross-links were reported mainly involving membrane proteins. Finding suitable reaction conditions for cross-linking of native proteins matching the physicochemical properties of complex multifunctional cross-linkers is a major difficulty.

Until recently the XL-MS cross-linking chemistry targeted primary amines. Also, it should be noted that, of the theoretically possible cross-links, only a small fraction are experimentally observed. Recently, a chemistry to cross-link *acidic residues* that generates structural information *complementary* to that obtained by aminespecific cross-linking has been reported. This significantly extends the scope of XL-MS analyses.

Aspartic and glutamic acid residues with carboxyl-terminating side-chains are attractive targets because of their prevalence in most proteins. Of all residues 5.5% are Asp and 6.7% Glu,[338] compared with 5.8% Lys. But since the intrinsic chemical reactivity of carboxylic acids is *low*, it presents practical challenges for cross-linking reactions and requires the use of a coupling reagent.

In a recent work[325] reported from Aebersold's group, a cross-linking chemistry has been introduced that connects proximal *carboxyl* groups — *acidic* cross-linking (AXL) — where side-chains of Asp and Glu residues are cross-linked with *dihydrazides* using the coupling reagent 4-(4,6-dimethoxy-1,3,5-triazin-2-yl)-4-methylmorpholinium chloride (DMTMM).[339] In contrast to 1-ethyl-3-(3-dimethylaminopropyl) carbodiimide hydrochloride (EDC), DMTMM is able to couple carboxylic acids with hydrazide-based cross-linkers at neutral pH (7–7.5); it has biocompatibility and also provides good reaction yields. Numbers of cross-linked peptides obtained with model proteins were observed to be in the same range as of those generated Lys-specific cross-linking using the reagent disuccinimidyl suberate (DSS). Additionally, a second set of cross-linking restraints was observed in the form of zero-length cross-links between Lys and Asp or Lys and Glu residues. To demonstrate the practical utility of the method, *they* applied it to multisubunit complexes in the megadalton range that have been recently probed with lysine-specific cross-linking and for which structural information is available. Both for the chaperonin TRiC/CCT from *Bos taurus* and the *Schizosaccharomyces pombe* 26S proteasome, cross-links were identified that are in agreement with the known structures of these complexes. Complementary to cross-linking chemistry targeting lysines, acidic and zero-length cross-links provide orthogonal sets of structural restraints.

In cross-linking work, compared to unreacted protein, cross-linked products appear in low or very low concentration; chromatographic or electro-

---

[338] In the most recent release of the SwissProt database, version 2014_04.

[339] Kunishima, M., Kawachi, C., Monta, J., Terao, K., Iwasaki, F., and Tani, S. (1999) 4-(4,6-Dimethoxy-1,3,5-triazin-2-yl)-4-methyl-morpholinium chloride: An efficient condensing agent leading to the formation of amides and esters. *Tetrahedron*, **55**, 13159–13170.

phoretic separation is necessary to increase relative concentration of cross-linked peptides and lessen interference of unmodified proteins/peptides.

In analysis by mass spectrometry, unambiguous charge state information for both precursor and fragment ions and increased mass accuracy are important for interpretation of cross-link data; these can come from high mass accuracy MS/MS. Mass accuracies < 10 ppm are essential; this is more so when information only at MS^1 level is used. The MS^1 level data remain important because cross-link data as it involve two peptides entails enormous search space — these numbers can reach millions or even billions for larger databases. For analysis/validation, *manual analysis* has been used. There have been innovative approaches to handle this.[322]

The MS analysis algorithms mostly are for specific (chemistry-specific, such as amine-reactive, or cleavable) cross-linkers. Differential comparison of cross-linked and control samples to identify cross-linked peptides is used. To identify cross-linked peptides, labeling strategies such as tryptic digestion in labeled ^{18}O, ^{15}N, thus generating characteristic isotopic signatures, have been used. Still, MS^1 methods are very limited where there is sample complexity. They yield a large number of ions that cannot be used to assign to a tryptic peptide.

In identification of cross-linked peptides from LC-MS/MS data, there is dramatic combinatorial explosion of the search space with increasing sample complexity. Leitner *et al.*[323] calculated the search space for proteome-level amine-reactive cross-linking using data taken from the UniProt/Swiss-Prot database (version 15.10). By selecting only fully tryptic peptides, considering a maximum of two missed cleavages and a length of 5–45 amino acids for each peptide but without any variable modifications such as phosphorylation or oxidation, possible combinations amounted to 7.3×10^9 for *Escherichia coli*, 8.2×10^{10} for *Saccharomyces cerevisiae*, and 7.5×10^{11} for *Homo sapiens*. Even for small proteomes such as *Escherichia coli* (4367 proteins), there is search-space explosion — the search space is $\sim 60,000$ times larger; for the human proteome (20,333 proteins), the increase in complexity is $\sim 600,000$-fold.

With superior computational power, it may become feasible to enumerate all possible combinations also from large sequence databases and to search them with the presently available, and also with to be in the near future available, search tools. Notwithstanding this, unconstrained matching in such a huge search space will result in many *false positives*, especially when *low resolution* MS/MS spectra are probed against all possible peptide combinations. Verification and *validation* of the results that the different algorithms provide without relying on *manual validation* remain at present the biggest unresolved issue.

In molecular modeling of proteins and protein complexes, cross-linking data can provide additional information about the spatial distance (though not highly accurate) between the amino acids on the *surface* of *folded* proteins; those can be used as distance constraints for *molecular modeling* of protein folds and complex topologies. Regarding this, Havel *et al.*[340] put forward the following *three* rules: *many imprecise* distance constraints (e.g., cut-off distances or residue-residue contact information) are to be preferred over *few* constraints with *precise* distance information; distance constraints from residues that are widely separated in the primary sequence are to be preferred over sequentially nearby residues; and distance constraints should involve as many different residues as possible. The above rules were extended[341] by including the following three observations made on the myoglobin fold: distance information should be accurate to within a few Angstrom; that modeling software is to be preferred which produces native-like conformations; and dissimilarity between native and non native structural models should be as large as possible.

"How many cross-links can be observed in native protein complexes? Are distance constraints from cross-linking experiments useful for filtering out false-positive predictions of protein complexes? Which length should the ideal intermolecular cross-linker have? How many cross-links are needed for a reasonable prediction of the complex topology?" These would be some relevant questions to ask regarding role of cross-linking data in protein structure studies. Aebersold *et al.*[323] asked precisely these questions and carried out a theoretical analysis on the applicability of cross-linking-derived distance constraints on the 54 crystal structures of the first Protein-Protein Docking Benchmark set[342] of Chen and Weng. These analyses addressed several important questions for cross-linking studies. Their conclusion: the incorporation of cross-link data into computational topology prediction of protein complexes *significantly increases* the likelihood of predicting *native-like* complex structures.

4.10 Manual *de novo* sequencing of peptides

Sequencing peptides to ascertain positions of constituent amino acids is basic to protein primary structure determination. At one time there was only Edman degradation; now mass spectrometry is the major, expeditious alternative. In mass spectrometry the sequencing can be done by a manual *de novo* method, or, now almost overwhelmingly, by applying software on

---

[340]Havel, T. F., Crippen, G. M., and Kuntz, I. D. (1979) Effects of distance constraints on macromolecular conformation. 2. Simulation of experimental results and theoretical predictions. *Biopolymers*, **18**, 73–81.
[341]Cohen, F. E. and Sternberg, M. J. (1980) On the use of chemically derived distance constraints in the prediction of protein structure with myoglobin as an example. *J. Mol. Biol.*, **137**, 9–22.
[342]Chen, R. and Weng, Z. (2002) Docking unbound proteins using shape complementarity, desolvation, and electrostatics. *Proteins*, **47**, 281–294.

the CID fragments. Reconstitution of the amino acid sequence in a peptide from CID fragment ions is effectively done by identifying the fragments and considering probable ways they were bound together before fragmentation.

There are a number of popular procedures[343,344,345,346 347,348,349] for proceeding step by step from CID spectra to deduce the origin of the ions observed in mass spectra and from that the sequence. These are all quite satisfactory; their differences are in some details of their approach. Here we summarize major points in the *de novo* procedure and illustrate them through a few examples. There has also been a graph theory approach[350] reported in the literature.

Nomenclature of the fragments—by Roepstorff and Fohlman,[196] later modified by Biemann *et al.*[197]—has been given earlier in this chapter. There are *sequence ions* and *satellite ions*; the satellite ions are formed from loss of small molecular fragments from the sequence ions and often are useful in recognizing the latter in the spectra. It is the sequence ions which provide the peptide structure.

4.10.1 Procedure in manual sequencing

First, a few general observations. Above all we should remember that all the rules we discuss evolved from *experimental observations* on mass spectra of known peptides. As we have said earlier, peptide fragmentation in MS/MS can be effected in a number of ways: the most common method is CID (collision-induced dissociation); ETD/ECD comes next, then there

[343]*De novo* peptide sequencing tutorial at IonSource.com http://www.ionsource.com/ tutorial/DeNovo (this "leans heavily" on a *de novo* sequencing course presented in 1991 at the University of Virginia taught by Donald F. Hunt and Jeffrey Shabanowitz).

[344]Chapter 4 of Kinter and Sherman's book, Kinter, M. and Sherman, N. E. (2000) Protein sequencing and identification using tandem mass spectrometry. Wiley-Interscience.

[345]Mandal, A., Naaby-Hansen, S., Wolkowicz, M. J., Klotz, K., Shetty, J., Retief, J. D., Coonrod, S. A., Kinter, M., Sherman, N., Cesar, F., Flickinger, C. J., and Herr, J. C. (1999) FSP95, a testis-specific 95-kilodalton fibrous sheath antigen that undergoes tyrosine phosphorylation in capacitated human spermatozoa. *Biol. Reproduct.*, **6**, 1184–1197.

[346]Mehta, A., Kinter, M., Sherman, N., and Driscoll, D. M. (2000) Molecular cloning of apobec-1 complementation factor, a novel RNA-binding protein involved in the editing of apolipoprotein B mRNA. *Molec. Cell. Biol.*, **20**, 1846–1854.

[347]Medzihradszky, K. F. (2005) Peptide sequence analysis. *Methods in Enzymology*, **402**, 209–244.

[348]Medzihradszky, K. F. and Chalkley, R. J. (2013) Lessons in *de novo* peptide sequencing by tandem mass spectrometry. *Mass Spectrom. Rev.*, DOI: 10.1002/mas.21406 published online October 29, 2013.

[349]MatrixSciencePeptideSequencingMS.pdf.

[350]Yan, B., Pan, C., Olman, V. N., Hettich, R. L., and Xu, Y. (2005) A graph-theoretic approach for the separation of *b* and *y* ions in tandem mass spectra. *Bioinformatics*, **21**, 563–574.

Table 4.3: Amino acid residue names, masses, and immonium ions.

Residue	1-letter code	3-letter code	Residue mass monosiotopic	Residue mass average	Immonium ion (m/z)
Alanine	A	Ala	71.03712	71.079	44.05 (w)
Arginine	R	Arg	156.10112	156.188	129 (w)
Asparagine	N	Asn	114.04293	114.104	87.09 (m)
Aspartic Acid	D	Asp	115.02695	115.089	88.04 (m)
Cysteine	C	Cys	103.00919	103.144	76 (w)
Glutamic Acid	E	Glu	129.04260	129.116	102.06 (m)
Glutamine	Q	Gln	128.05858	128.131	101.11 (m)
Glycine	G	Gly	57.02147	57.052	30 (w)
Histidine	H	His	137.05891	137.142	110.07 (s)
Isoleucine	I	Ile	113.08407	113.160	86.1 (s)
Leucine	L	Leu	113.08407	113.160	86.1 (s)
Lysine	K	Lys	128.09497	128.174	101.11 (s)
Methionine	M	Met	131.04049	131.198	104.05 (m)
Phenylalanine	F	Phe	147.06842	147.177	120.08 (s)
Proline	P	Pro	97.05277	97.117	70.07 (s)
Serine	S	Ser	87.03203	87.078	60.04 (m)
Threonine	T	Thr	101.04768	101.105	74.06 (m)
Tryptophan	W	Trp	186.07932	186.213	159.09 (m)
Tyrosine	Y	Tyr	163.06333	163.170	136.08 (s)
Valine	V	Val	99.06842	99.133	72.08 (s)

is PSD (post-(ion)source decay), and finally, the much less used photofragmentation. In CID, HCD, and also in IRMPD, dissociation produces b, y, a ions, weakest peptide bonds breaking first; in ECD and ETD, c, z, a ions are produced, except where proline present — phosphorylation and glycosylation entities remain unaffected. Proline's unusual structure however causes b- and y-ions not to appear — presence of a cyclic side chain at its secondary amine as well as α-carbon atom position does not provide a site for protonation and therefore of subsequent rupture. Thus, unlike the general case of fragmentation of tryptic peptides at the C-terminal side of a lysine or arginine, proline, when present at the C-terminal side of a lysine or arginine, prevents cleavage there. Another aspect to be noted is the energy deposited in the dissociation process; this can radically alter the ions that are observed. A broad categorization in this regard would be low-energy dissociation and high-energy (as in MALDI) dissociation. The b- and y-ions are observed in all the methods; v-, w-, z-ions appear in high-energy fragmentation.

There is no unique way to proceed in *de novo* manual sequencing; in a way the task is somewhat like solving a jigsaw puzzle. The peptide fragmented in a certain way; identities of the fragments or most of the fragments are known; the purpose is to arrive at the amino acid sequence of the un-

Table 4.4: Immonium and related ions characteristic of the 20 standard amino acids.

Amino acid	Immonium and related ion masses		Comments
Ala	44		
Arg	129	59, 70, 73, 87, 100, 112	129, 73 usually weak
Asn	87	70	87 often weak, 70 weak
Asp	88		Usually weak
Cys	76		Usually weak
Gly	30		
Gln	101	84, 129	129 weak
Glu	102		Often weak if C-terminal
His	110	82, 121,123, 138, 166	110 very strong
			82, 121, 123, 138 weak
Ile/Leu	86		
Lys	101	84, 112, 129	101 can be weak
Met	104	61	104 often weak
Phe	120	91	120 strong, 91 weak
Pro	70		Strong
Ser	60		
Thr	74		
Trp	159	130, 170, 171	Strong
Tyr	136	91, 107	136 strong, 107, 91 weak
Val	72		Fairly strong

Falick, A. M., Hines, W. M., Medzihradszky, K. F., Baldwin, M. A., and
Gibson, B. W. (1993) Low-mass ions produced from peptides by
high-energy collision-induced dissociation in tandem mass spectrometry.
Reproduced with kind permission of Springer Science and Business Media
from *J. Am. Soc. Mass Spectrom.* **4**, 882–893.

fragmented peptide from a knowledge of the fragment ion masses. It should be mentioned that all fragments do not appear with comparable intensities; some are of such low intensities that they are not even observed. Physicochemical reasons exist why a certain ion is not observed with sufficient intensity, but consideration of such loss processes is sufficiently complicated that for practical purposes it is all but untractable and not worth the *de novo* sequencer's time. From certain characteristic signs, and checking of internal consistencies, a conclusion is reached. The amino acid residue data are once again provided here (Table. 4.3) for convenience.

Some more general observations: low m/z ions do not usually show up in spectra taken with ion traps (mostly trapping conditions used are such that low m/z ions even if they are formed are not retained — "the 1/3 rule" — the cut-off point is at the 1/3 of the precursor MH$^+$ ion m/z).[351] Immonium

---

[351]See, also, Kiyonami, R., Sclabach, T., Schwartz, J., Miller, K., and Taylor, L. Analysis of low mass ions in peptide fragmentation spectra with a linear ion trap. www.thermo.com.cn/Resources/201008/11154640729.pdf.

Table 4.5: Combinations of amino acid residues with equal/nearly equal masses.

Amino acid residue	Residue mass (Da)	Δ mass (Da)
Leucine	113.08407	
Isoleucine	113.08407	0
Hydroxyproline	113.04768	0.03639
Asparagine	114.04293	
Glycine + Glycine	114.04294 (2×57.02147)	0
Ornithine	114.07931	0.03638
Glutamine	128.05858	
Glycine + Alanine	128.05859 ($57.02147 + 71.03712$)	0
Lysine	128.09497	0.03639
Oxidized methionine	147.03540	
Phenylalanine	147.06842	0.03302
Glycine + Valine	156.08989 ($57.02147 + 99.06842$)	
Arginine	156.10112	0.01123
Glycine + Glutamic Acid	186.06407 ($57.02147 + 129.04260$)	
Alanine + Aspartic Acid	186.06407 ($71.03712 + 115.02695$)	0
Tryptophan	186.07932	0.01525
Valine + Serine	186.10045 ($99.06842 + 87.03203$)	0.03638
Proline + Threonine	198.10045 ($97.05277 + 101.04768$)	
Valine + Valine	198.13684 (2×99.06842)	0.03639

ions appear at low m/z; it is helpful to search for them *first*; they provide indications about the amino acids present in the peptide. Several ions show satellite lines owing to neutral loss (or addition) of H_2O, of NH_3. Besides, there could be side chain losses. These ions are a, b, y, and z. So once the latter ions are recognized, it is helpful to check if their satellite ions are present. Additionally, there are internal a-type and internal b-type ions.

At the outset, one needs to be aware of what the precursor peptide mass MH^+ is. This would be known since this is the ion that was subjected to fragmentation. The relationship between the y-ion, b-ion mass and the mass of the peptide mass MH^+ that would be repeatedly used to correlate b-ion y-ion pairs is $y_i + b_{n-i} = MH^+ + 1$. If M is the peptide, MH^+ is that peptide protonated. Then the m/z of the protonated peptide can be written in terms of the y and the b ions as $(M+H)^+ = b + y - 1$; the y-ion and b-ion masses make up the precursor peptide m/z, the 1 on the right side is to account for the proton that appears on the left side. Rearranging the above equation, y and b ion m/zs can be written in terms of the other: $b = (M+H)^+ + 1 - y$, and $y = (M+H)^+ + 1 - b$. Whenever a b- or y-ion is suspected or recognized, it is good practice to check for the existence of its complement y- or b-ion.

One way of starting the analysis of the spectrum is to examine first the low m/z region. The low energy region is the one where immonium ions would be expected to show up (as in Examples 1 and 2 given below); in ion-trap based experiments they would not be (Example 3). Note in Table 4.4 that all immonium ions are of low mass; they immediately give indication to residue(s) present in the peptide.

The C-terminus of a *tryptic* peptide is either lysine or arginine. Since in protein mass spectrometry trypsin is almost invariably used as the reagent in proteolysis, all such resulting peptides are tryptic peptides. m/z of singly charged y_1 ion is given by $y_1 =$ residue masses of the (C-terminus) amino acid + C-terminus $[17(OH)/16(NH_2)] + 2$ Da [1 for N-terminus hydrogen– to form NH_2, 1 for protonation of this NH_2] = residue mass + 19 Da. With -COOH at the C-terminus we should expect for lysine K-OH = 128.09 + 19 = 147.09 and for arginine R-OH = 156 + 19 = 175 (for NH_2 in place of OH it would be 1 Da less). So, presence of 147 ion or 175 ion in the spectrum (that would correspond to y_1 ion) would give a strong indication about the residue at the C-terminus. With y_1 mass known, the position of b_{n-1} (the penultimate b-ion) can be calculated, and also the presence of this peak in the spectrum checked. On the basis of residue masses, the following relationship can be written for m/z of the *penultimate* b ion. $b = (M+H)^+ - 128$ (if K was the terminus or -156 here if R was the terminus) $- 18$ (because H_2O that was lost to generate the peptide bond must be recognized here). Tryptic peptides yield a relatively more abundant y-ion series.

While b_1 cannot be observed, b_2 can be, and with that satellite a_2 (not always appears) arising from a loss of 28 (CO) from b_2. Such b_2/a_2 pair is often present, and this is a good place to proceed from on the b-series ions. The next step is to look for the b_2 ion. b_2 would appear at the low end of the spectrum. b ions also produce neutral loss ions losing NH_3 (-17 Da) or H_2O (-18 Da). Using the relationship $y_i + b_{n-i} = MH^+ + 1$ one can calculate the corresponding y_{n-2} ion (this will be at the high end).

This y_{n-2} information is to be retained for use a little later. After b_2, larger-size b ions can be traced in steps. The amino acid residue masses can be *added* one by one — this is best done by starting with the lowest mass amino acid — and by checking if that ion has appeared in the spectrum. When such an ion is found, it is important to be sure that the mass increase is because of a *single* amino acid and *not a sum* of two amino acids (there are several such accidental coincidences — see Table 4.5). One needs to also watch for isobaric amino acids. The masses of fragments, when there is a combination of residues, are given in Table 4.5.

With y_1 identified (K or R), b_{n-1} can be identified by subtracting K or R mass from MH^+. That is, the entire b series is then identified. (It should be mentioned that with tryptic peptides, b ion series may not be prominent; the y ion series is, because the C-terminus is usually more basic.)

Table 4.6: Characteristic side-chain losses of the 20 standard amino acids from the molecular ion.

Amino acid	Characteristic losses (Da) from MH^+
Ala	-
Arg	-100
Asn	-58
Asp	-59
Cys	-47
Gly	-
Gln	$-59, -72$
Glu	$-36, -60, -63, -73$
His	-81
Ile/Leu	-57
Lys	$-59, -72$
Met	$-47, -48, -62, -75$
Phe	$-91, -92$
Pro	-
Ser	-31
Thr	-45
Trp	-130
Tyr	$-107, -108$
Val	-43

Medzihradszky, K. F. and Burlingame, A. L. (1994) *Methods: A companion to methods in Enzymology*, **6**, 284–303.
Reproduced by permission of Elsevier Limited.

With y_{n-2} identified earlier, the y-ion series can now be extended to *lower* masses. Here, one needs to keep on *subtracting* amino acid masses from y_{n-2} to generate one by one the remaining members of the y-ion series. It is important to check at the end that the b-ion series and the y-ion series tally with each other. There could be additional lines from doubly charged ions and some high-energy ions — though all of them may not be observable — beside the b-ion and the y-ion series. For triply charged ion, a series of doubly charged ions may be expected. The mass change from side-chain loss is given in Table 4.6. Fragment ion masses are given in Table 4.7. (Specific rules are there, such as for P.)

It must also be borne in mind that there could be (one or more) post-translational modifications which would add to residue mass(es), as a result of which overall ion masses of fragments would get modified.

At the final step, of course, all the residue masses have to be added up to check that they amount to the mass of MH^+.

Table 4.7: Rules for the calculation of fragment ion masses.

Fragment	Mass calculation using residue weights	Mass calculation from other fragments
a_i	Σ residue weights $- 27$	$b_i - 28$
b_i	Σ residue weights $+ 1$	$MH^+ + 1 - y_{n-i}$
c_i	Σ residue weights $+ 18$	$b_i + 17$
d_i	Σ residue weights $- 12 -$ side chain	$a_i - (R_i - 15)$
for Ile	Σ residue weights $- 55$ or $- 41$	$a_i - 28$ or $- 14$
for Thr	Σ residue weights $- 43$ or $- 41$	$a_i - 16$ or $- 14$
for Val	Σ residue weights $- 41$	$a_i - 14$
v_i	Σ residue weights $+ 74$	$x_{i-1} + 29$
w_i	Σ residue weights $+ 73$	$x_{i-1} + 28$
for Ile	Σ residue weights $+ 87$ or $+ 101$	$x_{i-1} + 42$ or $+ 56$
for Thr	Σ residue weights $+ 87$ or 89	$x_{i-1} + 42$ or $+ 44$
for Val	Σ residue weights $+ 87$	$x_{i-1} + 42$
x_i	Σ residue weights $+ 45$	$y_i + 26$
y_i	Σ residue weights $+ 19$	$MH^+ + 1 - b_{n-1}$
Y_i	Σ residue weights $+ 17$	$y_i - 2$
z_i	Σ residue weights $+ 2$	$y_i - 17$
Internal fragments		
b-type	Σ residue weights $+ 1$	
a-type	Σ residue weights $- 27$	

Medzihradszky, K. F., Sequence determination of peptides.
Reproduced by permission of Elsevier Limited from
Methods of Enzymology, Vol. 402, 209–244, 2005.

4.10.2 Examples of manual sequencing

In the following pages, we discuss *de novo* sequencing of three samples.

Example 1: Sequencing a peptide of m/z **1187.6 (+1) – (1188.3 MH$^+$ average).**

The MALDI PSD mass spectrum of a tryptic peptide in the first example[347] is shown in Fig. 4.28.

Several immonium ions can be unambiguously identified by the presence of Arg (m/z 70 – Arg/Pro 70, 87, 112), Ile/Leu (m/z 86 – Ile/Leu 86), His (m/z 110 – His 110 – and 166), Phe (m/z 120 - Phe 120), and Tyr (m/z 136 – Tyr 136). MH$^+$ is at 1187.6. This was a tryptic peptide, featuring y_1 ion at m/z 175, which implies a C-terminal Arg and the corresponding b_{n-1} ion was also detected at m/z 1014. At m/z 1073, there is an abundant ion that corresponds to a 115 Da loss from the molecular ion. Considering the correlation given above, this mass difference is consistent with b_1 mass of 116 Da, which identifies the N-terminal residue as Asp. Thus, we obtain the two terminii of the sequence: (Asp-   - Arg).

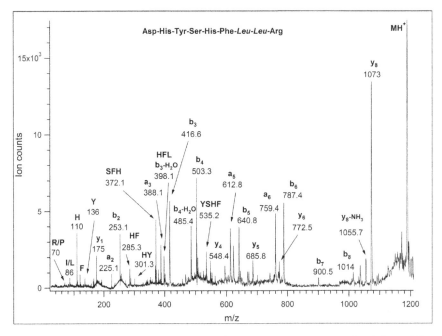

Figure 4.28: MALDI PSD spectrum of m/z 1187.6, a tryptic peptide of the 50 kDa subunit of DNA polymerase α from *S. pombe*. Reproduced by permission of Elsevier Limited from *Methods in Enzymology*, Vol. 402, 209–244, 2005.

Carbonyl- and water-losses from sequence and internal fragments can help identify b-type ions. The spectrum shows some abundant ions that are good b-type candidates, such as m/z 787, 640, 535, 503, 416, 398, 285, and 253. The ions m/z 503.3 and 416.6 are likely to be b ions; they seem to lose 18 (H_2O) to produce 485.4 and 398.1 respectively. Similarly the ion m/z 787.4 which seems to lose 28 (CO) to produce 759.4, 640.8 which seems to lose 28 (CO) to produce 612.8, and 253 which seems to lose 28 (CO) to produce 225.1. The ions 787 and 640 are 147 Da apart, corresponding to a Phe residue, and the y-ion complementary to the 640 b-ion occurs at m/z 548. The ion at 535 does not fit in the series as there is no amino acid with a 105 Da residue mass (640 − 535 = 105). With phenylalanine, the sequence now is (Asp- -Phe- -Arg).

The abundant ion at 503.3 seems to lose water and give 485.4 (observed), therefore, 503.3 is likely to be a b ion. The difference between the b ion at m/z 640 and 503 is 137, corresponding to a His. Its complementary ion (1188.3 − 503 = 685) is observed. 668 also is observed (not marked in the spectrum), seemingly from NH_3 loss (−17). This leads to the sequence: (Asp- -His-Phe- -Arg).

The next b-ion is at m/z 416, 87 Da apart, indicating a Ser-residue. There is water loss from this b-ion, generating the ion m/z 398. The corresponding y ion ($1188.3 - 416 = 772.3$) is detected at 772.5. This takes us to the sequence: (Asp- -Ser-His-Phe- -Arg).

The ion m/z 398.1, following 416.6, does not fit in the series; it is formed most probably via water loss from 416. The ion m/z 285.3 is 131 Da lower than the b-ion m/z at 416, which could be owing to Met (131.04), but no other ion — immonium ion, side-chain loss, y-ion support this conclusion. It is likely an internal fragment (His-Phe is 285.3). The mass difference between 416 and 253 is 163, which corresponds to Tyr. With this we obtain some more of the sequence: (Asp- -Tyr-Ser-His-Phe- -Arg).

Since the N-terminal residue has already been established as Asp, and $253 - 116(b_1) = 137$ (His is 137), the second amino acid must be a His residue. Therefore: (Asp-His-Tyr-Ser-His-Phe- -Arg).

If we consider MH$^+$ 1187.6 and Asp-His-Tyr-Ser-His-Phe- -Arg, the mass gap is 226. From immonium ion lines, the series contains a Leu/Ile with a residue weight of 113; the other missing amino acid must also be either a Leu or Ile. Indeed there is an y_2 ion at m/z at 288 (not marked, small peak), corresponding b-ion at m/z at 900.5, and $y_2 - NH_3$ ion at m/z 271 (not marked). Thus the sequence of this peptide is Asp-His-Tyr-Ser-His-Phe-*Leu/Ile*-*Leu/Ile*-Arg. With PSD fragmentation it is not possible to differentiate between isomeric Leu/Ile residues.

The b-type ions listed above but not identified as sequence ions, such as m/z 285, 398, and 535, are internal fragments corresponding to His-Phe, His-Phe-Leu/Ile, and Tyr-Ser-His-Phe sequences, respectively. Similarly, ions at m/z 301 (His-Tyr), 372 (Ser-His-Phe), and 672 (His-Tyr-Ser-His-Phe) are b-type internal fragments, only the a-type fragments formed by the carbonyl-loss were not detected. The presence of basic His residues explains why more abundant b- and internal ions than C-terminal fragments are observed.

Example 2: Sequencing a peptide with precursor ion at m/z 557.8 (2+)

The low energy CID spectrum[348] of m/z 557.8(2+) peptide is shown in Fig. 4.29.

The immonium ions (60, 72, 120) indicate the presence of Ser, Val, and Phe residues; the presence of 87 indicates Asn, 101 Lys/Gln; the presence of 147 indicates y_1 for lysine (tryptic peptide; 129 — Arg immonium ion — is not seen). The paucity of peaks in the high mass region presents a fortuitous advantage. A series of ion *pairs* which are separated by m/z 17 can be discerned: m/z 1,027–1,010; 970–953; 856–839; 709–692; 622–605, and 475–458. As this was a spectrum obtained via collision-cell-derived CID, one would guess that these are y-ions, since such spectra contain *few*

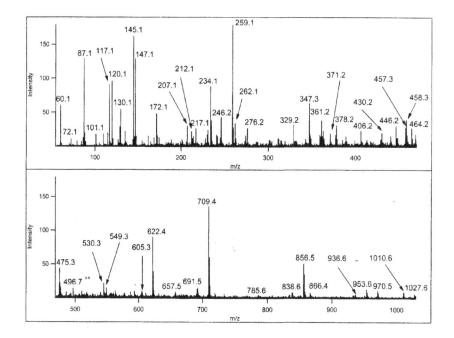

Figure 4.29: Low-energy CID spectrum of m/z 557.8 (2+). The corresponding sequence determined from this spectrum is Ser-Gly-Asn–Phe-Ser-Phe-Gln-Thr-Val-Lys. Reproduced by permission of Wiley Periodicals, Inc. from *Mass Spectrometry Reviews*, DOI: 10.1002/mas.21406 published online October 29, 2013.

high-mass b-ions while y-ions are observed in abundance. The m/z differences 57, 114, 147, 87, and 147 can be interpreted to suggest the presence of the sequence Gly–Asn–Phe–Ser–Phe. (y-ions extend from the C-terminus; so the loss of 57 from 1,027 means loss of Gly from the N-terminal side, and so on.)

The next y-ion is probably at 347.3, through a loss of 128 from the ion at 475.3. The 128 loss could be due to a loss of either Gln or Lys. Since Gln (128.06) and Lys (128.09) are isobaric (Table 4.5), this point cannot be immediately resolved owing to the limitation of mass accuracy.

We can now look at the spectrum to identify some b-ions. Using the equation $b_1 + y_{n-1} = MH^+ + 1$, and applying it to the y 1027.6 ion, we obtain for the N-terminal segment 88.1 $((557.8 \times 2)-1 +1 - 1027.6)$ which implies Ser (plus H^+); but b_1 cannot be formed. However, with this, the working sequence would be Ser-Gly-Asn-Phe-Ser-Phe-Gln/Lys-   -Lys.

Usually b_2 and a_2 are abundant. b_2 should be 145 $(87 + 57 + +1)$, b_3 259 $(b_2 +114)$, b_4 406 $(b_3 + 147)$, b_5 493 $(b_4 +87)$. All these ions are observed. The ions at m/z 262 and 172 could be internal fragments Asn-Phe (262) and

Gly-Asp (172). Loss of CO from 262 would produce 234, which is observed. However, a potential y_2 fragment, for a Ser-Lys C-terminus with Leu/Ile completing the sequence, too would produce 234. With monoisotopic mass substitution the m/z for Asn-Phe $-$ CO would be 234.1237 and that for this potential y_2 would be 234.1448. The measured m/z was 234.1387, meaning $+64$ ppm and -26 ppm mass deviations for these two cases respectively, too close to be resolved by the mass spectrometer (QSTAR) used here. If the ninth residue is Ser, then the presence of Ile/Leu (113) at the eighth position would meet the total mass requirement, but it would not explain the *absence* of immonium ion for Ile/Leu and the *presence* of the immonium ion for Val. So, Ser is *not* suitable for the ninth position. However, if we take valine for the ninth residue, then by difference between the mass of the lowest y mass fragment assigned so far (347) and y_1 (147) we obtain 200; and from this if we subtract 99 (valine mass), we obtain 101, which is the residue mass for threonine Thr. Presence of valine is supported by the presence of the 246 ion in the spectrum. The complete sequence therefore is Ser-Gly-Asn-Phe-Ser-Phe-Gln/Lys-Thr-Val-Lys.

The y_4 ion with Gln or Lys at the seventh position implies a mass difference of 0.036 ($\sim$75 ppm); this could be resolved with current high resolution instruments. Using *MS-Product* in Protein Prospector, a listing of all Gln/Lys-containing fragments can be obtained. The observed ion at 259 is a good possibility (Phe-Gln/Lys $-$ NH$_3$) but this overlaps with b_3. The fragment at 212.110 representing Gln/Lys-Thr $-$ H$_2$O gives a better lead; here the mass error is $+33$ ppm for Gln versus -137 ppm for Lys. Therefore, for the sequence we have Ser-Gly-Asn-Phe-Ser-Phe-Gln-Thr-Val-Lys— SGNFSFQTVK.

Example 3: Sequencing a 898.4 m/z (+1) peptide

This example involves sequencing of the peptide whose MS/MS spectrum[352] is shown in Fig. 4.30; the m/zs of the major ions observed in its ion-trap-based experiment are 881.4, 880.4, 862.3, 836.3, 784.4, 766.3, 752.3, 735.4, 718.2, 687.3, 669.3, 639.3, 622.2, 604.2, 586.3, 540.2, 523.1, 529.3, 426.2, 370.2, 359.2.

M for this peptide sample is 897.4; $M + 2$ therefore is 899.4. Since this is a tryptic peptide, its C-terminus is either Lys or Arg. Both 147 and 175 are below the lower limit of the spectrum (ion trap-based experiment, 1/3 cutoff rule applies, no ion below 1/3 of 899 remains trapped and can appear), so the residue at the C-terminus cannot be easily ascertained. Were 784.4 a b ion, we would expect the m/z of its potential complementary y ion to

[352]Nambi, S., Gupta, K., Bhattacharyya, M., Ramakrishnan, P., Ravikumar, V., Siddiqui, N., Thomas, A. T., and Visweswariah, S. S. (2013) Cyclic AMP-dependent protein lysine acylation in mycobacteria regulates fatty acid and propionate metabolism. *J. Biol. Chem.*, **288**, 14114–14124.

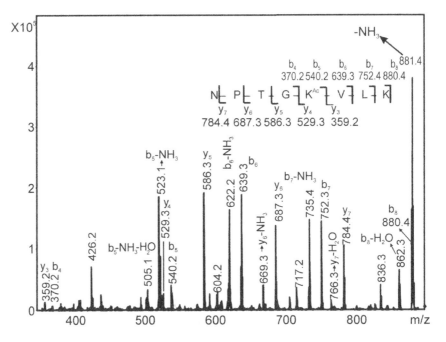

Figure 4.30: CID MS/MS spectrum of the tryptic peptide containing the acetylated Lys residue for FadD2 (m/z 898.5). Reproduced by permission of the American Society for Biochemistry and Molecular Biology, Inc. from *J. Biol. Chem.*, Vol. 288, 14114–14124 (2013).

be of m/z 899.4 − 784.4 = 115. This m/z is much lower than the lowest m/z C-terminus y ion (147 or 175) that can be expected; 784.4 therefore is *not* a b ion; it is a y ion.

We select a high intensity peak near the center of the spectrum; let us take 586.3. By subtracting 586.3 from 899, we obtain 312.7. This is not observed (lowest m/z ion observed is 359.2), so 586.3 has no complementary ion on the spectrum.

The difference between 604.2 and 586.3 (17.9) does not correspond to any amino acid. 639.3 and 622.2 are two peaks obviously separated by NH_3 loss (−17), so are 687.3 and 669.3 (−18). 639.3 − 586.3 = 53 does not correspond to any amino acid. However, 687.3 − 586.3 = 101 corresponds to threonine Thr 101.05. Further, 586.3 − 529.3 = 57.0 matches with glycine Gly. We seem to have recognized two adjacent amino acids. However, 529.3 − 426.2 = 103.1 does not match with anything. Going to the other direction, next, 718.2 − 687.3 = 30.9 does not correspond to any amino acid. 735.4 is obviously a NH_3-loss (− 16.9) peak from 752.3. After that 752.3 − 687.3 = 65.0 does not correspond to any amino acid. 784.4 and 766.3 are two peaks differing by H_2O (-18). $784.4 - 687.3 = 97.1$ corresponds to proline Pro. 880.4 and 862.3 obviously are two peaks differing by H_2O. $880.4 - 784.4 = 96$ does not correspond to any amino acid. So, Pro-Thr-Gly are part of one side (N-terminus side) of the series, or sequence, giving y-ions at 784.4, 687.3, 586.3.

Continuing, the peaks 880.4, 752.3, 639.3 therefore should arise from the other side (the C-terminus side) — of the sequence. There, we have $880.4 - 752.2 = 128.1$ (lysine). $752.3 - 639.3 = 113.0$ (113.08 leucine L, isoleucine Ile I). 540.2 and 523.1 are obviously related by NH_3 loss — difference of 17. Continuing, $639.3 - 540.2 = 99.1$ would be valine V. ($586.3 - 540.2 = 46.1$ does not match with any amino acid residue, therefore 540.2 peak does not belong to this side of the series.) $540.2 - 529.3 = 10.9$ does not match any amino acid, therefore 529.3 does not belong to this series.

Further, $540.2 - 426.2 = 114.0$ does not quite fit with the b-series (were 426.2 a b-ion peak we would have expected a NH_3-loss peak at around 409). $540.2 - 370.2 = 170.0$ does not match with any amino acid; it could be, however, a derivative. $529.3 - 426.2 = 103.1$ does not match with any amino acid. $529.3 - 359.2 = 170.1$. Note that the 170.0 appears in the other series as well, this may mean that this is a *modified* amino acid at the center of the peptide; one possibility is Lys(Ac).

So, on the N-terminus side, we have Pro-Thr-Gly-Lys(Ac), and, on the C-terminus side Lys(Ac)-Val-Leu-Lys. Since it is a tryptic peptide, one could check for the possibility of Lys at the C-terminus and the two sides connected at Lys(Ac). In that case the sequence would be Pro-Thr-Gly-Lys(Ac)-Val-Leu-Lys. $M = \Sigma$ amino acid masses $+ 1 + 17$; $MH^+ = \Sigma$ amino acid masses $+ 1 + 17 + 1$. Checking with the MH^+ mass ($898.4 = 897.4 + 1$) and subtracting the series PTGK(Ac)VLK mass ($97.1 + 101.1 + 57.1 + 170.1 + 99.1 + 113.1 + 128.1) + 2 + 17$, one obtains ($899.4 - 784.7$) 113.7; the remaining amino acid therefore seems to be asparagine. Thus, the full sequence is Asn-Pro-Thr-Gly-Lys(Ac)-Val-Leu-Lys.

CHAPTER 5
EXAMPLES FROM BIOLOGICAL APPLICATIONS

In this chapter, we present in some detail a few examples of applications of mass spectrometry to protein science. We begin with an example of intact protein analysis, which would show the present status of the top-down proteomics field. This is followed, in order, by an example of differential mass spectrometry in analyzing apolipoproteins in HDL samples, an application of the *de novo* approach to structure derivation of conotoxins, and then a mass spectrometric study of ribosomes. After that are a couple of examples from neuroscience: identification of neural proteins from a single CNS synapse type — Parallel Fiber/Purkinje cell synapse, and neurexin-LRRTM2 interaction in synaptogenesis. Then we present an application of ion mobility mass spectrometry to human plasma proteome. An example of time-dependent mass spectrometric study to low quantities of mixtures of noncovalently bound biomolecules comes next in a study of viral capsid formation; that is followed by an analysis of the structure of the enzyme ATPase. Finally, we present a couple of examples from mass spectrometry imaging.

5.1 Mapping intact protein isoforms using top-down proteomics

The lack of efficient intact protein fractionation methods has been a significant limiting factor for the past inability of top-down proteomics to go into a proteome scale. In the work of Kelleher *et al.*[51] presented here, they, using a *four-dimensional* separation system along with the latest version of Swiss-Prot database, identified 1,043 gene products from human cells that are dispersed into more than 3,000 protein species created by PTMs, RNA splicing, and proteolysis. Identification of proteins up to 105 kDa and those with up to 11 transmembrane helices was feasible with greater than 20-fold increases in both separation power and proteome coverage. The work provided precise correlations to individual genes, and it mapped a number of previously undetected isoforms of endogeneous human proteins which included some where multiple modifications had taken place owing to accelerated cellular aging induced by DNA damage. At the time of writing, this level of proteome coverage represented the most comprehensive

220

Introduction to Protein Mass Spectrometry
http://dx.doi.org/10.1016/B978-0-12-805123-8.00005-9

Copyright © 2016 Elsevier Inc.
All rights reserved.

implementation of top-down mass spectrometry so far, with an approximate tenfold increase in identifications of intact proteins for any microbial system and a greater than 20-fold increase over any previous work in mammalian cells. The work portends emerging possibilities of large-scale interrogation of whole protein molecules. They called large-scale identifications of the human proteome with the highest content of molecular information preserved for primary structures of endogenous proteins the discovery mode. This is a large body of work, what follows here is a summary description of the accomplishments; still, it is expected to show where the top-down proteomics field stands at the present.

In this work, using solution isoelectric focussing (sIEF)[50] and gel-eluted liquid fraction entrapment electrophoresis (GELFrEE)[49] in tandem for fractionation by protein IEF and size, respectively — together comprising *two*-dimensional *liquid* electrophoresis — the two-dimensional *gel* electrophoresis is bypassed and hence the associated low recovery and extensive workup steps. That is followed by nanocapillary LC and MS — the *third* and the *fourth* dimensions of the four-dimensional separation, for both low[353] and high[354] molecular mass proteins, before MS/MS steps. A 12 T linear ion trap Fourier-transform LTQ FT Ultra or the Orbitrap Elite, Thermo Fisher Scientific, mass spectrometer was used. For detection of precursor ions ~ 30 kDa, and for increased sensitivity, an ion trap was employed. The mass spectrometer RAW files were processed with in-house software called PTMCRAWLER to assign masses; the intact masses and fragment masses determined were searched against a human proteome database (Swiss-Prot).

In their liquid chromatography platform, using 0.5-1 mg of input protein, a peak capacity of well over 2,000 ($\sim 2,500$) was obtained for separation of protein molecules in solution. Combining this with the separation power of the mass spectrometer (peak capacity ~ 75), the resulting peak capacity of the four-dimensional system was greater than 100,000 ($\sim 187,500$) for proteins below approximately 25 kDa, 20-fold higher than the peak capacity for high-resolution two-dimensional gels. Using fragmentation data acquired with < 10 ppm mass accuracy for searching with

[353]Lee, J. E., Kellie, J. F., Tran, J. C., Tipton, J. D., Catherman, A. D., Thomas, H. M., Ahlf, D. R., Durbin, K. R., Vellaichamy, A., Ntai, I., Marshall, A. G., and Kelleher, N. L. (2009) A robust two-dimensional separation for top-down mass spectrometry of the low-mass proteome. *J. Am. Soc. Mass Spectrom.*, **20**, 2183–2191.

[354]Vellaichamy, A., Tran, J. C., Catherman, A. D., Lee, J. E., Kellie, J. F., Sweet, S. M. M., Zamdborg, L., Thomas, P. M., Ahlf, D. R., Durbin, K. R., Valaskovic, G. A., and Kelleher, N. L. (2010) Size-sorting combined with improved nanocapillary liquid chromatography-mass spectrometry for identification of intact proteins up to 80 kDa. *Anal. Chem.*, **82**, 1234–1244.

a tailored software[355] and databases with highly annotated primary sequences,[356] identification and characterization of isoforms were achieved. This resulted in a deep consideration of known PTMs, alternative splice variants, polymorphisms, endogenous proteolysis, and diverse combinations of all these sources of molecular variation at the protein level. Together with the curation of the Swiss-Prot database, the result was used to map each given protein identification to a single gene (except in rare cases where multiple genes could produce the identical sequence).

A total of 1,043 proteins, originating from 1,045 human genes and having unique Swiss-Prot accession numbers, were identified. Over 400 species were core histones. Seventy-seven percent of the protein products of these genes displayed amino (N)-terminal acetylation. Further, arising from fragmentation evidence for 3,093 protein isoforms/species the following range of PTMs were detected: 645 phosphorylations, 538 lysine acetylations, 158 methylations, 19 lipid/terpenes, and five hypusines.

Detecting PTMs based on intact mass values, they came up with pairs of protein species showing characteristic mass differences (such as $\Delta 14\,\mathrm{Da}$ for methylation, $\Delta 80\,\mathrm{Da}$ for phosphorylation, etc.). For proteins less than $20\,\mathrm{kDa}$, with mass differences consistent within $0.05\,\mathrm{Da}$, they detected 225 pairs with monomethylation, 185 with di-methylation and 122 with tri-methylation/acetylation; they also detected 87 cases consistent with double acetylation, 140 with mono-phosphorylation, and 100 with diphosphorylation. Overall, 2,130 such mass shifts were found on the entire HeLa data set for all isotopically resolved proteins. The prevalence of uncharacterized mass shifting events in the human proteome is a notable finding.

Of the 1,043 proteins identified with intact mass information, 431 were from using isotope spacings, 54% of these matched within $2\,\mathrm{Da}$ the species identified from the database; whereas of the 331 that were identified by deconvolution of charge states, 130 were manually determined to be of high quality and 51% of these matched within $200\,\mathrm{Da}$. Those outside these windows were clearly identified by fragmentation, but had some mass discrepancies to be explained.

Precise identification of protein isoforms, because of functional differences in the latter, adds substantially to the information content of proteomic analyses in higher eukaryotes. For a gene-specific identification of protein isoforms, matching fragment ions from both termini besides

---

[355]Durbin, K. R., Tran, J. C., Zamdborg, L., Sweet, S. M. M., Catherman, A. D., Lee, J. E., Li, M., Kellie, J. F., and Kelleher, N. L. (2010) Intact mass detection, interpretation and visualization to automate top down proteomics on a large scale. *Proteomics*, **10**, 3589–3597.

[356]Roth, M. J., Forbes, A. J., Boyne, M. T. II, Kim, Y., Robinson, D. E., and Kelleher, N. L. (2005) Precise and parallel characterization of coding polymorphisms, alternative splicing, and modifications in human proteins by mass spectrometry. *Mol. Cell. Proteomics.*, **4**, 1002–1008.

the intact protein mass, are usually sufficient. In this work, nine of the approximately 15 isoforms of histone H2A, despite their greater than 95% sequence identity, were fully characterized in an automated fashion (including the H2A.Z and H2A.X variants) with an additional three having Δm greater than 1 Da (H2A type 1-D, 2-C and 2-B). They also identified nine S100 proteins, several α and β-tubulins, seven unique isoforms of human keratin, MLC20, BTF3 and their related sequences (which are 97% and 81% identical, respectively), and over 100 isoforms/species from the high-mobility group (HMG) family. In the 40–110 kDa range, this work included extensive characterization of GRP78, a 70.6-kDa heat-shock protein (more than 12 fragment ions mapping to each terminus), and identification of several proteins greater than 90 kDa, such as P33991 (97 kDa) and Q14697 (104 kDa).

Extensive use of SDS in the 2D liquid electrophoresis platform causes reduced bias against integral membrane proteins. Of the 1,043 total identifications from HeLa cells, 32% were membrane-associated proteins, with 62% of these annotated as integral membrane proteins. In a study of a mitochondrial membrane fraction using modified chromatographic procedures (A: 60% formic acid in water, B: 100% isopropanol; the gradient used: 0% B at 0 min, 25% B at 5 min, 60% B at 50 min, 95% B at 56 min) for enhanced separation of membrane proteins, they identified an additional 46 integral membrane proteins from a single three-dimensional experiment (without isoelectric focussing). The species that eluted from the column during LC-MS revealed proteins with a distribution of 1–11 transmembrane helices.

To bring out the DNA damage response, cells were treated with etoposide which was followed by four-dimensional fractionation and top-down tandem mass spectrometry. Using gene ontology analysis, all identifications in this work were annotated according to cell compartment (like nucleus, membrane, etc.) — the result is shown in Fig. 5.1(a) — or biological process (like regulation of biological processes, cell adhesion, etc.) — the result is shown in Fig. 5.2. They detected many proteins involved in cell cycle regulation and apoptosis, including nine that interact with proliferating cell nuclear antigen during repair of DNA damage. They identified several proteins involved in the Fanconi anaemia pathway including FANCE, RAD51AP1, RAD23B and RPA3. A number of cyclin-dependent kinase (CDK) inhibitors were identified, such as p27$^{\text{Kip1}}$ (CDKN1B) and p16$^{\text{INK4a}}$ (CDKN2A), T53G1 and the protein product from a target gene of p53 (Q9Y2A0, p53-activated protein 1).

To readout phosphphorylation stoichiometry closely, they monitored, using the three-dimensional fractionation approach (GELFrEE-nanocapillary LC-MS), 17 phosphoprotein targets across three time points at three different concentrations of etoposide. Increases in the occupancy of phosphorylation were found in H2A.X-pSer139 (γH2A.X) after treatment with

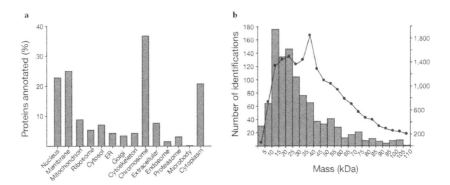

Figure 5.1: Proteome analysis metrics. (a) A gene ontology analysis for the identifications in this study. (b) Plot showing the molecular mass distribution for the unique identifications obtained. The line graph depicts the theoretical molecular mass distribution for the human proteome (Swiss-Prot, *Homo sapiens*, 20,223 entries). Reproduced by permission of Macmillan Publishers Limited from *Nature*, Vol. 480, 254–258, 2011.

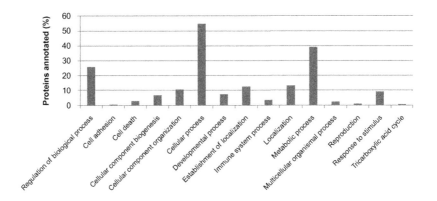

Figure 5.2: Gene Ontology analysis of the proteins identified in this study showing the proportion of proteins that are annotated in the specified biological processes. Reproduced by permission of Macmillan Publishers Limited from *Nature*, Vol. 480, 254–258, 2011.

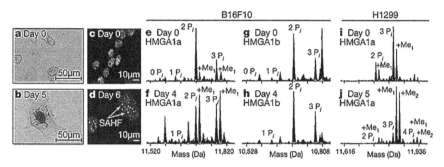

Figure 5.3: Monitoring dynamics of HMGA1 isoforms during senescence in B16F10 and H1299 cells. After induction of DNA damage by transient treatment with campothecin for H1299 cells or etoposide for B16F10, progression of accelerated senescence was monitored by SA-β-Gal (a,b) or DAPI staining to monitor formation of senescence-associated heterochromatic foci (SAHF) (c,d) over the specified recovery period. Changes in modification profiles on HMGAqa (e,f) and HMGA1b (g,h) from B16F10 showed mild increases to phosphorylation occupancy but a significant increase in methylation levels on multiply phosphorylated species. A more striking increase in both methylation and phosphorylation was observed in senescent H1299 cells (i,j). No such methylation was observed in the HMGA1b profiles for either cell line. Reproduced by permission of Macmillan Publishers Limited from *Nature*, Vol. 480, 254–258, 2011.

25 or 100 μM etoposide for 1 h. Consistent with engagement of the DNA-repair machinery, a return to basal levels of phosphorylation of γH2A.X was found after a 24 h recovery from treatment. A strong correlation was observed between the mass spectrometric determination of phosphorylation stoichiometry of γH2A.X with the results from immunofluorescence and western blotting run in parallel.

Using 3D fractionation, over 2,300 species (from 690 proteins) in H1299 cells and 2,300 species (from 708 proteins) in B16F10 melanoma cells were tracked in the days after a 24 h treatment with camptothecin or 5 h of etoposide, respectively. After induction of DNA damage, the classic hallmarks of stress-induced senescence in H1299 and B16F10 were monitored over several days, including cell enlargement and formation of senescence-associated heterochromatic foci. As both B16F10 and H1299 cells entered stress-induced senescence, although levels of γH2A.X remained the same as in control cells, a striking *up*regulation was observed of methylated forms of di- and tri-phosphorylated HMGA1a, but *not* of its *splice variant* HMGA1b as can be seen in Fig. 5.3.

Fragmentation data for two multiply modified species, HMGA1a and HMGA1b, are presented in Fig. 5.4. Regarding phosphorylations on HMGA1a, it was determined Ser 101 and Ser 102 occupied in the 2Pi form

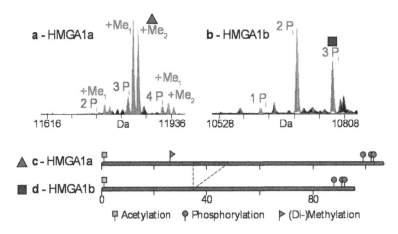

Figure 5.4: Characterization of two HMGA1 isoforms (indicated with a triangle, and a square). Isoform profiles are shown for (a) HMGA1a and (b) HMGA1b. Diagram highlighting the sites of alternative splicing and post-translational modification present on the (c) HMGA1a (tri-phosphorylated, di-methylated form) and (d) HMGA1b (tri-phosphorylated form). (e) Graphical fragment map showing precise localization of the di-methylation site at Arg25 and the addition of 11 amino acids present in HMGA1a but absent in (f) HMGA1b. Reproduced by permission of Macmillan Publishers Limited from *Nature*, Vol. 480, 254–258, 2011.

and there was evidence that the third site pointed predominantly toward
pSer 98. Some occupancy for pSer 43 was seen for the 3Pi and 4Pi forms,
a site only available in the splice region specific to the HMGA1a variant
(Fig. 5.4e). In senescent H1299 cells, the effect on methylation, for day
5, was particularly dramatic; both the mono- and di-methylated species
(which also contained multiple phosphorylations) increased, reproducibly,
to be greater than 80% of the total signal for species from the *hmga1* gene.
(This was also seen in biological replicates.) The methylation site was localized precisely to Arg 25, which is consistent with earlier work on HMGA1
proteins. Methylated HMGA species in damaged cancer cells undergoing
apoptosis also showed similar response, but the B16F10 and H1299 cells prepared here were clearly senescent (as measured by annexin V staining and
fluorescence-activated cell sorting analysis through day 6). Arg 25 is localized in the first AT-hook DNA-binding region (residues 21-31)(Fig. 5.4e); it
is possible, during accelerated cellular senescence, the R25me1 and R25me2
marks perturb DNA-kinking and allow HMGA1a to be preferentially incorporated into senescence-associated heterochromatic foci. Also noted were
other changes in bulk chromatin: hypoacetylation on all core histones, increased levels of H3.2K27me2/3, and decreased H3.2K36me3.

Success in large-scale proteome coverage foreshadows the feasibility for
interrogating the natural complexity of protein primary structures that exist within human cells and tissues, culminating in a definitive description
of protein molecules present in the human body. Also, full mapping of intact isoforms on a proteomic scale, detecting co-variance in modification
patterns, has the potential to unravel the post-translational logic of intra-
cellular signaling. This was the first time top-down proteomics was achieved
at this large scale.

In a later work from the same group which has been a far larger high-
throughput Top Down study,[357] identification of 1,220 proteins from the
transformed human cell line H1299 at a false discovery rate of 1% has been
reported with a significantly improved proteome coverage. Experimental
procedure was similar to their earlier work: sIEF, GELFrEE, LC/MS; CID,
HCD, ETD; data handling by ProSightPC 3.0.

In all, 347 mitochondrial proteins were identified (which is $\sim 23\%$ of all
the annotated human mitochondrial proteins — in a given cell type only
a fraction of the total is to be expected — including $\sim 50\%$ of the mitochondrial proteome below 30 kDa and over 75% of the subunits comprising
the large complexes of oxidative phosphorylation). Of the proteins iden-

[357]Catherman, A. D., Durbin, K. R., Ahlf, D. R., Early, B. P., Fellers, R. T., Tran, J. C.,
Thomas, P. M., and Kelleher, N. L. (2013) Large-scale top down proteomics of the human
proteome: membrane proteins, mitochondria, and senescence. *Mol. Cell. Proteomics*, **12**,
3465–3473.

tified, three hundred (301) were integral membrane proteins containing between one and 12 transmembrane helices; nuclear proteins represented the largest number of the total identified proteins; 42 histone proteins were identified, including macro-H2A (O75367) and five linker histone H1 proteins (H1.0, H1.2, H1.4, H1.5, and H1x); 82 of annotated ribosomal proteins were also identified ($\sim 65\%$ of the nonmitochondrial ribosomal proteins). More than 5,000 proteoforms were observed. Many of them showed post-translational modifications like phosphorylation and methylation; and over a dozen of them showed lipid anchors (such as myristoylation, geranylgeranylation), some previously unknown. By comparing untreated and senescent cells (senescence induced by DNA-damaging agent campothecin), hyperphosphorylation of HMGA2 (High-mobility group AT-hook 2 protein) was detected.

5.2 Quantitative analysis of intact apolipoproteins in human HDL

Differential mass spectrometry(dMS)[358,359,360] provides relative quantitation from full-scan mass spectrometry data. Using multiple LC-MS runs, at each time mass spectral peak intensities and m/z ratio between samples in two conditions are compared. In this label-free LC-MS method small but statistically significant differences in low-abundance peptides are detected; large but statistically insignificant differences in peptides present at much greater concentrations are ignored. Here we present an example[361] in which dMS is combined with *top-down* analysis of apolipoproteins isolated from human high-density lipoprotein HDL_3 — from patients having high and low HDL-cholesterol (HDL-c) levels. A high level of HDL-c (aka 'good cholesterol') is hypothesized to be cardioprotective in humans.

[358]Wang, W., Zhou, H., Lin, H., Roy, S., Shaler, T. A., Hill, L. R., Norton, S., Kumar, P., Anderle, M., and Becker, C. H. (2003) Quantification of proteins and metabolites by mass spectrometry without isotopic labelling or spiked standards. *Anal. Chem.*, **75**, 4818–4826.

[359]Wiener, M. C., Sachs, J. R., Deyanova, E. G., and Yates, N. A. (2004) Differential mass spectrometry: A label-free LC-MS method for finding significant differences in complex peptide and protein mixtures. *Anal. Chem.*, **76**, 6085–6096.

[360]Meng, F., Wiener, M. C., Sachs, J. R., Burns, C., Verma, P., Paweletz, C. P., Mazur, M. T., Deyanova, E. G., Yates, N. A., and Hendrickson, R. C. (2007) Quantitative analysis of complex peptide mixtures using FTMS and differential mass spectrometry. *J. Am. Soc. Mass Spectrom.*, **18**, 226–233.

[361]Mazur, M. T., Cardasis, H. L., Spellman, D. S., Liaw, A., Yates, N. A., and Hendrickson, R. C. (2010) Quantitative analysis of intact apolipoproteins in human HDL by top-down differential mass spectrometry. *Proc. Nat. Acad. Sci., U.S.A.*, **107**, 7728–7733.

The HDL_3 samples were analyzed by a ThermoFisher reverse-phase nano-HPLC coupled to a LTQ-FT hybrid mass spectrometer. High-resolution Orbitrap ETD-MS/MS spectra were acquired either by data-dependent acquisition or by targeting a specific m/z.

Six human subjects, healthy middle-aged donors, having relatively low HDL-c (avg. 44 mg/dL, $N = 3$) and high (avg. 74 mg/dL, $N = 3$) were chosen. HDL_3 subfraction ($d = 1.13$–1.21 g/mL) was chosen for its athero-protective potential, an attribute that has been extensively characterized.

Lipoprotein particles were isolated from human plasma by density gradient ultracentrifugation. LC-MS parameters were adjusted to accommodate intact proteins (C8 stationary phase used). Each of the six samples was split into three technical replicates to increase the statistical power of the analysis. Instrument resolving power was increased to 100,000 for higher charge-state analytes. Interspersed by quality assurance/quality control samples, all samples were analyzed nonstop by LC-MS in < 1.5 days — to minimize instrument drift and variability. The modifications allowed for a reproducible analysis with a median percent coefficient of variation (CV) of approximately 36%, acceptable for dMS analysis (CV = $[\sigma$ (standard deviation)$/\mu$ (mean)]).

To identify the protein and characterize post-translational modifications, ETD-MS/MS spectra were analyzed by ProsightPC v1.0 using 2,000 Da and 20 ppm tolerances for precursor and fragment ion masses, respectively. Spectra were deisotoped, a list of monoisotopic, neutral fragment masses generated, and then searched against a well-annotated human protein database (Uniprot) consisting of 130,248 protein sequences with known and potential post-translational modifications to generate a total of 2,985,079 distinct protein forms. Expectation values < 1×10^{-5} were considered true positive identifications.

The accurate mass determination of these ion species within 0.1 Da allowed prediction that the three primary analytes measured are apolipoprotein C-III and two post-translationally modified forms of apolipoprotein C-III, *O*-glycosylation by the addition of NANA-(2→3)-Gal-β (1→3)-GalNAc ($\Delta m = 656.2037$ Da) and branched [NANA-(2→3)-Gal-β(1→3)]-[NANA-(2→6)]-GalNAc ($\Delta m = 947.3580$ Da). By addition of purified apolipoprotein C-III standard, a linear instrument response in the 100 nM–100 μM range was ensured.

In the samples analyzed, several proteins are observed which are quantitatively different. Figure 5.5 (top) shows the amino acid sequence of apolipoprotein C-III with the c and z^* ions labeled. ProsightPC 1.0 yielded a single protein match (apolipoprotein C-III, accession no. P02656), with an expectation score 2×10^{-49}. Figure 5.5 also shows a high-resolution ETD spectrum of $(M + 9H^+)^{9+}$ ions at $m/z = 1{,}047.7253$. A total of 35 c and 17 z^* matching ions assigned to the spectral peaks, with an average

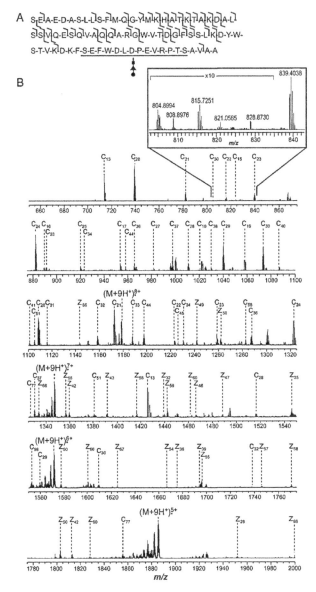

Figure 5.5: The amino acid sequence of apolipoprotein C-III with the matching c and z^* ions labeled (A). High-resolution electron-transfer dissociation spectrum of $(M+9H^+)^{9+}$ ions at $m/z = 1,047.7253$ (B). The NANA-$(2\rightarrow3)$-Gal-$\beta(1\rightarrow3)$-GalNAc- glycosylation, believed to be located at residue Thr[74], is localized to a portion of the protein between Asp[59]-Val[77]. Most of the z ions in the spectrum represent fragment ions containing sugar moiety (see the original for details). Reproduced by permission of the National Academy of Sciences, U.S.A. from *Proceedings of the National Academy of Sciences, U.S.A.*, Vol. 107, 7728–7733, 2010.

mass accuracy of 3 ppm, allowed definite identification of O-glycosylated form of apolipoprotein C-III [NANA-$(2 \rightarrow 3)$- Gal-β $(1 \rightarrow 3)$-GalNAc, $\Delta m = +656.2037$] ($m = 9{,}415.45$, $\Delta m = 656.2037$ Da), a post-translational modification of apolipoprotein C-III. The accurate mass fragment ions localized the O-glycosylation modification to the C-terminal end of this protein at Asp^{59}-Val^{77}, with Thr^{74} as the site of modification. Additional support for the presence of an O-glycosylation modification came from high-resolution CAD used in threshold dissociation mode which revealed seven b and three y matching fragment ions, the predominant fragmentation being the selective removal of the sugar side chain (-291.09 Da, -656.20 Da). This compound is believed to be associated with coronary artery disease.

To identify proteins of interest (as defined by m/z ratio, intensity and retention time) by dMS, expression profiles generated from the high-resolution LC-MS data were analyzed using the Rosetta Elucidator system (version 3.2). Features (observed isotope peaks) readily measured by the mass spectrometer — area under curves (AUC), greater than zero value in all 18 measurements were selected and the data were then exported from Elucidator. All feature peak areas were extracted from LC-MS data files by the Peak-Teller (R) algorithm. The statistical analysis was performed on log AUC of the features. To properly distinguish between biological replicates from technical replicates, the three technical replicates for each feature within each subject were first averaged. Two-sample t-tests were then performed on the three low HDL-c subjects vs. the three high HDL-c subjects, for each feature. Those features with p-values lower than 0.02 for two or more isotopes and charge state 2+ or greater were selected. To estimate %CV, features containing three or more zero peak area under curves values in individual samples were discarded. Those containing AUC values $< 5{,}000$ were taken as 5,000. The estimated platform %CV of 36% was obtained as the median of all individual feature %CVs.

The dMS analysis revealed multiple statistical significant differences in protein abundance. Of the 96,235 features detected, approximately 380 demonstrated statistically significant changes in protein abundance between high HDL-c and low HDL-c subject groups at $p < 0.02$ and filtering. Several features of ~ 9 kDa showed significant abundance differences across the HDL_3 samples. Figure 5.6 shows the dMS quantitation results of three forms of apolipoprotein C-III measured in the six patient samples. Interestingly, the three major forms of the protein showed higher abundances, on average, in samples of patients having low HDL-c concentrations.

The work quantitatively determined differences in abundances of protein and protein isoforms in the presence of a complex background of unchanging proteins. A notable interesting finding is the significantly high concentration of NANA-$(2 \rightarrow 3)$ Gal-β $(1 \rightarrow 3)$ GalNAc-ApoC-III in the low HDL-c samples.

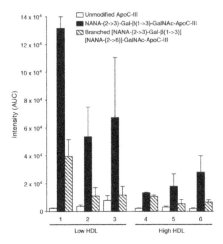

Figure 5.6: Quantitation of the three forms of apolipoprotein C-III. Error bars represent standard error of the mean for multiple ($n = 3$) replicate LC-MS injections. Subjects 1–3 low HDL; subjects 4–6 high HDL. Reproduced by permission of the National Academy of Sciences, U.S.A. from *Proceedings of the National Academy of Sciences, U.S.A.*, Vol. 107, 7728–7733, 2010.

5.3 Rapid sequence analysis of some conotoxins — combination of *de novo*, bottom-up methods

Venoms are mixes of neurologically active peptides. They block specifically diverse groups of ion channels; they can be used as research tool via isolation of a given ion channel, or as therapeutic, as ion channels are implicated in pain, heart arrhythmia, epilepsy, etc. In this section we present a work of Chait and coworkers[362] on toxins from the venom gland of cone snails — conotoxins. Their work revealed the complete amino acid sequence of 31 of the cone snails peptides, including King Kong toxin,[363] using less than seven percent of their subjects venom.

---

[362]Ueberheide, B. M., Fenyö, D., Alewood, P. F., and Chait, B. T. (2009) Rapid sensitive analysis of cysteine rich peptide venom components. *Proc. Nat. Acad. Sci. U.S.A.*, **106**, 6910–6915.

[363]'King Kong toxin, a component of the venom in some poisonous marine snails, has a peculiar power to go with its peculiar name. When injected into a meek little lobster in a tank full of superiors, the poison induces delusions of grandeur; the little guy starts marching around like he's king of the tank.' newswire.rockefeller.edu/2009/04/29/new-sequencing-technique-to-prod-benefits-from-killer-venom.

Conotoxins—the target peptides in this work, are usually about 10–40 amino acids long, are highly structured with extensive disulfide bonds. They act like miniature proteins, and are rich in post-translational modifications. About 500 different cone snails each producing 100–200 toxins—contributing to a library of approximately 100,000 different bioactive compounds—are known. The traditional characterization has been to undertake, after elaborate biochemical purification, N-terminal Edman sequencing of purified toxin—cDNA sequencing (the genome of these animals has generally not been sequenced; as a result there are only a limited number of cDNAs), and MS/MS of crude venom fractions. However, so far, the number of toxins characterized has remained small, mainly because of a lack of sufficient sequence-specific fragment ions. Chait and coworkers in sequence analysis of these disulfide-rich peptides use a combination chemical derivatization—ETD strategy for enhanced fragmentation, and apply the technique to obtain full sequences for 31 peptide toxins.

This work of their *de novo* sequencing strategy based on ETD, consisted of off-line (reverse phase) HPLC of 7% of the crude venom from the gland of a single cone snail, reduction to determine the number of peptide disulfide bonds, alkylation of individual fractions to yield information on the number of cysteines present, and generation of charged-enhanced precursors which was followed by MALDI TOF MS.

$(Me)_2NH^+$ converts cysteine into (dimethyl analog of) lysine.[364] The method has been called ETD *with* charge enhanced precursors—ETD with CEP; these charge-enhanced precursors are more amenable to bond cleavage (Fig. 5.7).

Figure 5.8 shows relative efficiencies of fragmentation in CAD, ETD, and ETD with CEP, for three toxins having M_r and z differences. For lower mass and lower charge state, CAD is effective (A); it can provide virtually full sequence coverage. However, with increasing mass or charge (D, G), the sequence coverage decreases. For ETD with low charge, sequence coverage is quite limited (B, H); however, with increasing charge almost entire sequence can be obtained (E). Large enhancements in fragmentation when ETD follows CEP (owing to large increase in charge state) are shown in C, F, I. In all the three cases there is substantial increase in fragmentation. In the third case of the 27-residue toxin—the so-called 'King Kong' toxin, the charge increase is from z = 3 to z = 6 via CEP, and the ETD results in almost total fragmentation, yielding nearly the full sequence. The order of the 2 C-terminal residues is not resolved, but it can be deduced that to satisfy the 6-cysteines requirement for the toxin, one of these two has to

---

[364]Reaction of cysteine converted into lysine (dimethyl analog)—adapted from Simon, M. D., Chu, F., Racki, L. R., de la Cruz, C. C., Burlingame, A. L., Panning, B., Narlikar, G. J., and Shokat, K. M. (2007) The site-specific installation of methyl-lysine analogs into recombinant histones. *Cell,* **128,** 1003–1012.

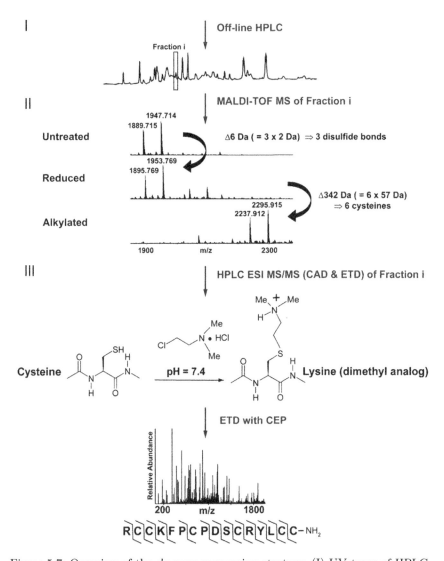

Figure 5.7: Overview of the *de novo* sequencing strategy. (I) UV trace of HPLC separation of crude venom extract from *C. textile*. (II) MALDI TOF MS of fraction I after no treatment, reduction, and alkylation. (III) On-line LC ESI-MS/MS using CAD and ETD on reduced and alkylated aiquots of fraction I. The final step shows the conversion of Cys residues to dimethylated Lys analogs followed by ETD fragmentation; MS/MS is shown for the $(M + H)^{+5}$ ion of the 1,889.715 Da species in II. Reproduced by permission of the National Academy of Sciences, U.S.A. from *Proceedings of the National Academy of Sciences, U.S.A.*, Vol. 106, 6910–6915, 2009.

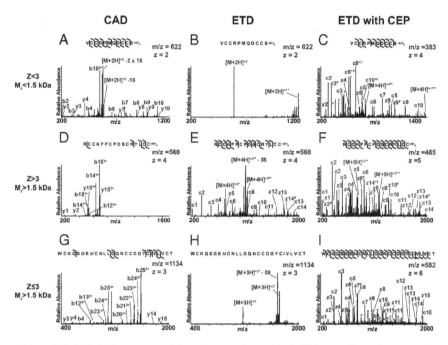

Figure 5.8: MS/MS spectra of 3 toxins studied with 3 different dissociation techniques [CAD (A, D, and G), ETD (B, E, and H), and ETD with CEP (C, F, and I)]. N-terminal fragment ions (b and c) are indicated by ⌐ and C-terminal fragment ions (y and z) are indicated by ⌐. Doubly charged ions are indicated with an asterisk. z ions resulting from cleavage at cysteine and loss of the cysteine side chain are indicated with #. Charge reduced species are labeled in the spectrum with a ·, indicating the number of electrons transferred to the precursor ion. Reproduced by permission of the National Academy of Sciences, U.S.A., from *Proceedings of the National Academy of Sciences, U.S.A.*, Vol. 106, 6910–6915, 2009.

be cysteine. Then the other one, to satisfy the mass requirement, has to be a threonine. An absence of z ion in the spectrum implies that a charged side-chain is absent in the terminal residue. Which means that the latter is likely to be threonine and not the dimethyl Lys analog (modified cysteine). This *de novo* sequencing result is in agreement with results from Edman degradation.

In their experiments, derivatized fractions were analyzed by nanoflow-HPLC interfaced to a microelectrospray ionization source on a Finnigan LTQ-ETD or LTQ-ETD-Orbitrap ion trap instrument. The charge-enhanced precursors from intact toxins were dissociated using ETD, and the resulting spectra acquired in high-resolution mode and subjected to *manual de novo* sequencing.

Interpretation of fragmentation spectra using the *de novo* CEP techniques requires considerable skill and time; one way to reduce the time required is to use sequence tag and any existing cDNA library. Using ETD and CEP it is possible to obtain from practically any toxin experimental sequence tags of ≈ 5 aa in a matter of minutes. Majority of cDNA sequences code for toxin *precursors*– prepropeptides, which consist of signal, pro- and mature-peptides. The active toxin sequence is located at the C-terminus of the prepropeptides. It is difficult, however, to predict precise start and stop sites of the mature peptide. The task is further complicated by presence of a number of isoforms and a very large number of post-translational modifications. Thus even after finding the experimental sequence tag in a cDNA library and the start and stop sites, it is necessary to ascertain the presence of any PTM or amino acid variations. For this, high quality MS/MS data are essential. From their work an example of a combined use of *de novo* sequencing and precursor cDNA database follows.

We describe here the steps they took in their sequencing of a +7 toxin (535 m/z) (+4 is the dominant charge state of the species in the ESI mass spectrum; +7 is the charge state in the mass spectrum of its dimethyl lysine analog — the absolute intensity is less in the latter). In this case the reduction resulted in a mass increase by 6 Da (= 3 × 2 Da) which implied there are three disulfide bonds; the alkylation with iodomethane resulted in a mass increase of 342 Da (= 6 × 57 Da) which indicated that there are six cysteines. The cysteine residues were then converted to their dimethyl lysine analogs using N,N-dimethyl-2-chloro-ethylamine to augment fragmentation in the ETD process. From doubly charged ion locations — on the mass spectrum, one could quickly conclude on the presence of the sequence tag R(307) G(463) Y(520) D(683) A(798) P(869) (six cysteines, 3302.092 Da). In the next step of cDNA database search was using their computer program Toxfinder, the inputs given were: $[M+H]^{+1}$, the number of cysteines (6), and the sequence tag RGYDAP which was obtained from manual inspection. Search was carried out over available cDNA precursor database using the search sequence tag in both directions and since the spectral resolution was low, considering Q/K and I/L as the same, respectively. Three types of output from such a search are possible: mass and sequence matches are reported, sequence matches are reported, and in the case the sequence tag is not found in the database, nothing is reported. In the present case Toxfinder returned the following 70-residue precursor sequence: [1] MEKLTILLLVAAVLTSTQALIQGGGDERQKAK-INFLSRSDRD [43] **CRGYDAP**C SSGAPCCDWWTCSARTNRCF [70]. It did not return a peptide candidate that fulfills both the mass and cysteine-content constraints. There could be several reasons for this. The tag could be simply a random match, or the sequence is not in the database, or the mature toxin is an isoform which is not in the database, or it contains post-translational modifications. They examined the latter alternatives.

The presence of six Cys residues which was inferred from iodomethane experiment followed by mass analysis means that the toxin should at least include the residues 43–69 in the above precursor sequence. The dominant singly-charged ions in the mass spectrum are consistent with the N-terminus starting at Asp-42. The remaining singly-charged ions in the low-mass region are consistent with the C-terminus ending at Phe-70. This would mean that the sequence starts at aspartic acid (D) and ends in phenylalanine (F). Since the measured mass is 3302.092 Da, and the calculated mass is 3224.186 Da, there remains a discrepancy of 77.906 Da. This 77.906 Da could be because of a PTM; it could also be a single bromination modification of a tryptophan residue (then the theoretical mass shift would be 77.910 Da). We note that there are two WW residues. These two Ws are adjacent in: DC⌉R⌉G⌉⌊Y⌉⌊D⌉AP⌉⌊C⌉⌊S⌉⌊S⌉⌊G⌉⌊AP⌉⌊C⌉⌊C⌉⌊D⌉⌊W⌉⌊W$_{Br}$⌉⌊ T⌉⌊C⌉⌊S⌉⌊A⌊R⌊T⌊N⌊RCF; from this c and z ions result. For the z10: 1285.6 and 1286.6 Da, for the z11: 1549.6, 1550.6, 1551.7, 1552.7 Da. Isotope effect tells which W is brominated. Inspection of the spectrum allows us to assign this putative bromination specifically to Trp-19. Which means that the toxin is: D [43] CRGYDAPCSSGAPCCDWW$_{Br}$ TCSARTNRCF [70].

5.4 Mass spectrometry of ribosomes

In this work of Robinson and coworkers,[365] the key to the mass determination of the very high molecular mass ribosomes is utilization of the fact that during electrospray ionization noncovalent adducts are formed, and whereas the adducts have higher masses than the original ribosomal constituent — that is, a noncovalent mass shift results, a peak broadening also takes place concurrently, owing to heterogeneity. A linear regression can be extracted between the two using some simulations. From this, when a certain peak broadening is observed, the accompanying mass shift can be ascertained using the regression, from which the original mass of the ribosomal constituent can be determined. In these experiments on ribosomes, a Waters Q-ToF2 mass spectrometer, modified for high mass operation and equipped with Z-spray nanoflow source, was used.

To ascertain the nature of the peak broadening and the mass shift effect, first a simulation of a hypothetical 800 kDa protein with 50+ charge state was carried out, and the peak structure of the species with addition of 200–205 Na^+ or Mg^{2+} ions studied. Na^+/Mg^{2+} adducts were chosen as they are readily identified in the charge state series of monomeric proteins released from complexes in tandem MS experiments. The peak intensities of adducts resulting from the model were required to conform to a Gaussian

---

[365]McKay, A. R., Ruotolo, B. T., Ilag, L. L., and Robinson, C. V. (2006) Mass measurements of increased accuracy resolve heterogeneous populations of intact ribosomes. *J. Am. Chem. Soc.*, **128**, 11433–11442.

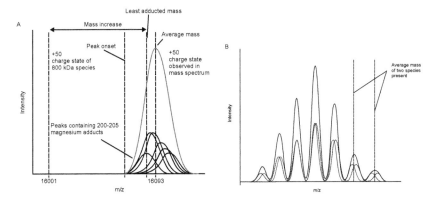

Figure 5.9: (A) The effect of sodium or magnesium adduct formation on the 50+ charge state of an 800 kDa species in an electrospray mass spectrum. The observed peak (shown in dashed line), is comprised of a series of unresolved Na/Mg adducts. It is possible to calculate an average mass of the complex by taking the m/z value of the centroid of the peak or a *least adducted mass* by estimating the m/z value near the onset of the peak. (B) Simulation of two overlapping charge state series and the resulting mass spectrum. It is clear that any attempt to measure the mass of this series using the centroid m/z will result in an average mass of the two species present. Reproduced by permission of the American Chemical Society from *Journal of the American Chemical Society*, Vol. 128, 11433–11442, 2006.

distribution of the type usually observed in electrospray spectra. At the centroid value of the peak an average mass over all the adducted species would be observed. Similarly, it may be expected that mass measurement at an appropriate point on the leading edge would correspond to the peak of the *least adducted mass* of the complex (Fig. 5.9 (a)). Heterogeneity (owing to incomplete binding) in protein composition also can produce unresolved peaks, then an average peak width is observed. Adduct formation and heterogeneity both can cause peak broadening. Figure 5.9 (b) shows the effects of overlapping charge state series in the mass spectrum.

Both the formation of adducts and any presence of heterogeneity in the protein assembly show up in observed spectra. Adducts of a given protein after desolvation cause a cluster of peaks owing to addition of mass and charge of the adherents retained. In a heterogeneous system each individual protein would similarly contribute its own cluster. The observed spectrum in a heterogeneous system therefore could be superposition of several clusters; they would need to be deconvoluted to discriminate between formation of adducts and heterogeneity in characterizing unknown protein assemblies. The extent of adducts retained after desolvation for several constituents of the protein assembly studied was estimated in this work.

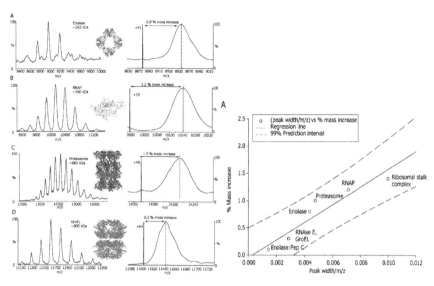

Figure 5.10: Relationship between mass increase and peak broadening. Left: Mass spectra of four multiprotein assemblies: octameric enolase, RNAP, 20S proteasome minus an α subunit, and GroEL. An expansion of the base peak including the isotope model calculated from the sequence mass reveals that each charge state has an associated mass shift that is independent of the mass of the complex. Right: Plot of % mass increase versus peak width (normalized to m/z) for a series of seven complexes. The dashed line indicates the 99% prediction interval for the regression line (line equation from regression analysis: % mass increase = 160 ((peak width/m/z) − 0.03). Reproduced by permission of the American Chemical Society from *Journal of the American Chemical Society*, Vol. 128, 11433–11442, 2006.

The protein composition of the 30S[366] ribosomal unit was determined — it was found to consist of four distinct species. In the study of intact 70S ribosomes, it was observed that the heterogeneity identified in 30S subunits is also present in the 70S.

Four noncovalent multiprotein assemblies of known composition: octameric enolase, RNAP, 20S proteasome minus an α-subunit, and GroEL were studied to assess the extent of adduct formation. Whereas all showed mass shifts to higher values (owing to adduct formation), the mass increase did not relate to the size of the complex (largest shift for the 390 kDa RNA polymerase complex, the smallest for the 800 kDa GroEL complex) (Fig. 5.10(left)(A-D)) — the requirement to desolvate ions and need to

---

[366]Sedimentation coefficient S (time) (The rate at which particles of a given size and shape travel under centrifugal force to the bottom of the container.) $= v(\mathrm{T}^{-1})/a(\mathrm{T}^{-2})$. $1\,\mathrm{S} = 1$ Svedberg $= 10^{-13}\,\mathrm{s}$ (100 fs).

retain noncovalent interactions in the gas phase (to provide stability) act in opposition. To assess the extent of adduct formation an approach that is independent of stability factor and extent of desolvation of the complex was needed.

Mass shifts and peak widths, however, were found to have a close direct relationship. Plots of percent mass increase (peak shift) against average peak widths (fwhm) from a series of seven complexes showed (Fig. 5.10 (right)) good linearity (correlation coefficient $R = 0.9556$). Further, the linear relationship was found to hold when examined on *E. coli* GroEL$_{14\text{-mer}}$ under a range of solvation conditions. The regression was tested also against three charge states from the GroEL$_{14\text{-mer}}$ (+68, +69, and +72) (figure not shown) with all points falling within 99% prediction interval ($R = 0.9507$), which meant that the method is charge-state independent. Overall it indicated that based only on the width of the peaks observed in the spectrum it is possible to estimate the mass increase.

The 30S ribosomal subunit has an *approximate* mass of 800 kDa and it consists up to 20 proteins and 1 RNA molecule (the 16S rRNA). To analyze the spectrum, Gaussian curves were first fitted to the apex of the peaks, and m/z values assigned to nine of the peaks Fig. 5.11 Left (top). Using these m/z values, mean molecular mass and standard deviation for 20 different possible charge state assignments were calculated. For the base peak (d) carrying +49 charge, the standard deviation was found to be the lowest (though absolute standard deviation value high). (A plot of residuals — difference between the mass obtained for each charge state and the mean value found — showed the mass errors of the g, h, i peaks were too high.) Disregarding then these three peaks, the standard deviation was replotted, and +52 charge state assigned for the base peak (d).

Using the best-fit assignment of (d) with charge as +52, a measured mass of 797,605 ± 119 Da was obtained. Average peak width in the mass spectrum was measured to be 85 m/z fwhm. Application of regression implied (with the peak width and mass available) 0.9% mass increase because of the formation of adducts. Since the anticipated mass of 30S using the established protein and RNA sequences is 818,784 Da, multiplying this by 1.009 one obtained 826,153 Da. This is larger than the measured value of 797,605 by 28,548 Da. (But this difference does not correspond to any 30S ribosomal protein; however, from biochemical analysis[367] it is known that preparations of *E. coli* ribosomes can contain less than stoichiometric quantities of protein S1.) Also known is the fact that 30S releases protein S6 in the mass spectrometer. This therefore suggests four possibilities: 30S, 30S−S1, 30S−S6, 30S−S1−S6. The mass for the peaks a–f is consistent

---

[367]Subramanian, A. R. and van Duin, J. (1977) Exchange of individual ribosomal proteins between ribosomes as studied by heavy isotope-transfer experiments. *Molec. Gen. Genet.*, **158**, 1–9.

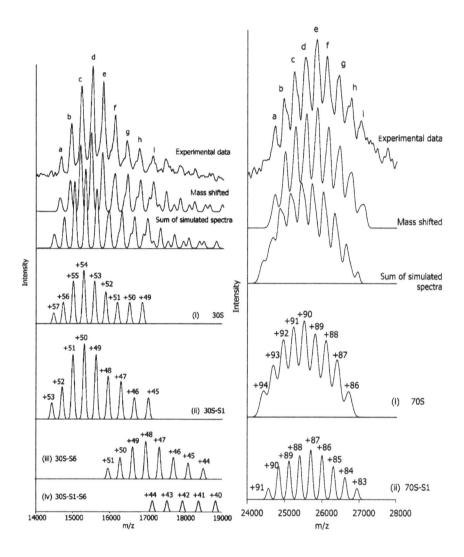

Figure 5.11: Left: (top) Electrospray mass spectrum of the small ribosomal subunit from *Thermus thermophilus*; (d) is the base peak in the charge state series. Charge states labeled (a–e) represent a well-defined peak shape consistent with one protein composition. Heterogeneity is evident for charge states g–i. Below the experimental spectrum are: simulations of the charge state distributions for the four possible compositions (i) the 30S subunit, (ii) the 30S subunit − S1, (iii) the 30S subunit − S6, and (iv) the 30S subunit − S1 and − S6. Right: (top) Experimental mass spectrum of 70S ribosomes from *Thermus thermophilus*. Below the experimental spectrum are simulations of the charge state distributions for two possible compositions of 70S ribosomes: (i) 70S ribosome with the full protein complement; (ii) the 70S ribosome − S1. Reproduced by permission of the American Chemical Society from *Journal of the American Chemical Society*, Vol. 128, 11433–11442, 2006.

with an average of the masses of intact 30S and 30S−S1, implying that both are present but are not resolved.

Considering the presence of S1 and S6, spectra of 30S, 30S−S1, 30S−S6, 30S−S1−S6 were then simulated, these individual spectra were summed, and the intensities were adjusted until the best-fit between the experimental and modeled mass spectra (the latter mass-shifted according to regression) were obtained. Good agreement was found; shoulders too were reproduced. Relative intensities of various simulations suggested that 50% of the 30S subunits contain the ribosomal protein S1.

Though it was anticipated that the 30S−S6 and 30S−S1−S6 intensities should have reflected the intensities of 30S and 30S−S1, but that is not observed. The discrepancy is attributed to the mean charge state difference (+54 and +50) between the two precursors. This is evident from the lower intensity of the simulated spectra of the 30S−S1−S6 compared to that of the 30S−S6. These simulated spectra illustrate how the four distinct populations (traceable to the ribosome preparation) combine to give the spectrum of 30S assembly.

Tandem MS-MS measurements were carried out (at 80V, where desolvation increases) to resolve overlapping populations of 30S ribosomes in the mass spectrum. In the 12,600–13,800 m/z region of the spectrum (a lower range than the 14,000–19,000 spectrum, apparently because desolvation has occurred) the presence of overlapping charge states is revealed within the unresolved peaks in the mass spectrum (in good agreement with simulations). This is taken as additional evidence for the overlapping 30S and 30S−S1 peaks (a–e) in the mass spectrum.

Then, isolating the *highest charge states* of the intact 30S and raising the collision cell voltage to 200 V — the spectrum shows that only one protein is released (facile loss of S6, it appears at about 1000 m/z, charge +12), and the stripped complex 30S−S6 appears at approximately 16,600 m/z. In the most intense peak of +51 charge, no contribution from peaks assigned to loss of both S1 and S6 could be detected. This is attributed to higher charge states of the 30S (62/63/64+) (S6 is lost from higher charge states of 30S compared to 30S−S1 (57/58+) and also because, in tandem with MS/MS isolation, predominance of the intact 30S charge states.

The intact 70S ribosomal particles are close to being 3 times larger than the 30S subunits with an approximate mass of 2.3 MDa; they consist up to 58 proteins and 3 RNA molecules (5S, 16S, and 23S ribosomal RNAs). Under carefully controlled experimental conditions, well-resolved peaks were obtained (Fig. 5.11 top-right (a–i). m/z values were estimated for all nine peaks and from the best-fit model the base peak (e) charge was obtained as +90. From this the molecular mass was calculated to be 2,325,148 ± 586 Da. The base peak width 294 m/z implied an 1.8% increase on the mass calculated from protein and RNA sequences, to give 2,337,213 Da, which is 12,065 Da greater than 2,325,148 Da. One possible explanation is that there is more than one population of 70S particles as was deduced for 30S subunits.

Two-component Gaussian fitting yielded, for two peak widths, 207 m/z and 110 m/z fwhm, which correspond to mass increases of 1.2% and 0.7%, respectively. Applying these to the sequence masses of intact 70S and 70S−S1 ribosomes, one obtains for their masses 2,323,439 and 2,251,569 Da, respectively. The 71,780 Da difference could be explained by any number of combinations of small ribosomal proteins. However, since the peaks are resolved in the spectra, indicating that there are not multiple combinations of different substoichiometric ribosomes, it is likely (like 30S) that this mass difference corresponds to incomplete binding of the individual ribosomal protein S1 (59,970 Da). Partial loss of the 96,000 Da stalk complex was also considered (but none of these simulations provided an adequate fit for the experimental data). From this, it was considered reasonable to assume that the intact protein 70S must also contain variant 30S subunits. The most likely explanation for the two masses determined for 70S particles is the presence and absence of S1, as deduced for 30S subunits.

The individual charge state distributions of 70S and 70S−S1 were simulated, intensities of the peaks adjusted to fit the experimental data, and summed. Mass adjustment from peak width was done, and compared with the experimental spectrum. Good agreement was observed. Although the shoulders observed in the experimental data for peaks b, c, and h (Fig. 5.11 top right) are not evident in the simulation, the broadening resulting from the overlapping signals is reproduced. Mass differences between experimental and adduct-adjusted masses of 70S and 70S−S1 are small — less than 2 and 5 kDa, respectively, < 0.1 and 0.3% — this simple model is the best fit to the experimental data. The major contribution to heterogeneity arises from the existence of 70S and 70S−S1.

5.5 Proteins in Purkinje cell post-synaptic densities

The formation and function of each type of synapse is controlled by a complex activation of signaling pathways through specific proteins. At present little is known about the biochemical composition of specific synapse types, though it is known certain individual proteins can localize to different classes of synapses. This is a brief note of a work from Chait's and Heintz's laboratories on identification of neural proteins of a *single* CNS synapse type: the parallel fiber/Purkinje cell synapse. The latter has unique physiological properties and it is known that it has involvement in neurological diseases. This mass spectrometric work of protein profiling at parallel

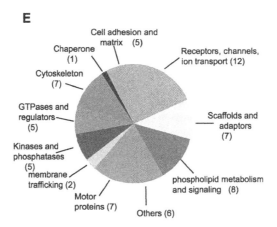

Figure 5.12: MS identified 65 different proteins in the complexes purified VGluδ2 mice. These proteins can be classified into 11 functional categories. The number of proteins from each category is indicated in parentheses. Reproduced by permission of PLoS Biology from *PLoS Biology*, Vol. 7, 2009.

fiber to Purkinje cell synapses identified a large number of postsynaptic proteins.[368]

In this study, genetically engineered mice were taken; they developed a transgenic line that expresses an affinity tag at the PF/PC synapse. Then, in order to separate PF/PC PSDs from other cerebellar synapses, using an anti-eGPF antibody, affinity purification was carried out, which consisted of biochemical purification (gel filtration) and subjecting to antibody containing magnetic material; the target PSD proteins adhered to the beads. The mass spectrometric procedure involved use of 1-D SDS-PAGE, MALDI (QqTOF) MS and MALDI-IT tandem MS (MS/MS). Data analysis was carried out using XProteo algorithm. 65 proteins were identified which contained MRCKγ not previously detected localized to synapses, and are shown in Fig. 5.12 grouped into 11 different functional categories. In this work a new group of proteins was recognized that can regulate or be regulated by phospholipid metabolism (Iptr1, synaptojanin 1 and 2, phospholipase B, ABCA12 and MRCKγ) or contain phospholipid-binding domains (Plekha7, annexin A6, and MRCKγ).

---

[368]Selimi, F., Cristea, I. M., Heller, E., Chait, B. T., and Heintz, N. (2009) Proteomic studies of a single CNS synapse type: the parallel fiber/Purkinje cell synapse. *PLoS Biology*, **7**, e1000083, 0001–0010.

5.6 Neurexin-LRRTM2 interaction effect in synapse formation

Here, we present a couple of more applications from neuroscience research. In these studies, mass spectrometry has been used to identify molecular interactions between pre- and postsynaptic neurons that mediate neuronal connectivity.

The role of Neuroligin-Neurexin interactions on synaptic development was known from earlier work,[369] and it was also known[370] that the leucine-rich repeat transmembrane neuronal protein — LRRTM1 — can induce synapse formation. Following the report from Linhoff *et al.* that LRRTM1 can induce excitatory synapse formation in culture, Südhof[371] and Ghosh[372] laboratories examined role of other LRR proteins in synapse formation. This led to the identification of LRRTM2 as a novel synaptogenic protein, and Neurexins as LRRTM2 receptors. Both studies employed the *bottom-up* approach to identify the LRRTM2 receptor.

Ko *et al.* demonstrated synaptogenic activity of LRRTM2 by expressing the gene in COS-7 cells and co-culturing the cells with hippocampal neurons. In that assay, hippocampal neurons formed presynaptic specializations onto the LRRTM2-expressing COS-7 cells. In addition, overexpression of LRRTM2 in hippocampal neurons led to an increase in the synapse number.

de Wit *et al.* carried out similar experiments using HEK293T cells to express LRRTM2 and demonstrated the synaptogenic activity of LRRTM2. To identify the domains of LRRTM2 that induce presynaptic differentiation, they expressed various deletion mutants of LRRTM2 in hippocampal cultures and showed that the extracellular LRR domain of LRRTM2 was required for its synaptogenic activity. To determine if LRRTM2 was required for synapse formation *in vivo*, an shRNA targeting LRRTM2 was cloned

[369]Craig, A. M. and Kang, Y. (2007) Neurexin-Neuroligin signaling in synapse development. *Curr. Opin. Neurobiol.*, **17**, 43–52. Also, Kang., Y, Zhang., X, Dobie., F, Wu., H, and Craig, A. M. (2008) Induction of GABAergic postsynaptic differentiation by alpha-neurexins. *J. Biol. Chem.*, **283**, 2323–2334.

[370]Linhoff, M. W., Lauren, J., Cassidy, R. M., Dobie, F. A., Takahashi, H., Nygaard, H. B., Airaksinen, M. S., Strittmatter, S. M., and Craig, A. M. (2009) An unbiased expression screen for synaptogenic proteins identifies the LRRTM protein family as synaptic organizers. *Neuron*, **61**, 734–749.

[371]Ko, J., Fucillo, M. V., Malenka, R. C., and Südhof, T. C. (2009) LRRTM2 functions as a neurexin ligand in promoting excitatory synapse formation. *Neuron*, **64**, 791–798.

[372]de Wit, J., Syewestrak, E., O'Sullivan, M. L., Otto, S., Tiglio, K., Savas, J. N., Yates III, J. R., Comoletti, D., Taylor, P., and Ghosh, A. (2009) LRRTM2 interacts with Neurexin 1 and regulates excitatory synapse formation. *Neuron*, **64**, 799–806.

into a lentiviral vector and injected into the hippocampus. Subsequent electrophysiological recordings from untransfected and sh-LRRTM2-expressing neurons revealed a marked decrease in synaptic response in sh-LRRTM2-expressing cells.

Both Ko *et al.* and de Wit *et al.* used similar strategies to identify the LRRTM2 receptor. Ko *et al.* fused the extracellular domain of LRRTM2 to human IG (Ig-LRRTM2) and produced recombinant protein in HEK293T cells that could be used as a bait to isolate LRRTM2-binding proteins. Ig-LRRTM2 was bound to protein A-sepharose and loaded on a column for affinity purification. Whole brain extracts were passed through the column, and bound proteins were initially analyzed by SDS polyacrylamide gel electrophoresis. This revealed that multiple proteins bound to Ig-LRRTM2. Proteins larger than Ig-LRRTM2 were excised from the gel and analyzed by mass spectrometry. This led to identification of 140 peptides, of which 31 were derived from neurexins.

The neurexins are part of a family of presynaptic proteins (Neurexin 1,2,3), which have multiple isoforms. Some neurexin isoforms include a specific sequence (SS4) that is absent from other forms of the proteins due to alternative splicing. To determine whether LRRTM2 binds to specific neurexins, various neurexins were expressed in HEK293T cells, and tested for binding to Ig-LRRTM2. These experiments revealed that LRRTM2 binds to both Neurexin 1α and 1β, but preferentially binds to neurexin isoforms that lack the SS4 insert. This contrasts with the binding specificity of neuroligins to neurexins, as neuroligin1 binds to both (S)S4 containing and lacking forms of neurexin 1α, but only the SS4 lacking form of neurexin 1β. Scatchard analysis indicated that the cell surface binding affinity of LRRTM2 and neurexin 1β was approximately 5 nM.

The approach of de Wit *et al.* was similar to that described above, with a few significant differences. In their case, Fc-LRRTM2 was used to affinity purify binding proteins from synaptosome fractions prepared from postnatal day-18 brains. LRRTM2-associated proteins were then analyzed by MuDPIT. They found about 300 proteins which copurified with the ecto domain of LRRTM2; in comparison, fewer than 100 proteins copurified with only Fc. This dataset was sorted based on the number of spectra that contained peptides corresponding to a specific protein (Fig. 5.13), which indicated that Neurexin family members were the most abundant proteins that preferentially associated with Fc-LRRTM2. Since many different Neurexin proteins have the same peptides, these mass spectrometric results by themselves do not identify a unique Neurexin protein, but strongly suggest that a Neurexin isoform is an LRRTM2 receptor.

A series of experiments were performed to examine the binding of Fc-LRRTM2 to the surface of 293T cells expressing different Neurexin isoforms. This showed that Fc-LRRTM2 does not show significant binding

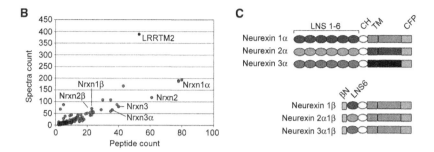

Figure 5.13: Mass spectrometric results show Neurexins as potential LRRTM2 receptors. The scatter plot is a graphical representation of a representative interaction screen. Each dot represents a protein that was identified by tandem mass spectrometry. The x axis represents the number of peptides and the y axis is the number of spectra in which the identified protein was found. In the diagram to the right, a schematic representation of Neurexin constructs is shown. Neurexin (Nrxn) 1α, 2α, 3α, and 1β refer to the wild type proteins. Nrxn $2\alpha1\beta$ and Nrxn $3\alpha1\beta$ are chimeric proteins of Nrxn1β with the LNS6 domain replaced with that of 2α and 3α, respectively. LNS, laminin, neurexin, sex-hormone-binding protein: CH, highly glycosylated region; βN, β specific leader; TM, transmembrane domain. All constructs have a C-terminal CFP (Citrine-Y(Yellow)FP) tag. Reproduced by permission of Elsevier Inc. from *Neuron*, Vol. 64, 799–806, 2009.

to N-cadherin or NGL2, but binds to neurexins. Among the neurexin isoforms, the highest level of binding is seen with Neurexin 1β, followed by Neurexin 1α. Neurexin 1β contains only one extracellular LNS domain that has previously been implicated in its interaction with Neuroligin. To determine whether the LNS domain of Neurexin 1β was specifically required for binding to LRRTM2, two additional constructs were tested in this work in which the LNS domain of Neurexin 1β was replaced by the corresponding domain from Neurexin 2 or Neurexin 3. The Neurexin $2\alpha1\beta$ construct showed significant but highly reduced binding compared to Neurexin 1β, whereas the Neurexin $3\alpha1\beta$ construct showed no specific binding. These results indicated that Neurexin 1α and Neurexin 1β are LRRTM2 receptors and that the LNS domain of Neurexin 1β is critical for its ability to bind LRRTM2.

Taken together, the observations of Ko *et al.* and de Wit *et al.* indicate that LRRTM2 can bind to Neurexins lacking the SS4 sequence. Additionally, interaction between these proteins requires the LNS6 domain of neurexin 1β.

5.7 Rapid analysis of human plasma proteome — an IMS-IMS-MS application

There is much need for state-to-state information on protein composition at physiological scale within living organisms. Because human plasma is in contact — either directly or indirectly — with all of the 10^{14}–10^{15} cells in our bodies (and perhaps as many non-human cells — e.g., those from bacteria or fungi) — analysis before and after applied perturbations may provide insight into why various changes occur. Similarly, the ability to monitor proteome dynamics within organisms would be invaluable. There have been efforts — such as attempts to characterize the proteome of *Drosophila melanogaster* (example of two proteins from a study of *Drosophila* proteome[373]) to track the temporal abundances of the protein complement across the organism lifetime. In the proteome of *Drosophila*, the concentration of the protein paramyosin remains relatively constant across the lifespan of the organism. The concentration of prophenyloxidase — involved in the immune response of the animal — is an indicator of the animal's age decreasing precipitously across its lifespan. Additionally, deletion of the black cells (bc) gene causes a reduction in average lifespan. A major hurdle in such studies is the time needed for the measurements. The main requirements are twofold: it should be possible to follow a maximum number of proteins from a large number of samples, and, the total time for analysis — in a high-throughput mode — should be as small as possible. In a work of Clemmer *et al.*[190] — a significant step toward developing multiplexed, two-dimensional ion mobility spectrometry (IMS-IMS) protocols — IMS combing, plasma proteome from 70 individuals were examined in duplicate in a single day (the instrument's present automated analysis capability can examine as many as 180 samples in duplicate over a 24 h period). For these experiments a time requirement of 4 min/sample was chosen and capability of tracking concentration changes for 60–70 (50–100) proteins across a large population. Commercial systems currently available and used with 2 LC separations and IMS-MS analysis yield information for a large number (2,928) of proteins but the data acquisition time is > 500 h. In comparison, in the present IMS-IMS-MS method, although information on fewer proteins is obtained, many more samples could be examined.

In the experimental procedure, from each plasma sample the six most abundant proteins — albumin, IgC, IgA, serotransferin, haptoglobin, and antitrypsin — were removed by introducing the plasma sample onto a MARS (Agilent's Multiple Affinity Removal System) LC column, and the

---

[373]Sowell, R. A., Hersberger, K. E., Kaufman, T. C., and Clemmer, D. E. (2007) Examining the proteome of *Drosophila* across organism lifespan. *J. Proteome Res.*, **6**, 3637–3647.

flow-through fraction having the lower abundance proteins—which eluted from $\sim 2\,\text{min}$–$\sim 5\,\text{min}$—was concentrated to $\sim 300\,\mu\text{L}$, for protein digestion. The proteins in each sample were denatured, reduced by dithiothreitol to reduce the disulfide bonds, and alkylated using iodoacetamide. Next the samples were digested with TPCK-treated trypsin for 24 h. The resulting peptides were desalted and dried. To each sample of digested peptide a mixture of angiotensin I and bradykinin was added as internal standards. For these two peptides—whose circulating/endogeneous concentrations/levels are expected to be negligibly low—three charge states, +1, +2, +3 are observed. Owing to differences in collision cross-sections, at least one of these ions is present in each IMS-IMS dataset frame. For ESI, a nanospray chip (Advion) containing 400 nozzles was used, and a $15\,\mu\text{L}$ aliquot of the plasma sample drawn into a new autosampler tip (Nanomate, Advion) was delivered to the chip. The operational procedure of the instrument has been described in Section 3.2.7.

A detailed analysis of 15 plasma samples was carried out side by side using an LTQ-FT mass spectrometer which yielded information on the peptides that are routinely observed in each of the samples. The MS/MS data were analyzed using the MASCOT suite of programs and Swiss Prot database.

The data output from the IMS-IMS-MS measurement consists of a list of peaks along with information on: IMS-IMS frame (comb pulse setting), drift time, m/z, and normalized intensity. Peptide assignments are made by comparing these with the corresponding dataset obtained from LC/MS analysis of the same plasma digest samples. Multiple drift time measurements with associated m/z values enhance data alignment capabilities and contribute to rapid comparisons of dataset features for multiple samples and is a particular advantage of this IMS-IMS-MS procedure.

Figure 5.14 shows results of a single frame (one comb pulse setting—that is, only one set of drift velocities for which the gate to the activation region opens—no data collection during intermediate values of drift velocities). In the frame for which data are shown, the centers of mobility selections were at 12.3, 14.9, 17.5, 20.1, 22.7 and 25.3 ms—approximately 2.6 ms apart. This value, ~ 2.6 ms, corresponds to the spacing obtained for the entire drift separation (both drift tubes) and is greater than the 1 ms mobility selection separation (the drift separation at the gate prior to ion activation). Between the comb teeth are the flat (lower intensity) regions; many ions which are selected at the comb teeth, after activation, show up in flat (between mobility selection) regions and thus contribute to resolving peaks (and adding features) which would not be possible without activation. A total ion drift time distribution obtained by integrating all m/z values for given drift times is shown above the 2D plot in the figure. A total ion mass spectrum obtained by integrating all drift time bins for given m/z values is shown to the left of the 2D plot. Over the four frames, the

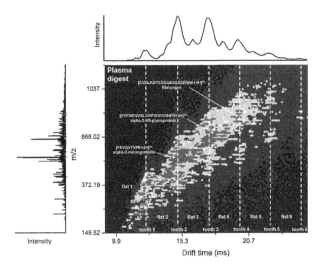

Figure 5.14: Two-dimensional drift time (m/z) dot plot obtained from the analysis of a typical plasma digest sample. A total ion drift time distribution is shown above the 2D plot; a total ion mass spectrum is shown to the left of the 2D plot. Three features with drift times 15.1, 18.1, and 19.9 ms and having m/z values 523.81, 694.35, and 807.05 have been assigned to peptide ions $[\mathrm{FEVQVTVPK+2H}]^{+2}$ α-2-macroglobulin), $[\mathrm{HTFMGVVSLGSPSGEVSHPR+3H}]^{3+}$ (α-2-HS-glycoprotein), and $[\mathrm{EVDLKDYEDQQKQLEQVIAK+3H}]^{3+}$ (fibrinogen), respectively. Reproduced by permission of Elsevier Inc. from *International Journal of Mass Spectrometry*, Vol. 283, 149–160, 2009.

number of features with intensities $> 4\%$ of the base peak height (≈ 40 ion counts) is ~ 3500 on average.

Tentative assignments are possible from comparison on m/z values of spectral features in LC-MS/MS and IMS combing datasets. Fig. 5.14 shows assignment to peptide ions from α-2-macroglobulin (2+), α-2-HS- glycoprotein (3+), and fibrinogen (3+) on the basis of peaks at drift times of 15.1, 18.1, and 19.9 ms having m/z values 523.81, 694.35, and 807.06, respectively. Determination of the numbers of assigned peptide ions and proteins observed in IMS combing experiments for 15 plasma samples shows that approximately half and three-fourths of the peptide ions and proteins, respectively, assigned from LC-MS/MS studies are matched to IMS-IMS-MS dataset features. From the IMS-IMS-MS analysis of 15 plasma samples overall 114 proteins were observed. Of these, 96 have been observed in at least 2 individual samples, 61 proteins in roughly half the samples, and 38 proteins in all the samples. The run-to-run reproducibility of the same sample was $\sim 4\%$ on the average. From sample workup to analysis, that is, for the complete experiment, the reproducibility was $\sim 15\%$.

By comparing intensities of ion-mobility peaks, protein concentrations can be estimated. For example, from comparisons of mobility distributions between two individuals — the ratios of peak intensities for three peptides [VSFLSALEEYYK +2H]$^{2+}$ (drift time 17.3 ms), [LLDNWDSVTSTFSK+ 2H]$^{2+}$ (drift time 19.2 ms), [DYVSQFEGSALGK +2H]$^{2+}$ (drift time 16.9 ms) (from the Apolipoprotein A1) showed ratios of 0.29, 0.33, 0.33, respectively. Since the experimental error margin is much less, this can be taken to represent the difference in the protein concentration between the two individuals.

A limitation of the IMS-IMS-MS method in this work is that MS/MS data are not available and the peptide ion assignment is based solely on a match of m/z values whereas a parallel CID-MS method would be superior. However, considering that the highest intensity m/z peaks are employed for identification by LC-MS/MS analysis, the assignments can be taken as reasonable. The strengths of the method as it stands now are: output of m/z values, high throughput, high dimensionality, ion mobility distributions (peaks and intensities), a comparison of two drift time measurements, and high reproducibility.

It is reasonable to assume on the basis of sample complexity that in ESI the relative ionization efficiency for a given peptide ion should be the same for individual samples; therefore the peptide ion intensities could be dependably related to protein concentration. The average difference between the intensity ratios for the mobility distributions of three peptide ions [VSFLSALEEYTK+2H]$^{2+}$, [LLDNWDSVTSTFSK+2H]$^{2+}$ and [DYVSQFE GSALGK+2H]$^{2+}$ from the protein Apolipoprotein A1 from the samples of two individuals is 0.32 ± 0.02 ($\sim 6\%$) [0.29, 0.33, and 0.33, respectively]. In another dataset comparison, for the peptides [VGFYESD-VMGR+ 2H]$^{2+}$, [FEVQVTVPK+2H]$^{2+}$, and [NQGNTWLTAFVLK+2H]$^{2+}$ from the protein α-2-macroglobulin sample of two other individuals the average peptide ion intensity ratio is 0.31 ± 0.06 ($\sim 19.4\%$). This error is still relatively low with respect to the propagation of error (using the experimental error of 15%) for a ratio.

That combing datasets of IMS can match about 50% of the peptides ions identified by LC-MS/MS analyses is to be considered particularly good, as the IMS method needs an order of magnitude *less* time. And this could be considered as the lower bound of the IMS-IMS-MS method. Though limited to high abundance proteins which may miss many biomarker candidates of low concentration, it is also true that currently many clinical assessments of health of individuals use many high abundance proteins — and the protein coverage afforded by IMS is not very different from that can be obtained using high-throughput LC-MS techniques. The 4-min analysis could find many applications in the monitoring of families of proteins such as the apolipoprotein family of specific classes of enzymes.

5.8 Topology of two transient virus capsid assembly intermediates

In this section we present a work of Ashcroft *et al.*[374,375] in which they follow *in vitro* the path of rigid viral capsid assembly of MS2 bacteriophage by ESI-TWIMS and determine the quaternary architecture of two transient, on-pathway intermediates of known stoichiometry but unknown topology.

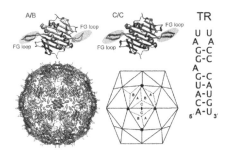

Figure 5.15: The molecular components used in the assembly assays. Coat protein dimer quasi-equivalent conformers are shown as ribbon diagrams, with the FG loops highlighted. Their relationships within a capsid are also shown; the A and C subunits have extended loops, while the B subunit loops fold back toward the globular core of the protein. In the complete capsid, A/B dimers surround the particle fivefold axes with a ring of B-type loops, while the A- and C-type loops alternate around the threefold axes. Also shown, the sequence of the TR RNA. For S2 TR stem-loop, the 5′ extension (beyond the 5′ A of TR) is : 5′CCGGCCAUUCAA (the difference in mass between TR and S2 is 4 kDa). Reproduced by permission of Elsevier Ltd. from *Journal of Molecular Biology*, Vol. 395, 924–936, 2010.

By combining ion mobility spectrometry-mass spectrometry with tandem mass spectrometry in a single experiment, they characterized — from cross-sectional area, stability — non-covalently bound macromolecular complexes in terms of mass, shape. Their results show that when a single subunit is removed from the complex such as by a process like CID, influence of the original quaternary structure of the complex bears upon the structure of the residual species. Complexes in which subunits are bound to several adjacent subunits, undergo significant collapse upon dissociation. In contrast, much of the original shape of the precursor complex is retained

[374]Knapman, T. W., Morton, V. L., Stonehouse, N. J., Stockley, P. G., and Ashcroft, A. E. (2010) Determining the topology of virus assembly intermediates using ion mobility spectrometry-mass spectrometry. *Rapid Commun. Mass Spectrom.*, **24**, 3033–3042.

[375]Basnak, G., Morton, V. L., Rolfsson, Ó., Stonehouse, N. J., Ashcroft, A. E., and Stockley, P. G. (2010) Viral genomic single-stranded RNA directs the pathway toward a T = 3 capsid. *J. Mol. Biol.*, **395**, 924–936.

if subunits in the latter have only a single neighboring subunit. Through IMS measurement of cross-sections of the intermediate species, theoretical calculations of cross-sections of possible intermediates with different topologies, and CID MS/MS, they conclude that MS2 bacteriophage assembly goes through two transient intermediates, one a 182.9 kDa hexamer with threefold symmetry (consisting of six coat protein dimer CP_2, three with bound RNA stem loops and three without) and the other a 304.8 kDa decamer with fivefold symmetry (consisting of ten coat protein dimer CP_2 five of which have bound RNA stem loops). To support their conclusions, they make use of IMS-derived results on mass and shape (particularly the change after monomer loss) of bimolecular complexes GroEL and MHC, and results of IMS-MS analysis of postdissociation products of the latter complexes.

Most viral capsids are spherical in shape with icosahedral symmetry. The MS2 structure protects and transports single stranded (ss) RNA. The coat protein monomer (CP) that forms the capsid in this case is of 13.7 kDa, its dimer (CP_2) is of 27.4 kDa, and the full MS2 capsid (T = 3 shell) is 2.5 MDa. The latter has a diameter of 28 nm, and its $\Omega_{IMS} \approx 616$ nm^2 (experimental value, uncertain; 606.3 in theory). The MS2 virus capsid icosahedral sphere scaffold is constructed of 180 coat protein monomers, based on 90 dimers. The dimers, whose structures are known from crystal structure and NMR studies, are of two types: A/B and C/C. A/B is an asymmetric dimer where the FG loop of the A subunit is extended and the B subunit is folded back; in C/C, a symmetric dimer, the FG loops are extended (Fig. 5.15).[376,377] Addition of a small nucleotide RNA stem loop (wild type genome sequence — 3,500 base pairs), denoted TR, to A/B or C/C conformations results in immediate formation of a 33.4 kDa intermediate initiation complex (RNA-protein sequence specific recognition event).[378] With the A/B, a symmetric dimer, the RNA binds in the lower region and causes an allosteric effect on the FG loop; as a result it assembles into the capsid structure.

In the mass spectrum of the dimer-RNA mix (CP_2 + TR) (2:1), taken at early stage, one observes at low m/z region (4,000 m/z) a few (4–5) peaks arising from the initiation complex. In the middle part (6,600–9,500 region) which is interesting because here the intermediates manifest, peaks from two species are observed; the first set of peaks because of a hexamer (at lower

[376]Valegard, K., Liljas, L., Fridborg, K., and Unge, T. (1990) The three-dimensional structure of the bacterial virus MS2. *Nature*, **345**, 36–41.

[377]Golmohammadi, R., Valegard, K., Fridborg, K., and Liljas, L. (1993) The refined structure of bacteriophage MS2. *J. Mol. Biol.*, **234**, 620–639.

[378]Stockley, P. G., Rolfsson, Ó., Thompson, G. S., Basnak, G., Francese, S., Stonehouse, N. J., Homans, S. W., and Ashcroft, A. E. (2007) A simple, RNA-mediated allosteric switch controls the pathway to formation of a T = 3 viral capsid. *J. Mol. Biol.*, **369**, 541–552.

m/z part — 182.9 kDa — tallest peak having 24+) which is followed by a set because of a decamer (at higher m/z part — 304.8 kDa — the tallest peak having 34+). A broad peak in the 20,000–30,000 region (full capsid-related signal) shows up at a later stage in time. Without further RNA if additional coat protein was added, formation of capsid assembly takes about 2 hours; an additional aliquot of CP_2 speeds up the assembly. For CP_2:TR (1:1) with only A/B dimers, the assembly is slow; for fast assembly, with 2:1, both A/B + C/C dimers are needed. Size-exclusion chromatography, light scattering/UV results are consistent with ESI-MS data; and electron microscopy results confirm capsid formation. Isotope chase experiment confirmed assembly-competency of higher-order species and that the capsid assembly is by CP_2 dimer addition. To confirm the stoichiometry of higher-order species, variant RNA and mutant CP2 were used. Driftscope plots indicated presence of various capsid assembly intermediates.

In order to standardize ESI-TWIMS-MS cross-sections Ω_{TWIMS}s of MS2 assembly intermediates with respect to a protein with known cross-sectional area, comparison was made with cross-section results on GroEL from an experiment containing both the species. The mass spectrum of GroEL (801.0 kDa) (Fig. 5.17) shows peaks in the m/z region $\simeq$ 12,000–14,000 with Ω_{TWIMS} 224.6 nm^2 (Ω_{Theory} 221.3 nm^2). Cross-sections of MS2 viral capsid assembly species were modeled using an in-house-developed Monte Carlo algorithm[379,380] based on a MOBCAL projection approximation method[381] tested earlier on large complexes.[382] Results Ω_{Theory} of those calculations for CP_2, CP_2 + TR, possible hexamer and decamer structures, and for the full capsid T = 3 are shown in Fig. 5.16 besides experimental Ω_{TWIMS}.

With GroEL, the starting assumption was that nothing is known about the intermediates. GroEL:14-mer consists of two 7-membered rings — two donuts one on top of the other — with a central cavity. The GroEL mass spectrum shows (Fig. 5.17) two sets of peaks — the low m/z peaks (13+, 14+) are from the unbound monomeric subunit expelled from GroEL. The

[379]Knapman, T. W., Berryman, J. T., Campuzano, I., Harris, S. A., and Ashcroft, A. E. (2010) Considerations on experimental and theoretical cross-section measurements of small molecules using travelling wave ion mobility spectrometry-mass spectrometry. *Int. J. Mass Spectrom.*, **298**, 17–23.

[380]Smith, D. P., Knapman, T. W., Campuzano, I., Malham, R. W., Berryman, J. T., Radford, S. E., and Ashcroft, A. E. (2009) Deciphering drift time measurements from travelling wave ion mobility spectrometry-mass spectrometry studies. *Eur. J. Mass Spectrom.*, **15**, 113–130.

[381]Mesleh, M. F., Hunter, J. M., Shvartsburg, A. A., Schatz, G. C., and Jarrold, M. F. (1996) Structural information from ion mobility measurements: effects of the long-range potential. *J. Phys. Chem.*, **100**, 16082–16086.

[382]van Duijn, E., Barendregt, A., Synowsky, S., Versluis, C., and Heck, A. J. (2009) Chaperonin complexes monitored by ion mobility mass spectrometry. *J. Am. Chem. Soc.*, **131**, 1452–1459.

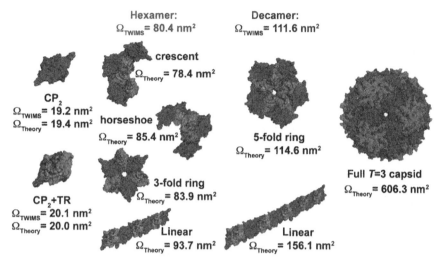

Figure 5.16: ESI-IMS-MS-measured Ω values (Ω_{TWIMS}) compared with computational calculations (Ω_{Theory}) for the CP_2, and $[CP_2:TR]$ and (CP, TR) building blocks, and potential hexamer and decamer capsid assembly intermediates. The structures were modeled on the final capsid topology taking into account the mass and stoichiometry discerned previously for these species, in addition to the interaction geometry observed in the intact capsid. Artificially extended, linear subunit arrangements were modeled for both the hexamer and decamer intermediates for comparison and were found, as expected, to yield calculated Ω values significantly larger than those measured by ESI-IMS-MS, while the threefold ring, crescent, horseshoe and fivefold ring modeled structures gave calculated Ω values in good agreement with their respective intermediates. Reproduced by permission of John Wiley and Sons, Ltd. from *Rapid Communications in Mass Spectrometry*, Vol. 24, 3033–3042, 2010.

higher m/z group of ions are from unfragmented GroEL. For CID study, the 66+ ion was chosen as the precursor. In the CID MS/MS spectra at three collision voltages 20 V, 70 V and 120 V, peaks owing to the 13-mer, 744 kDa appear at higher m/z (predominant ion 34+); in the low m/z region peaks from 57.2 kDa 1-mer appear (predominant ion 31+). Both the rings together (14-mer): Ω_{Theory} 221.3 nm^2 (in the experiment of Ashcroft *et al.* Ω_{TWIMS} 224.6 nm^2); only one ring (13-mer): Ω_{Theory} 217.1 nm^2, Ω_{TWIMS} 120.2 nm^2.[383,382] The value Ω_{TWIMS} 120.2 nm^2 obtained in the experiment for the 13-mer is remarkably low (45% lower), in comparison to the value 217.1 nm^2 from the calculation for GroEL 13-mer which was obtained by just removing a monomer and assuming nothing else happened in the

[383]Uetrecht, C., Versluis, C., Watts, N. R., Wingfield, P. T., Steven, A.C., and Heck., A. J. R. (2008) Stability and shape of hepatitis B virus capsid in vacuo. *Angew. Chem. IE*, **47**, 6247–6251.

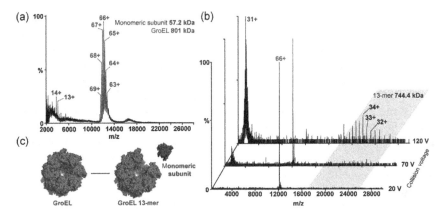

Figure 5.17: (a) ESI-IMS-MS spectrum of the GroEL chaperonin complex dominated by the intact 14-mer with an average of 66+ charge state ions (801 kDa; $\sim m/z$ 12,000), accompanied by a trace of unbound monomer (13+/14+ charge state ions; 57.2 kDa; $\sim m/z$ 4,400). (b) ESI-MS-CID-IMS-MS spectra with increasing collision voltage from the precursor 66+ charge state ions of intact GroEL, showing residual 13-mer (28+ to 38+ charges; $\sim m/z$ 22,000) and the unfolded, expelled monomer carrying a disproportionately high number of charges (average 31+ charge state ions; $\sim m/z$ 1,800). (c) Schematic model of GroEL dissociation assuming that the residual 13-mer and expelled monomer subunit do not undergo any structural rearrangement after dissociation. Reproduced by permission of John Wiley and Sons, Ltd. from *Rapid Communications in Mass Spectrometry*, Vol. 24, 3033–3042, 2010.

removal process. However, if a monomer is taken out it may be expected to disturb the ring structure substantially causing structural rearrangement. Because the removed subunit would be expected to have originally not only noncovalent interaction with neighbors on its own ring, but also with the subunits on the ring below.

Additionally they studied MHC-1 (major histocompatibility complex class 1, a linear system) which contains a heavy chain, a light chain, and a non-specific peptide subunit bound to the heavy chain (the peptide binding groove is on the heavy chain), substantially different in structure from the ring system of GroEL. The mass spectrum of MHC-1 in Fig. 5.18(a) shows the following peaks: peptide 1.07 kDa (1+), light chain 11.8 kDa (7+, 6+, 5+), MHC−peptide 43.6 kDa (12+, 11+), MHC 44.7 kDa (13+, 12+, 11+). The three CID spectra of MHC-1 12+ precursor at collision voltages 20, 60, and 100 V in Fig. 5.18(b) show: peptide 1+; light chain 11.8 kDa 5+ (tallest), 6+; heavy chain 31.9 kDa 11+, 8+, 7+. In Fig. 5.18c are models, from left to right: MHC, peptide, MHC−peptide, heavy chain, and light chain. Calculated and experimental cross-sections are: MHC-1 Ω_{TWIMS} 29.0 nm^2, Ω_{Theory} 27.8 nm^2; heavy chain, Ω_{TWIMS} 24.8 nm^2, Ω_{Theory}

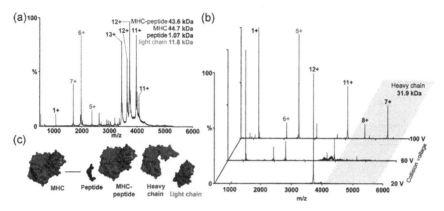

Figure 5.18: (a) ESI-IMS-MS spectrum of the major histocompatibility complex class 1 (MHC-1), showing the intact complex with an average of 12+ charge state ions (44.7 kDa;$\sim m/z$ 3,800) together with two unbound subunits, the light chain (average 6+ charge state ions; 11.8 kDa; $\sim m/z$ 2,000) and the peptide (MH$^+$; $\sim m/z$ 1,100), and the MHC-1 complex less the peptide (average 11+ charge state ions; 43.6 kDa; $\sim m/z$ 4,000). (b) ESI-MS-CID-IMS-MS spectra with increasing collision voltage from the precursor 12+ charge state ions of intact MHC-1 showing heavy chain (average 8+ charge state ions; 31.9 kDa; $\sim m/z$ 5,000), light chain (average 5+ charge state ions; 11.8 kDa; $\sim m/z$ 2,400) and peptide (MH$^+$; $\sim$ 1,100) product ions. (c) Schematic model of MHC-1 dissociation assuming that the residual heavy and light chain subunits do not undergo any structural rearrangement after expulsion from the intact complex. Reproduced by permission of John Wiley and Sons, Ltd. from *Rapid Communications in Mass Spectrometry*, Vol. 24, 3033–3042, 2010.

24.1 nm^2; light chain, Ω_{TWIMS} 12.6 nm^2, Ω_{Theory} 11.8 nm^2. There is little difference between calculation and experimental results. This seems to indicate that dissociation products came out with minimal unfolding.

The MS/MS CID spectra of the 24+ ion of the hexamer intermediate (182.9 kDa) and 34+ ion of the decamer intermediate (304.8 kDa), are shown, respectively, in Figs. 5.19(a) and 5.19(c). At 140 V collision voltage, for hexamer − [CP] (169.2 kDa) Fig. 5.19(a), the tallest peak is 17+. Other ions are 16+, 18+. In Fig. 5.19(c), decamer − [CP] (291.1 kDa), the tallest peak is 27+. Other ions are 26+, 28+. The fragmented representation of the hexamer after monomer expulsion, that results in a cavity, was generated by removal of atoms constituting one CP monomer from the threefold ring hexamer structure. In the case of decamer, however, expulsion of a monomer causes little difference in the structural size and shape. Here, the fragmented representation was generated by removal of atoms constituting one CP monomer from fivefold ring decamer structure model. The

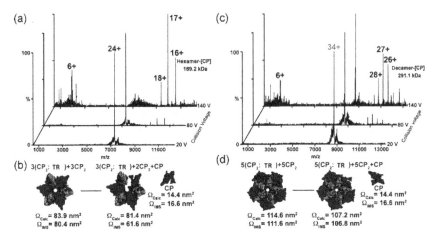

Figure 5.19: (a) ESI-MS-CID-IMS-MS spectra from the precursor 24+ charge state ions of the hexamer assembly intermediate showing the dissociated [3(CP$_2$:TR) + 2CP$_2$ + CP] product (i.e., loss of a CP monomer) (average 17+ charge state ions; 169.2 kDa; $\sim m/z$ 10,000) and the expelled, unfolded CP monomer (average 6+ charge state ions; 13.7 kDa; $\sim m/z$ 2,100). (b) Schematic model of hexamer assembly intermediate dissociation, assuming that the residual complex does not undergo any stuctural rearrangement after monomer expulsion. (c) ESI-MS-CID-IMS-MS spectra from the precursor 34+ charge state ions of the decamer assembly intermediate showing the dissociated [5(CP$_2$:TR) + 4CP$_2$ + CP] product (i.e., loss of a CP monomer) (average 27+ charge state ions; 291.1 kDa; $\sim m/z$ 11,000) and the expelled, unfolded CP monomer (average 6+ charge state ions; 137 kDa; $\sim m/z$ 2,100). (d) Schematic model of the decamer assembly intermediate dissociation, assuming that the residual complex does not undergo any structural rearrangement after monomer expulsion. Reproduced by permission of John Wiley and Sons, Ltd. from *Rapid Communications in Mass Spectrometry*, Vol. 24, 3033–3042, 2010.

calculated and TWIMS decamer cross-sections: decamer, [5(CP$_2$:TR) + 5CP$_2$] : Ω_{Theory} 114.6 nm^2, Ω_{TWIMS} 111.6 nm^2; 5(CP$_2$:TR) + 4CP$_2$ Ω_{Theory} 107.2 nm^2, Ω_{TWIMS} 106.8 nm^2, CP Ω_{Theory} 14.4 nm^2, Ω_{TWIMS} 16.6 nm^2. Here, measured cross-sections are close to the calculations. So this seems to fit with a decamer structure. As regards calculated and TWIMS cross-sections for the hexamer, threefold ring [3(CP$_2$:TR) + 3CP$_2$], Ω_{Theory} 83.9 nm^2, Ω_{TWIMS} 80.4 nm^2, Model [3(CP$_2$:TR) + 2CP$_2$] after monomer CP removal, but with the basic threefold form retained Ω_{Theory} 81.4 nm^2, Ω_{TWIMS} 61.6 nm^2 (-25%). Ejected CP: Ω_{Theory} 14.4 nm^2, Ω_{TWIMS} 16.6 nm^2. The experimental cross-section Ω_{TWIMS} 61.6 nm^2 in comparison to the calculated cross-section Ω_{Theory} 81.4 nm^2 is remarkably low (-25%), indicating a significant change in the structure and shape — more like a collapse with respect to its previous structure. It is reasonable to infer that if a

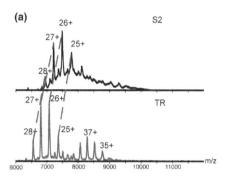

Figure 5.20: Mass spectrometry of assembly with S2 RNA. Stoichiometry of the threefold intermediate $[3(CP_2:RNA) + 3CP_2]$ using TR and S2 to initiate reassembly in the 2:1 CP_2:RNA stoichiometry at $t = 30$ min. The charge-state distributions corresponding to the intermediates formed during TR and S2 assembly, in the m/z range 6,000–12,000 are indicated. The difference in mass between TR and S2 is 4 kDa. The observed masses for the threefold intermediates are 182.7 and 194.7 kDa, yielding a mass difference of 12 kDa, as expected. The peaks corresponding to the extended fivefold intermediate with TR are clearly visible in the lower spectrum with TR (35+ to 38+) but are essentially undetectable in the upper one for S2. Reproduced by permission of Elsevier Ltd. from *Journal of Molecular Biology*, Vol. 395, 924–936, 2010.

monomer is taken from the ring, the ring is disturbed — it may collapse leading to smaller size and lower cross-section. Therefore with respect to various possibilities the preferred structure would be a threefold ring. The overall conclusion therefore is: there are two major higher-order intermediates, a hexamer: 182.9 kDa, $(3(CP_2 + TR) + 3CP_2)$, Ω_{TWIMS} 80.4 nm^2; topologywise it has a threefold axis of symmetry. And a decamer: 304.8 kDa, $(5(CP_2 + TR) + 5CP_2)$, Ω_{TWIMS} 111.6 nm^2; its topology has a fivefold axis of symmetry.

They also monitored the effect of coat protein variations and RNA mutants (different RNA aptamers) to confirm the stoichiometry of assembly intermediates, and also to see if the higher-order species and assembly pathways are the same. A comparison of mass spectra of TR clamp (23 nt, 7.3 kDa) and TR (19 nt, 6 kDa) show both hexamer and decamer (not shown in Figure here), but extended stem loop S2 (31 nt, 10.2 kDa) shows little evidence of the decamer (Fig. 5.20). They conclude RNA acts as an allosteric switch of coat protein conformation throughout the assembly process, and simplifying combinatorial possibilities immensely it determines the precise assembly pathway taken to the final T = 3 shell capsid.

5.9 Mass spectrometry of intact V-type ATPases reveals bound lipids and the effects of nucleotide binding

In this section, we present a work of Robinson *et al.*,[384] who show by ESI-MS that rotary adenosine triphosphatases (ATPases)/synthases (from *Thermus thermophilus* and *Enterococcus hirae*) can remain intact in vacuum along with membrane and soluble subunit interactions.

Rotary ATPases/synthases are molecular machines that are central to biological energy conversion. Both V-type and F-type complexes consist of two reversible motors: the ion pump/turbine in the membrane-embedded V_0/F_0 sector, and the chemical motor/generator in the V_1/F_1 sector; a central rotating shaft mechanically couples F_1-F_0, and V_1-V_0; peripheral stalks hold the two sections together. The ratio of protons and ATP — the two fuels that drive the two motors — determines the mode of operation. Movement of Na^+ or protons across the membrane is mediated by the V_0/F_0 domain, whereas interaction with nucleotides and inorganic phosphate either to produce or consume ATP, in the case of the eukaryotic F- and V-type families, respectively, involves the V_1/F_1 domain. For both functions, however, eubacteria and archaea typically have only one type of rotary ATPase/synthase. Most bacterial complexes are of the F-type, but some bacteria and all known archaea have complexes closely related to eukaryotic V-type ATPases. The physiological function of most prokaryotic complexes, whether of F- or V-type, is ATP synthesis; but, in many, regulatory functions have evolved that allow, if required, reversal into ATP-driven proton pumps. Despite the fact that a large body of structural information exists, there has been no high-resolution structures of any intact rotary ATPases/synthases. Thus, information on regulatory allosteric changes that involve both the soluble head and the membrane sector are not there. Similarly unavailable has been information on heterogeneous interactions with lipids and nucleotides. The first cryo-EM data revealing views of the entire membrane-embedded region has been available only recently.[385] A schematic diagram of V ATPase[386] is shown in Fig. 5.21 for reference.

---

[384]Zhou, M., Morgner, N., Barrera, N. P., Politis, A., Isaacson, S. C., Matak-Vinković, D., Murata, T., Bernal, R. A., Stock, D., and Robinson, C. V. (2011) Mass spectrometry of intact V-type ATPase reveals bound lipids and the effects of nucleotide binding. *Science*, **334**, 380–385.

[385]Lau, W. C. and Rubinstein, J. L. (2010) Structure of intact *Thermus thermophilus* V-ATPase by cryo-EM reveals organization of the membrane-bound V_0 motor. *Proc. Natl. Acad. Sci. U.S.A.*, **107**, 1367–1372.

[386]Forgac, M. (2007) Vacuolar ATPases: rotary proton pumps in physiology and pathophysiology. *Nat. Rev. Mol. Cell. Biol.*, **8**, 917–929.

Earlier LILBID mass spectrometry had demonstrated existence of intact rotary ATPases/synthases in vacuum.[111,112] Previous[387,388] well-resolved ESI-mass spectra of the V_1 domain of ATPase however did show lack of the V_0 domain, and also any V_0-V_1 interaction, apparently due to the dissociation of the complex in the absence of the protective micelle. The ES procedure employed in the present work enabled determination of subunit stoichiometries and the identity of tightly bound lipids within the membrane rotors, and also interrogation of subunit interactions within intact rotary ATPase/synthases. It permitted probing of synergistic effects of lipid and nucleotide binding.

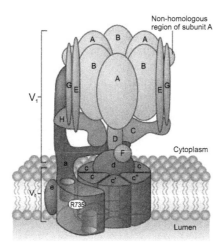

Figure 5.21: V ATPase. Reproduced by permission of Macmillan Publishers Limited from *Nat. Rev. Mol. Cell. Biol.*, Vol. 8, 917–929 (2007).

A mass spectrometer[389] designed to handle high-mass complexes was used in the present work. Assignment of mass spectral peaks to the intact particle that contained 26 subunits and nine different proteins is shown in Fig. 5.22 (a). The central group of ion peaks is from intact rotary TtATPases, the ion group to its immediate right arises from the loss of E peripheral stalk subunits, and the group further right is from subsequent loss of G subunits. To the left of the intact ATpase ion group are ions from membranous

[387]Esteban, O., Bernal, R. A., Donohue, M., Videler, H., Sharon, M., Robinson, C. V., and Stock, D. (2008) Stoichiometry and localization of the stator subunits E and G in *Thermus thermophilus* H^+-ATPase/Synthase. *J. Biol. Chem.*, **283**, 2595–2603.

[388]Kitagawa, N., Mazon, H., Heck, A. J., and Wilkens, S. (2008) Stoichiometry of the peripheral stalk subunits E and G of yeast V1-ATPase determined by mass spectrometry. *J. Biol. Chem.*, **283**, 3329–3337.

[389]Hernandez, H. and Robinson, C. V. (2007) Determining the stoichiometry and interactions of macromolecular assemblies from mass spectrometry. *Nat. Protoc.*, **2**, 715–726.

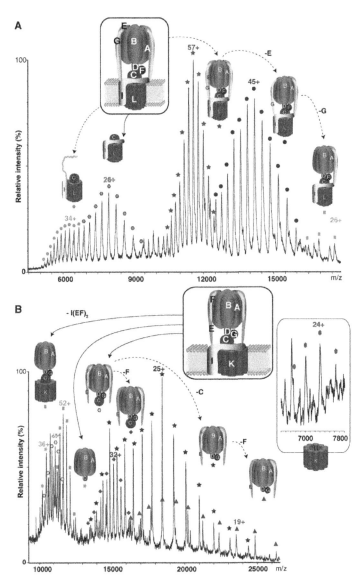

Figure 5.22: Mass spectra of the intact rotary ATPases from *T. thermophilus* and *E. hirae*. (A) Peaks are assigned to the intact *Tt*ATPase complex (stars), loss of membrane subcomplex (ICL_{12}) in solution and gas phases (at the left — peaks around 34+ and 26+, respectively) and dissociation of subunits E and G from peripheral stock (peaks around 45+ and 26+, at the right). (B) For *Eh*ATPase, the membrane subcomplex is observed in contact with the soluble head (squares, peaks around 52+). (Inset) Mass spectra of the K ring in aqueous solution. Reproduced by permission of the American Association for the Advancement of Science from *Science*, Vol. 334, 380–385, 2011.

subcomplex V_0 (ICL$_{12}$). The bimodal distribution of the peaks indicates that they originate both in solution and gas phases. Solution and gas phase loss of ICL$_{12}$ observed are 10% and 7%, respectively. Under the experimental conditions used, approximately 28% of the complex remains intact. The corresponding soluble V_1 is also observed (not shown here), which confirms spontaneous dissociation of the complex under these solution conditions. Assignment of all complexes was made using parallel simulations and an in-house software *Mass*ign. On the basis of MS-determined subunit masses, added to that masses of lipid and nucleotide binding and that due to incomplete desolvation, they obtain a total overall mass of 659,202 (±131) Da.

In *Eh*ATPase, the membrane-embedded rotor complex K subunit is larger than the corresponding L subunit of *Tt*ATPase; the former subunit contains four transmembrane helices (TMH) in comparison to the two in the latter. From the mass of the membrane ring, the stoichiometry in the K ring was determined to be 10 (175 kDa). Because of inadequate desolvation the spectrum is not well resolved when the complex emerges intact from the micelle. Better desolvation is achieved by increasing activation energy; under that condition, however, two effects are observed: either the membrane ring remains intact but the peripheral stalks and subunit I undergo dissociation, or, the stalk subunits, which are two in number, remain attached, the membrane region undergoes disruption (Fig. 5.22 (b)).

In the K_{10} ring (stoichiometry of the lipid binding is 10) a series of negatively charged cardiolipins was identified as the lipid components. Six cardiolipin isomers (see, Fig. 5.23) were identified; and from quantitative analysis, specific binding of one lipid per subunit was deduced (*Eh* K subunit 1: 1.2 ± 0.1 cardiolipins). Ten negatively charged symmetric cardiolipins were docked inside the K_{10} ring, in proximity to the conserved Lys[32]. The four hydrophobic chains positioned with two chains emanating from both sides of the polar head provided a hydrophobic lining to the inside of the ring (Fig. 5.23(a)).

For the L_{12} ring of *Tt*ATPase they identified phosphatidylethanolamine (PE) in the role of bound lipid (Fig. 5.23). Tandem MS of a subcomplex containing membrane subunits and peripheral stalks showed dissociated L subunits, some with (*apo*) some without (*holo*) the lipid. Undisrupted subunit was found to bind six lipids; this is supported by results of protein:lipid ratio quantification—one L subunit: 0.55 ± 0.1 PE (that is, one L subunit dimer per PE). Each L subunit has two transmembrane helices and one conserved Glu[63]. This is compatible with EM data, which shows that the enzyme has a sixfold symmetric membrane-embedded ring. Modeling results show that six Glu[63] residues are locked in an occluded position (proton transfer can take place in the remaining six active glutamate positions (Fig. 5.23(i)-(iii)); this effectively halves the proton:ATP ratio, keeping lipid binding at the dimer interface.

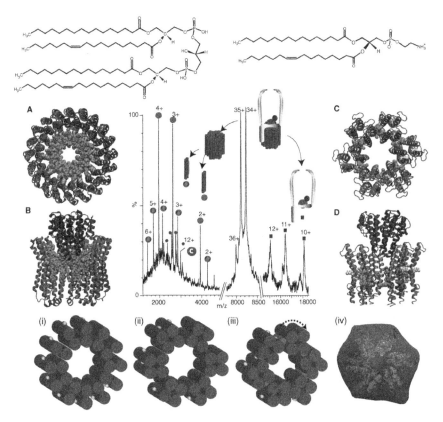

Figure 5.23: Lipid binding and its effect on the membrane rings. (Upper panel) Predominant lipid molecules, one of the six cardiolipins (left) and phosphatidylethanolamine (PE, right), found in intact *Eh*ATPase and *Tt*ATPase, respectively. (Central panel) Tandem MS of a subcomplex from *Tt*ATPase [$ICL_{12}E_2G_2F$ (282,740 daltons)] leads to disruption of the L_{12} ring, releasing proteolipids L ± PE (2+, 3+, 4+, 5+, 6+ at the left — 8,539 daltons; 2+, 3+, 4+ peaks — 7,849 daltons) and a stripped-complex ICE_2G_2F (squares, 10+, 11+, 12+ at the right — 184,242 daltons). Atomic structure of the K_{10} ring of *Eh*ATPase with docking of 10 cardiolipins to show reduction in the inner diameter (A) and after docking subunit C (B). Models for sixfold symmetry of the L_{12} ring with six PE molecules (dark gray shade, on the inner ring) (C) and the subunit C (central crest) docked onto the ring (D). (Lower panel) schematics of the rotor ring with 12 L subunits each having two TMHs (cylinders) and one conserved Glu^{63} (white spot) as seen in EM of two-dimensional crystals of isolated L ring (i). Transformation into a sixfold symmetric ring [(ii) and (iii)]. (iv) Comparison of the sixfold symmetrical model with EM data reported previously for the intact *Tt*ATPase (iv). Reproduced by permission of the American Association for the Advancement of Science from *Science*, Vol. 334, 380–385, 2011.

Whereas *E. hirae* has both F- and V-type ATPases, *T. thermophilus* has only one type of rotary ATPase; *in vitro* the latter operates in both proton pumping and ATP synthesis modes. Since each L subunit has two TMHs, it would mean switching between the 12-membered ring from a high proton:ATP ratio in position for synthesis and a lower one biased toward pumping. This picture imitates the eukaryotic V-type enzyme, where four TMHs are provided by gene duplication each with one active glutamate. For this dual-function rotary ATPase, one might expect this switching to be assisted by specific lipid binding.

The lipid-binding patterns in L_{12} (93.6 kDa) and K_{10} (160 kDa) rings preparing toward interaction with their respective C subunits (35.8 and 38.2 kDa) — which are likely conserved — were found to produce inner rings (orifices) of approximately the same size (38–39 Å); cardiolipin lining in the K_{10} ring reduces the orifice from 54 to 38 Å, whereas conversion of a 12- to a six-membered L-ring for *Tt*ATPase (with phosphatidylethanolamine) reduces the orifice from 47 to 39 Å.

In the presence of ATP the predominant species are the intact complex and the subcomplex ICL_{12} (Fig. 5.24 (a) bottom). Under this condition tandem MS of V_1 shows loss of subunit F (Fig. 5.24 (b) top), and then loss of subunit D. In the absence (or at low concentration) of ATP, the B subunit is lost from the intact complex (intact complex minus B is observed — Fig. 5.24 (a) top); also, CL_{12} is observed — from dissociation of I from ICL_{12} (Fig. 5.24 (a) top). Under this condition of ATP depletion, loss of subunit B is followed by extensive dissociation of the soluble head. Also, under this condition, loss of subunit D is possible before loss of subunit F, which leaves F in the subcomplex interacting directly with A_3B_3 hexamer. The latter interaction is supported by x-ray structure and cryo-EM reconstruction of, respectively, the isolated A_3B_3DF complex from *Tt*ATPase and yeast V-ATPase. X-ray structure of the auto-inhibited F_1 head of *Escherichia coli* F-ATPase shows[390] interactions in the soluble head between the subunits analogous to F_1 ε and β. Subunit ε was found to make similar direct interactions with the soluble head when the *E. coli* F_1 complex was subjected to the same tandem MS procedure. The braking mechanism to prevent unregulated consumption of ATP in *E. coli* is well established; similar MS dissociation patterns in *E. coli* F-ATPase and in *Tt*ATPase suggest a similar mechanism in prevention of ATP hydrolysis in the uncoupled V_1 complex of *Tt*ATPase.

At low ATP concentrations, loss of subunit I from ICL_{12} takes place to form CL_{12} (Fig. 5.24 (a) top). Expansion (to reveal fine structure) of

---

[390]Cingolani, G. and Duncan, T. M. (2011) Structure of the ATP synthase catalytic complex (F_1) from *Escherichia coli* in an auto-inhibited conformation. *Nat. Struct. Mol. Biol.*, **18**, 701–707.

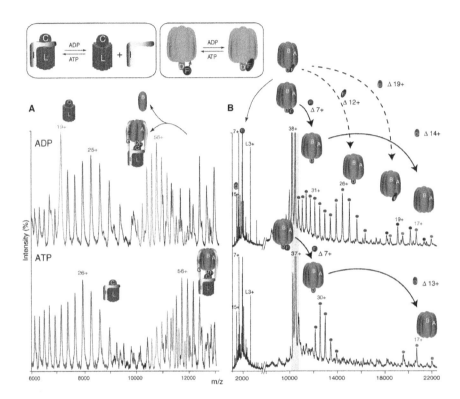

Figure 5.24: Nucleotide binding and its effects on intact TtATPase and the membrane-embedded and soluble head complexes. (A) Depletion of ATP leads to dissociation of the B subunit from the head and subunit I from ICL_{12}. (B) Tandem MS of the soluble head (A_3B_3DF) reveals sequential loss of subunits F and D in the presence of $50\,\mu M$ ATP. In TtATPase solutions containing $50\,\mu M$ ADP, two conformations of subunit F are evident from the bimodal distribution of charge states formed for the V_1 stripped of F ($\Delta 7 +$ and $\Delta 12+$) together with a direct loss of subunit D. The $\Delta 7+$ series is similar to that formed from the ATP-bound complex. The $\Delta 12+$ series is consistent with an extended conformation of subunit F. (Inset) Schematic representation of effects of ATP/ADP on the membrane ICL_{12} complex and the movement of an extended subunit F in the A_3B_3DF complex. Reproduced by permission of the American Association for the Advancement of Science from *Science*, Vol. 334, 380–385, 2011.

the mass spectral peaks (obtained under high energy activation) that are assigned to ICL_{12} when compared with simulated spectra reveals coexistence of the subcomplex with different numbers of PE lipid (689.5 Da) and ATP (507 Da) bound in solution; the experimental data are consistent with accommodation of six lipids and up to two nucleotides (ATP or ADP). This is in agreement with suggestions that a eukaryotic functional equivalent of subunit I can bind selectively to ADP, undergo conformational change, and thus sense cellular nucleotide levels.[391] Ion mobility measurements on TtATPase, ICL_{12}, and CL_{12} subcomplexes were carried out to look into this conformational change. Since broad arrival-time distributions (ATDs) would be expected for ions with multiple conformations, relatively compact ATDs for both the intact ATPase and CL_{12} (the range found is 10.62–11.49 ms in the latter case for ion range 23^+–21^+) were taken to indicate one predominant conformation. ATDs for the ICL_{12} complex, in contrast, are much broader (the range found is 16.05–19.48 ms for ion range 27^+–24^+), indicating multiple conformations of subunit I in the isolated V_0 complex. Tethering of subunit I by forces exerted from the peripheral stalks — subunits E and G — as suggested by Lau and Rubinstein[385] is taken as possible reason for the lack of conformational heterogeneity in the intact complex. This constraint disappears when subunit I in V_0 is released from the intact complex; flexibility of the hinge domain between the soluble and transmembrane domains then leads to conformational heterogeneity.

Robinson *et al.* find their IM data for the V_0 complex, based on modeling of the ICL_{12} and CL_{12} complexes, are consistent with a range of conformational states with the soluble domain of I of the intact TtATPase in the range $90°$–$135°$ to the proton channel. The corresponding number from EM density map is $90°$. From this, and considering preferential binding of ADP to the proposed site near the hinge region[391] they propose that interaction between I and CL_{12} are destabilized resulting in facile loss of subunit I under low-ATP conditions.

To test if the proton gradient would affect proton translocation in the isolated V_0 complex like cellular nucleotide levels do, they increased the pH of the ATP-containing TtATPase solution to 9.0 to effect reduction of proton concentration. After this incubation at pH = 9.0, mass spectra showed formation of subcomplexes following loss of IGE from the intact complex; from loss of IGE, the subcomplex $A_3B_3DFEGCL_{12}$ is formed. With increasing incubation time the membrane subcomplex ICL_{12} also undergoes dissociation of subunit I, to form CL_{12}. This observation and the dissociation of subunit I from V_0 together suggest a regulatory role for the

---

[391] Armbrüster, A., Hohn, C., Hermesdorf, A., Schumacher, K., Börsch, M., and Grüber, G. (2005) Evidence for major structural changes in subunit C of the vacuolar ATPase due to nucleotide binding. *FEBS Lett.*, **579**, 1961–1967.

subunit I wherein both proton and ATP concentrations are sensed. Because in this work facile loss of subunit I is observed under both low-$[H^+]$ and low-ATP conditions, and because cryo-EM data do not show extensive interaction between I and the membrane ring, the present work proposes a mechanism in which subunit I moves away from the ring and the resulting gap sealed with membrane lipids. Earlier proposals invoked locking together of subunit I with the CL_{12} membrane ring causing prevention of relative movement which explained the absence of passive H^+ translocation in isolated V_0.[392,393]

In the present work it is seen that lipids with two and four hydrophobic chains associate with subunits with two and four TMHs — subunits L and K, respectively. The fact that in both ATPases the lipids selected from the available pool are not the most common ones in the cell signifies that their selection is for their specific structural roles and metabolic regulation. This supports the view[394] that membrane proteins possess specific lipid binding sites and the lipids, by defining the conformations and inner dimensions of the membrane rings, have the ability to fine tune subunit interactions. The nucleotide binding experiments show that V_1 senses a decrease in cellular ATP concentrations both by the movement of subunit F and changes in interactions at the A:B interface. V_0 is also sensitive to low [ATP] and $[H^+]$, both of which promote displacement of subunit I. A suggestion from the present workers is that membrane lipids afterwards seal the proton-conducting channel. As a consequence, when reversible dissociation takes place *in vivo*, both ATP and proton/ion gradients are conserved.

5.10 Mass spectrometric imaging of biological material

In this section, we present two examples of mass spectrometric imaging studies; one that used high-speed MALDI TOF imaging technique, and the other that employed the laser ablation electrospray ionization (LAESI) technique. Imaging studies have also been undertaken by using the DESI-MS technique.[395]

---

[392]Zhang, J., Myers, M., and Forgac, M. (1992) Characterization of the V_0 domain of the coated vesicle (H+)-ATPase *J. Biol. Chem.*, **267**, 9773–9778.

[393]Beltrán, C. and Nelson, N. (1992) The membrane sector of vacuolar H(+)-ATPase by itself is impermeable to protons. *Acta Physiol. Scand. Suppl.*, **607**, 41–47.

[394]Hunte, C. and Richers, S. (2008) Lipids and membrane protein structures. *Curr. Opin. Struct. Biol.*, **18**, 406–411.

[395]Eberlin, L. S., Ifa, D. R., Wu, C., and Cooks, R. G. (2010) Three-dimensional visualization of mouse brain by lipid analysis using ambient ionization mass spectrometry. *Angew. Chem. Int. Ed. Engl.*, **286**, 873–876.

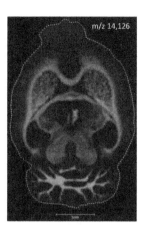

Figure 5.25: Right panel: The ion image of myelin basic protein isoform 8 in a rat brain section. After the image acquisition was finished, matrix was removed from the section with an acetone wash and the section was stained with H&E and an optical image obtained (left panel) of the same section. The scale bar in the image is 5 mm. Courtesy: R. Caprioli, Mass Spectrometry Research Center, Vanderbilt University. Reproduced by permission.

In MALDI imaging mass spectrometry technology, a thin slice of the tissue sample is mounted on an energy-absorbing matrix target which moves through the path of a stationary laser beam. Spot by spot (pixel by pixel), ablation material is detected by mass spectrometry; from a single scan hundreds of images are acquired from which a two-dimensional—and in turn a three-dimensional—image of the tissue can be constructed.[396] An image of myelin basic protein isoform 8 in a rat brain's section obtained using this method is presented in Fig. 5.25.

The second example of imaging is one that uses the LAESI technology. In LAESI, water molecules present in a biological sample are made to absorb energy supplied by a focussed mid-infrared laser beam; the ablated material is then subjected to electrospray postionization followed by mass spectrometric analysis. There have been a number of studies using

---

[396]Seeley, E. H. and Caprioli, R. M. (2012) 3D imaging by mass spectrometry: a new frontier. *Anal. Chem.*, **84**, 2105–2110.

this method.[397,398,399,400,401] Here we present an example[402] of rat-brain imaging; specifically, we highlight a part of the work in which by deconvoluting two mass peaks of nominally the same mass, separate distributions of two m/zs in the sample are obtained.

In this experiment, the mass spectra were externally calibrated with CAD fragments of the doubly protonated human [Glu^1]-fibrinopeptide B peptide, providing a mass accuracy of $\sim 5\,\text{mDa}$, and a mass resolution $m/\Delta m > 6000$ in the m/z range 50–1000. The ultimate spatial resolution was determined by the diameter of the ablation craters, which was $\sim 200\,\mu\text{m}$. For the rat-brain section experiment, images from which are shown in Fig. 5.26, the brain was cut into $100\,\mu\text{m}$ sections; selected tissue sections collected on to microscope slides were kept at $-80°\text{C}$ until LAESI analysis, and were stained with cresyl violet. Assignment of the brain regions analyzed was done using a rat brain atlas.

The deconvoluted distributions of the two images in Fig. 5.26 show substantial qualitative differences. In low resolution, overlaps of the peaks result in skewed overall distribution. The difference between the accurate masses of the 20-times protonated α-1 chain of rat hemoglobin and that of PC(34:1) is 228 mDa. Although the former peak is much broader, a Gaussian deconvolution using a 20 mDa window resulted in an acceptable separation between the distributions. The resulting distributions are compatible with the tissue accumulation in the brain of other lipids for PC(34:1) and with that of the heme unit for the hemoglobin.

[397] Nemes, P., Barton, A. A., and Vertes, A. (2009) Three-dimensional imaging of metabolites in tissues under ambient conditions by laser ablation electrospray ionization mass spectrometry. *Anal. Chem.*, **81**, 6668–6675.

[398] Parsiegla, G., Shreshtha, B., Carrrière, F., and Vertes, A. (2012) Direct analysis of phycobilisomal antenna proteins and metabolites in small cyanobacterial populations by laser ablation electrospray ionization mass spectrometry. *Anal. Chem.*, **84**, 34–38.

[399] Vaikkinen, A., Shreshtha, B., Kauppila, T. J., Vertes, A., and Kostianen, R. (2012) Infrared laser ablation atmospheric pressure photoionization mass spectrometry. *Anal. Chem.*, **84**, 1630–1636.

[400] Nemes, P. and Vertes, A. (2012) Ambient mass spectrometry *in vivo* local analysis and *in situ* molecular tissue imaging. *Trends Anal. Chem. (TrAC)*, **34**, 22–34.

[401] Shreshta, B., Sripadi, P., Walsh, C. M., Razunguzwa, T. T., Powell, M. J., Kehn-Hall, K., Kashanchi, F., and Vertes, A. (2012) Rapid, non-targeted discovery of biochemical transformation and biomarker candidates in oncovirus-infected cell lines using LAESI mass spectrometry. *Chem. Commun.*, **48**, 3700–3702.

[402] Nemes, P., Woods, A. S., and Vertes, A. (2010) Simultaneous imaging of small metabolites and lipids in rat brain tissues at atmospheric-pressure by laser ablation electrospray ionization mass spectrometry. *Anal. Chem.*, **82**, 982–988.

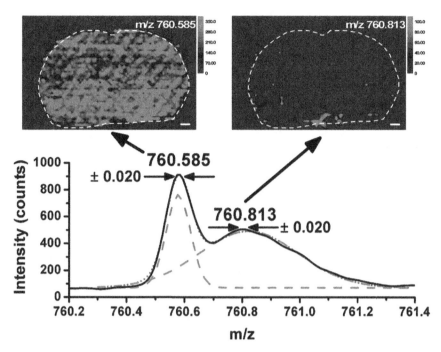

Figure 5.26: Separation of images of positively charged species with identical nominal mass by application of high resolving power. PC(34:1) (m/z 760.585) and the multiply charged α-1 chain of the rat hemoglobin (m/z 760.813) were separated by 228 mDa. Deconvolution of these peaks (solid black line) enabled rendering the separated species (dashed line) into distinct images. Reproduced by permission of the American Chemical Society from *Analytical Chemistry*, Vol. 82, 982–988, 2010.

CHAPTER 6
MASS SPECTROMETRY-BASED BIOINFORMATICS

The purpose of this chapter is to present a brief summary on the state of data analysis in protein identification and quantification. In a rapidly developing field, this necessarily means it is a snapshot in time, because what is in vogue at a given time soon gets outdated. Further, mass spectrometric analyses of biological samples take place under two special system conditions. The first one is that most of the mass spectrometers used in analyses are commercial; their operational modes and data output format are entirely determined by the manufacturers of those instruments. And with the enormous size of the data output with thousands of peptide MS/MS spectra, the second one is the requirement of software in analyses. Here the two major groups — proprietary software and open-source software — contend, and all these are constantly evolving. Depending on the nature of algorithm used there are several types. In the *de novo* method, sequences are obtained directly from the spectrum.[403,404,405] In another method,[406,407,408] unambiguous *peptide sequence tags* derived from spectra are used to search known sequences. In cross-correlation methods[409,410] also experimental spectra are compared against theoretical spectra. In the

---

[403] Johnson, R. S. and Taylor, J. A. (2002) Searching sequence databases via de novo peptide sequencing by tandem mass spectrometry. *Mol Biotechnology*, **22**, 301–315.

[404] Shevchenko, A., Sunyaev, S., Loboda, A., Shevchenko, A., Bork, P., Ens, W., and Standing, K. G. (2001) Charting the proteomes of organisms with unsequenced genomes by MALDI-quadrupole time-of-flight mass spectrometry and BLAST homology searching. *Anal. Chem.*, **73**, 1917–1926.

[405] Ma, B., Zhang, K., Hendrie, C., Liang, C., Li, M., Doherty-Kirby, A., and Lajoie, G. (2003) PEAKS: powerful software for peptide de novo sequencing by tandem mass spectrometry. *Rapid Commun. Mass Spectrom.*, **17**, 2337–2342.

[406] Mann, M. and Wilm, M. (1994) Error-tolerant identification of peptides in sequence database by peptide sequence tags. *Anal. Chem.*, **66**, 4390–4399.

[407] Sunyaev, S., Liska, A. J., Golod, A., Shevchenko, A., and Shevchenko, A. (2003) MultiTag: multiple error-tolerant sequence tag search for the sequence-similarity identification of proteins by mass spectrometry. *Anal. Chem.*, **75,**, 1307–1315.

[408] Tabb, D. L., Saraf, A., and Yates, J. R., III. (2003) GutenTag: high-throughput sequence tagging via an empirically derived fragmentation model. *Anal. Chem.*, **75**, 6415–6421.

[409] Eng, J. K., McCormack, A. L., and Yates, J. R. III (1994) An approach to correlate tandem mass spectral data of peptides with amino acid sequences in a protein database. *J. Am. Soc. Mass Spectrom.*, **5**, 976–989. Algorithms described in US patents 6,017,693 and 5,538,897. SEQUEST is distributed by Thermo Finnigan [from Yates/Eng SEQUEST home page].

[410] Pevzner, P. A., Dančik, V., and Tang, C. L. (2000) Mutation-tolerant protein identification by mass spectrometry. *J. Comput. Biol.*, **7**, 777–787.

Introduction to Protein Mass Spectrometry
http://dx.doi.org/10.1016/B978-0-12-805123-8.00006-0

Copyright © 2016 Elsevier Inc.
All rights reserved.

probability-based matching method,[411,412,413,414,415] a score is calculated based on the match between an observed peptide fragment and those calculated from a sequence search library incorporating judgment on the statistical significance of the match. An additional element is that often instrument manufacturers have their own inclination toward some specific software. Increasingly, this is now software that provides a platform to consider multiple diverse algorithms.[416] There have been reviews on software used for protein identification and quantification.[417,418] Several recent algorithms have been reviewed by Law and Kim.[419]

6.1 Peptides to proteins

6.1.1 Peptide mass fingerprinting

In *peptide mass fingerprinting* (PMF) (protein fingerprinting — developed by several groups independently), only the peptide masses need to be known; experimentally obtained peptide masses (after proteolysis and MS analysis) are compared against peptide masses obtained by *in silico* fragmentation of proteins — organisms' known genome first translated into proteins — into peptides. The masses of the peptides obtained experimentally from the unknown protein are then compared with the theoretical peptide masses of each protein obtained from the genome code. Through statistical analysis, the best match is then obtained. First suggested by

[411]Perkins, D. N., Pappin, D. J., Creasy, D. M., and Cottrell, J. S. (1999) Probability-based protein identification by searching sequence databases using mass spectrometric data. *Electrophoresis*, **20**, 3551–3567.

[412]Field, H. I., Fenyö, D., and Beavis, R. C. (2002) RADARS, a bioinformatics solution that automates proteome mass spectral analysis, optimises protein identification, and archives data in a relational database. *Proteomics*, **2**, 36–47.

[413]Clauser, K. R., Baker, P., and Burlingame, A. L. (1999) Role of accurate mass measurement (±10 ppm) in protein identification strategies employing MS or MS/MS and database searching. *Anal. Chem.*, **71**, 2871–2882.

[414]Fenyö, D., Qin, J., and Chait, B. T. (1998) Protein indentification using mass spectrometric information. *Electrophoresis*, **19**, 998–1005.

[415]Zhang, N., Aebersold, R., and Schwikowski, B. (2002) ProbID: a probabilistic algorithm to identify peptides through sequence database searching using tandem mass spectral data. *Proteomics*, **2**, 1406–1412.

[416]Searle, B. C. (2010) Scaffold: a bioinformatic tool for validating MS/MS-based proteomic studies. *Proteomics*, **10**, 1265–1269.

[417]Fenyö, D., Eriksson, J., and Beavis, R. (2010) Mass spectrometric protein identification using the global proteome machine. *Methods Mol. Biol.*, **673**, 189–202.

[418]Zhang, G., Ueberheide, B. M., Waldemarson, S., Myung, S., Molloy, K., Eriksson, J., Chait, B. T., Neubert, T. A., and Fenyö, D. (2010) Protein quantitation using mass spectrometry. *Methods Mol. Biol.*, **673**, 211–222.

[419]Law, K. P. and Lim, Y. P. (2014) Data independent analysis and hyper reaction monitoring. *Expert Rev. Proteomics*, **10**, 551–566.

Henzel *et al.*,[420] it was pursued almost simultaneously by several
others.[421,422,423,424,425]

6.2   MS/MS fragments to peptides to proteins

The following subsection divisions are based on different data acquisition
strategies.

6.2.1   Peptide sequence tag

Mann and Wilm demonstrated a method of peptide identification[406] using MS/MS *fragmentation peaks* — in which just two or three amino acid
residues are adequate for identification. The basis of their method is: a set
of easily identifiable series of *sequence ions* yields a *partial sequence* of the
peptide. Using this partial sequence, a peptide can be divided into three
regions: the first region characterized by the added masses (m_1) of region 1,
the region 2 having the *partial sequence*, and region 3 having added masses
(m_3) (Fig. 6.1). Thus, by definition the peptide mass M = m_1 + m_2 +
m_3. They called the three-segment construct 'm_1-partial sequence-m_3' the
peptide sequence tag, and showed that it is a unique signature and highly
specific identifier of that peptide, and, in comparison to *partial* sequence
information alone, is up to 1 million time *more* discriminating, and is applicable in the presence of an unknown post-translational modification as well
as an amino acid substitution between an entry in the sequence database
and measured peptide. These modifications can rapidly be located in a sequence using the sequence tag method; an erroneous identification of a tag
will not result in a false positive identification. The partial sequence by
itself, in comparison to the peptide sequence, however, is too short to be
searched in a database.

---

[420]Henzel, W. J., Billeci, T. M., Stults, J. T., Wong, S. C., Grimley, C., and Watanabe,
C. (1993). Identifying proteins from two-dimensional gels by molecular mass searching
of peptide fragments in protein sequence databases. *Proc. Natl. Acad. Sci. U.S.A.*, **90**,
5011–5015.

[421]Pappin, D. J., Højrup, P., and Bleasby, A. J. (1993). Rapid identification of proteins
by peptide-mass fingerprinting. *Curr. Biol.*, **3**, 327–332.

[422]Mann, M., Højrup, P., and Roepstorff, P. (1993). Use of mass spectrometric molecular weight information to identify proteins in sequence databases. *Biol. Mass Spectrom.*,
22, 338–345.

[423]James, P., Quadroni, M., Carafoli, E., Gonnet, G. (1993). Protein identification by
mass profile fingerprinting. *Biochem. Biophys. Res. Commun.*, **195**, 58–64.

[424]Yates, J. R., Speicher, S., Griffin, P. R., and Hunkapiller, T. (1993). Peptide mass
maps: a highly informative approach to protein identification. *Anal. Biochem.*, **214**,
397–408.

[425]Profound is a tool for searching a protein sequence collection with peptide mass
maps. A Bayesian algorithm is used to rank the protein sequences in the database according to their probability of producing the peptide ions (http://prowl.rockefeller.edu).

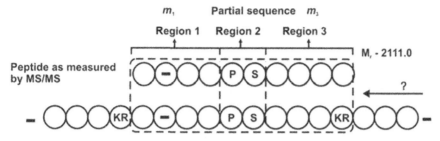

Figure 6.1: Principle of matching peptide sequence tags to a proposed sequence. The upper chain of amino acids represents the *peptide sequence as measured by MS/MS*, and the lower chain represents amino acids *in the sequence database* that the tag is compared to. Note that the partial sequence divides the peptide into three regions. The added mass m_1 of the residues in region 1, together with the N-terminus, is a match criterion as is the added mass in region three, m_3. In region 2, the sequence is known. Furthermore, it can be required that the peptide obeys the cleavage condition of the proteolytic enzyme, marked by KR for trypsin. The left point arrow indicates that both search directions may have to be considered. Reproduced by permission of the American Chemical Society from *Analytical Chemistry*, Vol. 66, 4390–4399, 1994.

We show the application of this method using a hypothetical tandem mass spectrum obtained on tryptic peptides (enzyme specificity: trypsin). Here the measured peptide molecular mass (M) is 2111±0.4 (masses are monoisotopic and in daltons); the run of sequence ions (ions observed) is 977.4, 1074.5, 1161.5. Using this information and from residue mass differences, presence of Pro and Ser as *partial sequence* (PS) is obtained: $1074.5-977.4=97.1$ (P), and $1161.5-977.4-97.1=87.0$ (S).

We first assume that the ion series is a B series. The added molecular mass m_1 of region 1 is defined as the N-terminus of the peptide plus the sum of the amino acid residue masses up to the partial sequence. Since the structure of b ions is H-(NH-CHR$_i$)-CO$^+$, R$_i$, being the amino acid side chains, m_1 would be equal to the lowest molecular mass in the ion series, 977.4 Da in the example (the lowest mass b ion would be to the N-terminus.); m_3 is the sum of the amino acid molecular masses in region 3 plus the molecular mass of the C-terminus. For b ions, m_3 therefore would simply be the difference between the measured molecular mass of the peptide M and the highest mass in the ion series, which gives $m_1 = 2111.0$ Da $-$ 1161.5 Da $= 949.5$ Da. Thus the complete sequence tag is: $m_1 = 977.4$ Da; PS; $m_3 = 949.5$ Da. We know, the peptide has been derived by tryptic cleavage of a protein. Therefore, in addition to matching the partial sequence, m_1 and m_3, we require that the amino acid N-terminal to region

1 and the C-terminal amino acid in the peptide be cleavage sites for trypsin. (Of course, the N- and C-terminus of the protein always match.)

The search produces in a single match, the peptide ASQSSTETQG*PS*-SESGLMTVK of Serine/Threonine-Protein Kinase from which the data for this example was constructed. Since the direction of the sequence is not known, the search is repeated by assuming that the ions are of y'' type. Since the structure of y ions is H_2-(NH-CHR_i-CO)-OH^+, m_1 is the peptide molecular mass M minus the largest mass in the ion series plus the mass of two hydrogens. In the example, $m_1 = 2111.0\,\text{Da} - 1161.5\,\text{Da} + 2m_H = 951.5\,\text{Da}$. m_3 is equal to the lowest mass in the ion series minus the mass of two hydrogens; in this case, $m_3 = 975.4\,\text{Da}$. The partial sequence has to be reversed for y'' before searching. However, there is no match found for the sequence tag $m_1 = 951.5$ Da; SP; $m_3 = 975.4$ Da, which demonstrates the discriminating capability of the algorithm. All these calculations are carried out automatically by the software PeptideSearch. The user only enters a "search string" consisting of the lowest mass of the ion series, the partial sequence, the highest mass, and the peptide molecular weight. Searches by other fragment series types, such as a, c, x, z series can be handled.[406]

The database can be searched on all five criteria (C-terminal cleavage site of preceding peptide (Arg or Lys); mass of region 1 (at unit mass accuracy); tag sequence (Pro Ser); mass of region 3 (at unit mass accuracy); C-terminal cleavage (Arg or Lys)) or just some of these five. The first approach (all five criteria) yields only exact matches while the second approach (some of those criteria) is error tolerant. For example, the program can be set to match peptides that have not been cleaved in a sequence-specific manner. More importantly, the algorithm can retrieve all peptides that match just two of the three regions. In the original implementation of the algorithm, any mismatch could be tolerated that is localized to one of the three regions.

Using this method, amino acid substitution can be rapidly determined and, in principle, even unsuspected post-translational modifications mapped to a single location. For example, if there is a substitution, say in the region 1, M would be changed, and searching by all the five criteria would not provide a match. Asking then to match regions 2 and 3 along with C-terminal specificity, would report tryptic peptides with the partial sequence PS match (which can be elongated to yield the correct mass m_3). In SWIS-SPROT, there are 25 such peptides. Then, by comparing the fragment ions to be expected from regions 2 and 3 with those that are observed, the list can be made short. In the next step, the fragment ions from region 1 can be compared to determine where the difference is.

Whereas the work emphasized extremely short partial sequences, every additional amino acid in region 2 increases the search specificity by a factor of about 20. With a partial sequence length of three amino acids, in the above example only two matches would be obtained instead of 25. With longer partial sequences it becomes more important to consider amino acid mismatches also in the partial sequence region, region 2.

There have been other work on sequence tags. Tanner *et al.*[426] describe a tool called InsPecT to identify post-translational modifications using MS/MS data. Given an MS/MS spectrum and a database, a database filter selects a small fraction of the database that with high probability is guaranteed to contain the peptide that produced the spectrum. InsPecT uses peptide sequence tags as filters that efficiently reduce the size of the database by a few orders of magnitude while retains with high probability the correct peptide. Modified peptides are identified by InsPecT with better or equivalent accuracy than other database search tools while being two orders of magnitude faster than SEQUEST, and substantally faster than X! Tandem on complex mixtures.[427] Tabb *et al.*[428] developed an open-source algorithm DirecTag to infer partial sequence tags directly from observed fragment ions. They used three separate scoring systems to evaluate each tag on the basis of peak intensity, m/z fidelity and complementarity. In datasets from several types of mass spectrometers; they claimed DirecTag reproducibility exceeded the accuracy and speed of InsPecT and GutenTag (GutenTag was developed by Tabb *et al.*[408] earlier).[429]

6.3 Data-dependent acquisition

6.3.1 SEQUEST

SEQUEST algorithm was introduced by Eng *et al.*[409]for correlating an uninterpreted, peptide fragment mass spectrum with amino acid sequences derived from a (protein or nucleotide) database. In SEQUEST, the character-based representations of amino acid sequences in a protein database are first converted into fragmentation patterns that can be matched to fragment ions in an MS/MS spectrum. Then, by matching the predictions to those observed in the experimental mass spectrum, a cross-correlation score X_{corr} is calculated, and the highest-scoring sequences are reported.

---

[426]Tanner, S., Shu, H., Frank, A., Wang, L. C., Zandi, E., Mumby, M., Pevzner, P. A., and Bafna, V. (2005) InsPecT: Identification of post-translationally (*database filters* that InsPecT constructs proved to be very successful in genomics searches) modified peptides from tandem mass spectra. *Anal. Chem.*, **77**, 4626–4639.

[427]A web interface to the InsPecT tool is available at http://peptide.ucsd.edu.

[428]Tabb, D. L., Ma, Z. Q., Martin, D. B., Ham, A. J., and Chambers, M. C. (2008) DirecTag: accurate sequence tags from peptide MS/MS through statistical scoring. *J. Proteome Res.*, **7**, 3838–3846.

[429]The source code and binaries for DirecTag are available from http://fenchurch.mc.vanderbilt.edu.

SEQUEST determines the amino acid sequence and thus the protein(s) and organism(s) corresponding to the MS/MS spectrum being analyzed. In their first work they demonstrated the approach applying to fragmentation data on peptides of known amino acid sequence, those generated by proteolytic digestion of proteins from whole cell lysates, antigenic peptides released from class II MHC molecules, and peptides obtained from the heavy chain of glycoasparaginase.

In developing the computer algorithm, the only information used is the mass of the peptide, which enables analysis of unrelated peptides generated with proteases of unknown specificity, and an interpretation of the MS/MS spectrum if the amino acid sequence is present in the protein database.

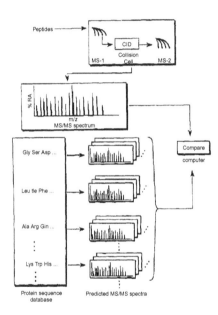

Figure 6.2: Flow chart that depicts the algorithm for searching protein databases with tandem mass spectrometry data. Reproduced with kind permission from Springer Science and Business Media from *J. Am. Soc. Mass Spectrom.*, Vol 5, 976–989, 1994.

The approach is shown schematically in Fig. 6.2. In the first step, computer reduction of data is carried out (such as m/z ratios converted to the rounded nominal values — nearest integer — which increases computational speed by a factor of ≈ 2). Only 200 most abundant ions in the spectrum are considered. In step 2, by matching molecular weight of the peptide to a linear sequence, sequences are identfied in a protein database for comparison against processed tandem mass spectrometric data. In step 3, the mass spectral information and the predicted fragment ions are then compared to yield a list of best-fit 500 sequences. The latter are then subjected to a correlation-based analysis which generates a final score and ranking of the sequences. The correlation-function[430] basis used in SEQUEST is shown in Fig. 6.3.

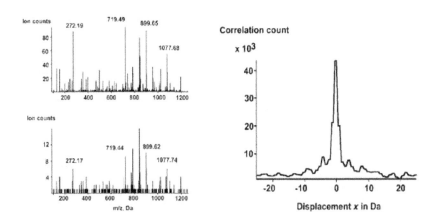

Figure 6.3: Fragment of the triple-charged ion of the peptide GVDLDQLLDMPN-NQLVELMHSR. The two spectra had been identified as being from the same precursor using the correlation approach on the entire spectrum. The correlation function shows a clear and absolute maximum at a relative mass shift of 0 Da. The intensity ratios of the fragments vary considerably between the two spectra, in contrast to the peak locations. Reproduced by permission of John Wiley and Sons from *Proteomics*, Vol. 3, 1597–1610, 2003.

It was observed that the cross-correlation score X_{corr} showed bias for larger peptides; therefore some normalization was required. To assess a match, SEQUEST uses the difference between the first- and second-ranked sequences (ΔC_n). This value depends on the database size, search parameters, and sequence homologies. MacCoss *et al.* later showed[431] that the use of a new scoring routine (SEQUEST-NORM) normalizes X_{Corr} values to be independent of peptide size and the database used to perform the search. This new scoring routine objectively calculates the percent confidence of

---

[430]Gentzel, M., Köcher, T., Ponnusamy, S., and Wilm, M. (2003) Preprocessing of tandem mass spectrometric data to support automatic protein identification. *Proteomics*, **3**, 1597–1610.

[431]MacCoss, M. J., Wu, C. C., and Yates III., J. R. (2002) Probability-based validation of protein identifications using a modified SEQUEST algorithm *Anal. Chem.*, **74**, 5593–5599.

protein identifications and post-translational modifications based solely on the X_{Corr} value.

Present-day tandem mass spectrometry instruments with much faster scan speeds generate even up to an order of magnitude more data in a given time; on a personal computer, SEQUEST™ database searches identifying proteins and their post-translational modifications can take several days; SEQUEST Cluster with parallel virtual machine (PVM) programming drastically reduces database search times. This is achieved by Thermo Electron Corporation/IBM by dividing the database search process into separate tasks and performing them in parallel on such as the IBM Cluster *e*server 1350 (IBM *e*server BladeCenter); at the end of the computer nodes completing analysis the results are integrated on the management node and a united output is generated.[432]

6.3.2 MASCOT

The MOWSE (MOlecular Weight SEarch) algorithm (public domain software) for identifying proteins by peptide mass fingerprinting was initially developed by Pappin and Bleasby for PMF searches (PMF paper[421] published in 1993), using a probability-based score for identification. This was the forerunner of MASCOT[411] that Matrix Science now develops and distributes. Along the way MOWSE was restructured so as to compute mass values "on the fly" from FASTA sequence databases, and new algorithms were coded for parallel execution on multiprocessor platforms. It supported, besides peptide-mass fingerprint and MS/MS fragment ion search, searches which combined mass data with amino acid sequence or composition. When in mid-1998 Matrix Science secured a license from ICRF (Imperial Cancer Research Fund) to develop and distribute MOWSE, with the name MAS-COT, it was made compatible with a number of platforms like SGI, SUN, DEC and Windows NT. Matrix Science website provides *free access* to MASCOT. For users who wish to run MASCOT on their own server, a license fee is required. Following are the various facilities available.[433]

MASCOT *Server* is live on the Matrix Science website for both peptide mass fingerprint and MS/MS database searches. A selection of popular sequence databases are online, including SwissProt, NCBInr, and the EST divisions of EMBL. This free service is ideal for evaluation and searching smaller data sets. For large scale in-house routine work running, use of licensed MASCOT Server is suggested. MASCOT *Distiller* offers a single, intuitive interface to native (binary) data files from AB Sciex, Agilent,

---

[432]Thermo's free demo version of ProteomeDiscoverer is available at: http://portal.thermo-brims.com/ Tutorial videos for the software are at: http://proteomicsnews.blogspot.com/p/proteome-discoverer-20-videos.html.

[433]Based on information from MatrixScience website. Accessed: September 29, 2014.

Bruker, Shimadzu, Thermo, and Waters. Raw data can be processed into high quality, de-isotoped peak lists. Optional toolboxes support MASCOT search result review, *de novo* sequencing, batch processing and quantitation. If not registered, MASCOT Distiller can still be used as a free project viewer, ideal for distributing search and quantitation results to colleagues. MASCOT *Daemon* is a batch automation utility, bundled with MASCOT Server. In combination with MASCOT Distiller, every step of a quantitation experiment can be fully automated. MASCOT *Parser* provides an API (Application Programming Interface) to MASCOT Server result files, making it easy to access search results from programs written in C^{++}, Java, Python or Perl. This is *free* for non-commercial use. MASCOT *Integra* is the complete solution for proteomics data management. It will scale to the largest projects, yet has a very affordable entry level, making it a practical choice for laboratories of all sizes.

6.3.3 PEAKS DB

PEAKS DB has been developed[434] by incorporating the *de novo* sequencing results into the database search. It also includes a new result validation method, *decoy fusion*, to solve the issue of over-confidence in the target-decoy method for certain types of peptide identification software. *De novo* sequencing was thought to be slow and required spectra with higher mass accuracy. (It had been mostly used when the protein database was unavailable.) With recent development of computer algorithms and computers, the speed is no longer an issue; high mass accuracy is also available. PEAKS DB takes advantages of fast *de novo* sequencing results along with several new features. The outcome is an increase in both sensitivity and accuracy. This is particularly true for mass spectral data obtained by ETD fragmentation, which makes the software a particularly useful tool for identifying peptides with PTMs. The spectra that produce highly confident *de novo* sequencing tags but no database hits are likely from novel or modified peptides. In an analysis purely based on database research those are currently rejected.

Whereas PEAKS DB belongs to the database search category of peptide identification software, it employs *de novo* sequencing as a subroutine, and utilizes the *de novo* sequencing results to improve the speed and accuracy of the database search. The main algorithmic steps of the software are: (1) *De novo* sequencing: The PEAKS algorithm is used to perform *de novo* sequencing for each input spectrum. (2) Protein shortlisting: The *de*

---

[434]Zhang, J., Xin, L., Shan, B., Chen, W., Xie, M., Yuen, D., Zhang, W., Zhang, Z., Lajoie, G. A., and Ma, B. (2012) PEAKS DB: *De novo* sequencing assisted database search for sensitive and accurate peptide identification. *Mol. Cell. Proteomics*, **11**, M111.010587.

novo sequence tags are used to find approximate matches in the protein sequence database, and all peptides of the protein shortlist are used to match the MS/MS with a rapid scoring function; only the 512 highest-scoring peptide candidates (including those with PTM) are kept for each MS/MS spectrum. (3) Peptide scoring: From the 512 candidates calculated in the peptide shortlisting step, a precise scoring function is used to find the best peptide for each spectrum. The similarity between the *de novo* sequence and the database peptide is an important component in the scoring function. (4) Result validation: A modified *target-decoy* approach is used to determine the minimum peptide-spectrum matching score threshold to meet the user's FDR requirement. (5) Protein inference and grouping: The high-confidence peptides identified through the above steps are used to infer the proteins. Those proteins that share the same set of peptide hits are grouped together for reporting purpose.

A modified target-decoy approach — *decoy fusion* — is used to estimate the FDR at any given score threshold. The use of the decoy fusion method is necessary for PEAKS DB's result validation as the standard target-decoy approach may underestimate the FDR of PEAKS DBs results. However, in PEAKS DB, the target and decoy sequences are not treated as separate entries in the database; instead, they are *concatenated* together *for each protein*. The software searches this newly generated database which contains the same number of protein entries, but each protein's length is doubled. After the search, the target and decoy identifications are separated by checking whether they are from the first or the second half of each concatenated sequence. The FDR is calculated, for each user-specified score threshold, as the ratio between the number of decoy hits and the number of target hits above the score threshold.

Two public datasets, one fragmented with CID (trypsin digest of *Pseudomonas aeruginosa*, http://www.marcottelab.org/MSdata/Data_12/-DATA/20090115_SMPA14_2.RAW.gz) and the other with ETD (Lys-C digest of a yeast lysate following strong cation-exchange fractionation prior to LC-MS; the raw data from fraction 10 — which was also used in a 2011 study by the Proteomics Informatics Research Group (iPRG) of the Association of Biomolecular Resource Facilities (ABRF)), were used to evaluate the performance of PEAKS DB. Both datasets were generated with LTQ-Orbitrap instruments.

In all experiments involving decoy sequences, the decoy sequences were produced by randomly shuffling the amino acids of each protein. Decoy peptides were removed before FDR calculation. That is, FDR = number of decoy hits/number of target hits. When a target-decoy method was used to estimate the FDR, the target and decoy databases were searched together.

Performance of PEAKS DB was compared by FDR curves with MASCOT 2.3 and SEQUEST (in Proteome Discoverer 1.2). For each peptide-spectrum match (PSM), SEQUEST outputs two scores, X_{Corr} and ΔC_n. In this experiment, $X_{Corr} + 5\Delta C_n$ was used as SEQUEST scores, since this combination produced the optimal FDR curve for SEQUEST. A comparison with the combination of MASCOT and Percolator was also conducted. At 1% FDR, the numbers of identified target PSMs are PEAKS DB (10,668) > MASCOT+Percolator (9,969) > SEQUEST (8,236) > MASCOT (7,515) from the CID dataset; and PEAKS DB (3,652) > MASCOT+Percolator (2,702)> MASCOT (2,398)> SEQUEST (2,233) from the ETD dataset. PEAKS DB also outperformed MS-GFDB (generating function database search tool) by approximately 58% and 8% in such a special comparison for CID and ETD, respectively.

PEAKS DB could confidently identify significantly more PSMs than MASCOT and SEQUEST. In comparison to MASCOT, at a 1% FDR, PEAKS DB could identify 42% more PSMs for the CID dataset, and 52% more PSMs for the ETD dataset. In fact, PEAKS DB identified more PSMs (9,494 for CID and 3,299 for ETD) at 0.1% FDR than MASCOT (7,515 for CID and 2,398 for ETD) at 1% FDR. While Percolator significantly improved the performance of MASCOT, PEAKS DB still outperformed MASCOT + Percolator by 7% for CID data and by 35% for ETD data at 1% FDR on these datasets.[435]

6.3.4 X! Tandem

X! Tandem algorithm was introduced by Craig and Beavis.[436] The central axiom of X! Tandem is: For each identifiable protein in the original protein mixture, there will exist at least one detectable tryptic peptide with $n \leq n_{R1}$.

In X! Tandem the ranking is done on the basis of *expectation values*. An expectation value can be defined for any valid scoring system that has the characteristic of the score being a maximum when a valid match is obtained.[437] If x represents a score for a mass spectrum $\mathbf{S}$, then the *survival function* $s(x)$ (the probability that a given random variable will be greater than a given quantity), for a discrete stochastic score probability distribution $p(x)$, can be defined as $s_{j(x)} = \Pr(X > x) = \sum_{i=j(x)}^{\infty} p_i$, where $\Pr(X > x)$ is the probability that the spectrum's score will have a value greater

---

[435]Bioinformatics Solutions Inc., besides PEAKS DB, markets PEAKS PTM which allows users to search with all of 600+ published PTMS, and SPIDER which is specifically designed to detect peptide mutations and perform cross-species homology searches. http://www.bioinfor.com accessed December 26, 2014.

[436]Craig, R. and Beavis, R. C. (2003) A method for reducing the time required to match protein sequences with tandem mass spectra. *Rapid Commun. Mass Spectrom.*, **17**, 2310–2316.

[437]Fenyö, D. and Beavis, R. C. (2003) A method for assessing the statistical significance of mass spectrometry-based protein identifications using general scoring schemes. *Anal. Chem.*, **75**, 768–774.

than x by random matching with sequences in a particular database, $\mathbf{D}$, and $p_{j(x)}$ is the discrete probability that best corresponds to a score of x. The expectation value $e(x)$ for $\mathbf{S}$ on $\mathbf{D}$ is then defined as $e_{j(x)} = n s_{j(x)}$, where n is the number of sequences scored. The discrete valued frequency function $f(x)$ can be converted into $p(x)$ by simple normalization, $p_i = f_i/N$, where N is the number of peptide sequences used to create the histogram (number of peptide spectrum matching per score bin): $N = \sum_{i=0}^{\infty} f_i$.

Applying the equations above on $s_{j(x)}$ and N, it is possible to estimate $p(x)$ and $s(x)$ for any scoring system with any combination of data, search parameters, and sequence. Once $s(x)$ is known, it can then be fit to an analytical function and its value extrapolated for any value of x.

In a plot of $p(x)$ against $\ln x$, any peptide corresponding to a score *within* the body of the stochastic distribution *cannot* be confidently assigned as being a valid identification. A score *higher* than the right hand boundary of the stochastic distribution may be assigned as potentially valid, with an associated expectation value.

The following steps may be used to reduce any well-behaved protein identification scoring system to an expectation value formulation for comparison with other systems. 1. Create the distribution $p(x)$ for each specific identification experiment, preferably experimentally. 2. Calculate $s(x)$ from $p(x)$. 3. Fit the high-scoring portion of $s(x)$ to a model distribution. 4. Use the fitted function to extrapolate $s(x)$ to the desired score and convert the extrapolated value to $e(x)$.

In practice, as sequences are being scored, it is very simple to construct the scoring frequency histogram "on the fly." The computational overhead is very little and the overall performance of a practical scoring engine is not affected.

Operationally, the experimentally acquired spectrum is first simplified to only those peaks that are similar to the peaks in the model spectrum. In X! Tandem the *preliminary score* is a dot product of the acquired and model spectra. Because only similar peaks are considered, this is the sum of the intensities of the matched b and y ions. Preliminary score (correlation score) = Dot product (by-Score)= $\sum_{i=0}^{n} I_i P_i$. Possible ions are represented by $P_i = 1$, and all other values are taken as 0. Hyperscore in X! Tandem is defined by HyperScore $= (by\text{-}Score).N_y!.N_b!$, that is modifying the preliminary score by multiplying by N factorial for the number of b and y ions assigned. The confidence that a score represents a valid identification increases as the distance between the high-scoring end of the stochastic distribution and the score for a valid peptide sequence. Tests with sample tandem spectrum and the NCBI nr protein sequence database show that in comparison to the preliminary score, the hyperscore application yields, in $p(x)$-$\ln x$ plot, the score for the valid sequence much farther right with respect to the right boundary of the stochastic distribution. Hence superiority of the hyperscore method. The hyperscore is much faster to calculate than SEQUEST's X_{Corr}.

In order to obtain E-values from hyperscores, X! Tandem makes a histogram of all the hyperscores for all the peptides in the database that might match the spectrum.[438] It then assumes that the peptide with the highest hyperscore is correct, and all others are incorrect. If the data on the right side of the histogram (from the peak of the histogram; this histogram is calculated independently for each spectrum) is taken and log-transformed (log of number of results on y-axis, hyperscore on x-axis), the data fall on a *straight line*. A straight line is the expected result from a statistical argument that assumes the incorrect results are random. In X! Tandem's assumption the top hypersocre is the only possible correct match. This match is significant if it is greater than the point at which the straight line through the log data intersects the log (number of results) = 0 line. Any hyperscore greater than this is unlikely to have arisen by chance. The E-value expresses just how unlikely a greater hyperscore is. X! Tandem calculates the E-value by extrapolating the straight line of the log histogram mentioned above. A hyperscore of 83 (taking a number for example) would occur by chance where the straight line crosses 83 (on x-axis). On the log (number of results) scale (that is, the y-axis), this corresponds to -8.2; so the E-value is -8.2. (E-value $= e^{-8.2}$.)

One of the X! Tandem's strengths is its automatic search for modified peptides — but only on proteins it has otherwise identified. Speed: ~ 200 times faster for nonspecific searches; ~ 1000 times faster if one also looks for oxidation, deamidation, and phosphorylation. It considers semi-tryptic peptides, and sequence polymorphisms.

X! Tandem can match tandem mass spectra with peptide sequences. This software has a very simple, sophisticated application programming interface: it simply takes an XML file of instructions on its command line, and outputs the results into an XML file, which has been specified in the input XML file. This format is used for all of the X! series search engines, as well as the GPM and GPMDB.[439]

Unlike some earlier generation search engines, all of the X! Series search engines calculate statistical confidence (expectation values) for all of the individual spectrum-to-sequence assignments. They also reassemble all of the peptide assignments in a data set onto the known protein sequences and assign the statistical confidence that this assembly and alignment is nonrandom. Therefore, separate assembly and statistical analysis software, e.g., PeptideProphet and ProteinProphet, do *not* need to be used.[439]

---

[438]Searle, B. (Proteome Software, Inc.) X! Tandem explained. https://proteome-software.wikispaces.com/file/.../XTandem-explained.ppt

[439]From http://www.thegpm.org/tandem December 31, 2014.

The latest release is SLEDGEHAMMER (2013.09.01). This release updates the E-value estimation algorithm and corrects several issues associated with using very high accuracy fragment ion mass tolerances; notably, the new E-value algorithm deals more effectively with malformed protein sequence lists, particularly sequence lists that definitely have very large numbers of protein sequences having very similar sequences.[439]

Parallel Tandem is a parallel driver for X! Tandem. These applications consist of a PVM (Parallel Virtual Machine) or MPI (Message Passing Interface) parallel program and several utility programs which reduce severalfold the time it takes to compare tandem mass spectra to amino acid sequences in a protein database.[440,441]

X!!Tandem is a parallel, high performance version of X! Tandem that has been parallelized via MPI to run on clusters or other non-shared memory multiprocessors running Linux. With the exception of the details related to MPI launch, it is run exactly as X! Tandem, and produces exactly the same results. It *differs* from Parallel Tandem in that the parallelism is handled internally, rather than as an external driver/wrapper.[442,443,444]

6.3.5 OMSSA

The Open Mass Spectrometry Search Algorithm (OMSSA)[445,446] is a probability-based matching algorithm, the type of statistical model used in BLAST,[447] and assumes a Poisson distribution. It compares the m/z values of experimental MS/MS spectra — after filtering noise peaks — with those calculated from peptides generated (the pepides' mass within a user-

[440]http://www.thegpm.org/parallel.

[441]Duncan, D. T., Craig, R., and Link, A. J. (2005) Parallel tandem: a program for parallel processing of tandem mass spectra using PVM or MPI and X! Tandem. *J. Proteome Res.*, **4**, 1842–1847.

[442]Bjornson, R. D., Carriero, N. J., Colangelo, C., Shifman, M., Cheung, K-H., Miller, P. L., and Williams, K. (2008) X!!Tandem, an improved method for running X! Tandem in parallel on collections of commodity computers. *J. Proteome Res.*, **7**, 293–299.

[443]Fenyö, D. (2000) Identifying the proteome: software tools. *Curr. Opin. Biotechnol.*, **11**, 391–395.

[444]Eriksson, J., Chait, B. T., and Fenyö, D. (2000) A statistical basis for testing the significance of mass spectrometric protein identification results. *Anal. Chem.*, **72**, 999–1005.

[445]Geer, L. Y., Markey, S. P., Kowalak, J. A., Wagner, L., Xu, M., Maynard, D. M., Yang, X., Shi, W., and Bryant, S. H. (2004) Open mass spectrometry search algorithm. *J. Proteome Res.*, **3**, 958–964.

[446]OMSSA MS/MS search engine: http://Pubchem.ncbi.nlm.nih.gov/omssa Return: OMSSA was a tandem mass spectra search engine. This search engine is no longer available. Due to budgetary constraints NCBI has discontinued OMSSA. Historical binaries are available from their ftp server ftp.ncbi.nih.gov/pub/lewisg/omssa/CURRENT/. Accessed October 15, 2014.

[447]Altschul, S. F., Madden, T. L., Schäffer, A. A., Zhang, J., Zhang, Z., Miller, W., and Lipman, D. J. (1997) Gapped BLAST and PSI-BLAST: a new generation of protein database search programs. *Nucl. Acids Res.*, **25**, 3389–3402.

specified tolerance of the precursor mass) from an *in silico* digestion of a protein sequence library. The resulting hits are then statistically recorded, ranked by E-values. The latter remain valid even when variable PTMs are included. The multiple theoretical spectra with variable PTMs can be considered non-redundant since they do not have the same precursor m/z — only a subset of the product ions are shared — thus it becomes *unnecessary* to explore the effect of redundancy on the E value. Variable PTMs increase the value of number of theoretical spectra generated by modified peptides.

Interfacing OMSSA with Percolator — the latter a post-search machine learning method for rescoring database search results — generates OMSSAPercolator.[448,449] It performs better than standard OMSSA; programmed using JAVA, it can be used as a standalone tool or can be used after integrating into existing data analysis pipelines.

6.3.6 Scaffold

Scaffold[416,450] is a meta-analysis tool that implements a pipeline of several statistical methods designed to validate results from database-search engines. Database-search results can be used from a variety of search engines, including several commercial engines (Mascot, Sequest, Phenyx, Spectrum-Mill, IdentityE, and Byonic) and open source engines (X!Tandem, OMSSA, MS-Amanda, MS-GF+, and MaxQuant). Scaffold infers protein identifications from peptide-to-spectrum matches (PSM)s across several samples grouped into biological and technical replicates to enable label-free quantitation using spectrum counting or integrated precursor intensities. Multiple modules are available to produce stable isotope labeled quantitative reports (with Scaffold Q+), post-translational modification site localization (with Scaffold PTM), and analysis of large scale data sets (with Scaffold perSPECtives).

Scaffold converts PSM scores to peptide probabilities, which essentially normalizes between search engines and allows for level comparisons of different data sets. Peptide probabilities are determined using a local false discovery rate based strategy similar to Percolator,[451] where discrimination minimization is done using linear discriminant analysis rather than support vector machines. Alternatively, if decoy search results are not available,

[448]Wen, B., Li, G., Wright, J. C., Du, C., Feng, Q., Xu, X., Choudhary, J. S., and Wang, J. (2014) The OMSSAPercolator: an automated tool to validate OMSSA results. *Proteomics*, **14**, 1011–1014.

[449]OMSSAPercolator is freely available and can be downloaded at http://sourceforge.net/ projects/omssapercolator/.

[450]Comments of Brian C. Searle on this Section are gratefully acknowledged.

[451]Käll, L., Canterbury, J. D., Weston, J., Noble, W. S., and MacCoss, M. J. (2007) Semi-supervised learning for peptide identification from shotgun proteomics datasets. *Nat. Methods*, **4**, 923–925.

PSMs can be probabilistically interpreted using the PeptideProphet[452] algorithm. Scaffold then scores the agreement between search engines and that is re-factored into a combined peptide probability.[453] The authors show that probabilities combined in this way are frequently more accurate and more sensitive than probabilities derived from individual search engines alone.

Scaffold uses a re-implementation of the ProteinProphet[454] algorithm to combine peptide probabilities into protein identifications using a statistical model that considers protein homology and the distribution of the number of peptides assigned to each protein. Due to the nature of MS/MS acquisition, correctly identified PSMs accumulate and several peptides are assigned to the same proteins. On the other hand, incorrect PSMs are randomly assigned across the FASTA database to proteins with only a single peptide identification. This fact is exploited by ProteinProphet, which attempts to learn the distribution of correct identifications in multiply assigned proteins compared to so-called one-hit-wonder proteins.

Most analysis pipeline programs separate out peptide assessment from protein grouping so that different sub-tools can be easily manipulated or interchanged. One disadvantage of that approach is that multiple peptide sequences can be equally assigned to the same MS/MS spectra and peptide sequence-based protein grouping algorithms choose one of those sequences at random. As larger protein databases are used the sequence search space increases and magnifies this issue. Scaffold uses a three-level graph structure that maintains multiple peptide assignments to the same MS/MS spectrum. In this scheme, the original mass spectral evidence is assigned to multiple possible peptides as a 'peptide group'. That peptide group is linked to multiple proteins by their associated peptide sequences. Using this approach, protein groups are constructed across the entire experiment. The authors show that despite the increasing number of protein sequence entries in different databases, Scaffold never identifies more than the correct number of protein groups, which indicates that false-positive rates are largely unaffected by redundancy in databases.

Quantification in the Scaffold Q+ module is performed on label-free or stable isotope labeled multiplex data using a non-parametric statistical model. Experiments are normalized using an iterative median polish algorithm, which takes a non-parametric approach similar to ANOVA. The main advantage of ANOVA-like strategies is that calculations are performed

---

[452]Keller, A., Nesvizhskii, A. I., Kolker, E, and Aebersold, R. (2002) Empirical statistical model to estimate the accuracy of peptide identification made by MS/MS and database search. *Anal. Chem.*, **74**, 5383–5392.

[453]Searle, B. C., Turner, M., and Nesvizhskii, A. I. (2008) Improving sensitivity by probabilistically combining results from multiple MS/MS search methodologies. *J. Proteome Res.*, **7**, 245–253.

[454]Nesvizhskii, A. I., Keller, A., Kolker, E., and Aebersold, R. (2003) A statistical model for identifying proteins by tandem mass spectrometry. *Anal. Chem.*, **75**, 4646–4658.

in intensity-space, rather than on ratios.[455] For each quantitative intensity value, Scaffold Q+ estimates an intensity-weighted confidence and standard deviation, and these are used to make up a kernel distribution. Intensity values converted to kernel distributions are aggregated across multiple peptides to form a kernel density estimate, which estimates a cumulative intensity for each protein. Protein changes across multiple biological samples are validated using permutation tests.

6.4 Targeted acquisition covering SRM and PRM

In PRM (parallel reaction monitoring) — a new targeted proteomics method — the third quadrupole of triple quadrupole (QQQ) is substituted by a high-resolution, high-mass-accuracy analyzer. Such a high-resolution analyzer can be used for parallel detection of all target product ions in one concerted step. In SRM, transitions are monitored one at a time; in PRM, parallel detection of transitions takes place in a single analysis.

Michalski *et al.*[456] (also Peterson *et al.*[457] and Tsuchiya *et al.*[458]) use for PRM a Q-Exactive instrument. Because of its parallel filling and detection modes it permits fast high-energy collision-induced dissociation peptide fragmentation. An enhanced Fourier transformation algorithm processes the image current from the detector that doubles mass spectrometer resolution. With fast isolation and fragmentation, the instrument achieves overall cycle times 1 s for a higher energy collisional dissociation method. More than 2,500 proteins could be identified in standard 90-min gradients of tryptic digests of mammalian cell lysate. The quadrupole-Orbitrap analyzer combination enables multiplexed operation at the MS and tandem MS levels; in a multiplexed single ion monitoring mode, the quadrupole rapidly switches among different narrow (0.4 Th at m/z 400) mass ranges that are analyzed in a single composite MS spectrum. The quadrupole similarly allows fragmentation, in rapid succession, of different precursor masses,

[455]Oberg, A. L., Mahoney, D. W., Eckel-Passow, J. E., Malone, C. J., Wolfinger, R. D., Hill, E. G., Cooper, L. T., Onuma, O. K., Spiro, C., Therneau, T. M., and Bergen, H. R. 3rd. (2008) Statistical analysis of relative labeled mass spectrometry data from complex samples using ANOVA. *J. Proteome Res.* **7**, 225–233.

[456]Michalski, A., Damoc, E., Hauschild, J.-P., Lange, O., Wieghaus, A., Makarov, A., Nagaraj, N., Cox, J., Mann, M., and Horning, S. (2011) Mass spectrometry-based proteomics using Q-Exactive, a high-performance benchtop, quadrupole Orbitrap mass spectrometer. *Mol. Cell. Proteomics*, **10**, 10.1074/mcp.M111.011015.

[457]Peterson, A. C., Russell, J. D., Bailey, D. J., Westphall, M. S., and Coon, J. J. (2012) Parallel reaction monitoring for high resolution and high mass accuracy quantitative, targeted proteomics. *Mol. Cell. Proteomics*, **11**, 1475–1488.

[458]Tsuchiya, H., Tanaka, K., and Saeki, Y. (2013) The parallel reaction monitoring contributes to a highly sensitive polyubiquitin chain quantification. *Biochem. Biophys. Res. Communications*, **436**, 223–229.

followed by joint analysis of the higher energy collisional dissociation fragment ions in the Orbitrap analyzer.

Peterson *et al.*[457] report that in the presence of a yeast background matrix PRM yielded quantitative data over a wider dynamic range than selected reaction monitoring because of PRM's high sensitivity in the mass-to-charge domain. Their PRM experiments yielded, on average, quantitative information between 2-4 concentration orders-of-magnitude (93%), with the majority of experiments resulting in quantification across three orders of magnitude (54%, 0.2 to 200 nM). The wider PRM isolation width, ± 1 *versus* ± 0.2 Th, resulted in improved dynamic range – neat, $10^{3.3}$ *versus* $10^{2.6}$, and matrix, $10^{2.7}$ *versus* $10^{2.2}$.

Tsuchiya *et al.*[458] have applied the PRM method to quantify polyubiquitin chains in biological matrices. They succeeded in the quantification of all the linkages from 50 amol to 100 fmol. Using the ubiquitin PRM, they dissected the ubiquitylation of ubiquitin-proline-β-galactosidase (Ub-P-βgal), and unambiguously identified that K29-linked ubiquitin chains are attached to Ub-P-βgal *in vivo*.

Post-acquisition reconstruction of precursor-fragment ion transitions has been referred to as *pseudo*-multiple reaction monitoring (p-mSRM); the term *pseudo* refers to the computer-calculated post-acquisition reconstruction of the transition. Full fragment ion spectrum is measured, in contrast to classical mSRM. Pak *et al.* show,[459] using LTQ-Orbitrap-Velos, that label-free MS2-based quantification is possible over at least five orders of magnitudes in very complex matrices, e.g., digested human plasma.

6.5 Data-independent acquisition

6.5.1 MS^{E}

In MS^{E},[460] after separation of sample components (UPLC®(UltraPerformance LC)/MS^{E} or HDMS^{E} (ion mobility)) in Stage 1, in Stage 2 (Generation of a complete MS dataset) the sample-component peaks arrive at the mass analyzer in very narrow time windows which requires the mass spectrometer to generate spectra very quickly while maintaining high spectral resolution, sensitivity, exact mass measurement, and in-spectrum dynamic range.

Since the identities of the peaks of interest are unknown at the beginning of the experiment, data is required on every detectable sample component. In the MS^{E} method, both precursor and fragment ions are generated by the mass spectrometer simultaneously while securing enough data points across each component peak to ensure correct peak integration

---

[459]Pak, H., Pasquarello, C., and Scherl, A. (2011) Label-free protein quantification on tandem mass spectra in an ion trapping device. *J. Integr. OMICS*, **1**, 211–215.

[460]Waters (White Paper, October 2011 720004036EN LB-AP).

and quantitative accuracy. With few parameters to adjust, MS^E requires a minimum of method development, and needs no prior knowledge of the sample components. No pre-selection of ions occurs prior to fragmentation, so no data is lost.

The ToF mass spectrometer rapidly and continuously cycles between two states. In state 1, all the ions from the ion source, are transmitted through the collision cell (collision energy is kept low so that no fragmentation occurs) to the mass analyzer and recorded as a precursor ion spectrum. In state 2, all the ions from the ion source are transmitted through the collision cell (collision energy *ramped* to generate maximum information from fragment ions) to the mass analyzer and recorded as a fragment ion spectrum.

In Stage 3 for alignment of spectra and data interpretation, with software algorithms that profile each chromatographic peak and determine their retention times, fragment ion spectra are assigned to their associated precursor ion peaks so that all the information necessary to identify each compound of interest is collated and are readily available. Precursor and fragment spectra are then aligned to retention times and linked together.[461] The software separates spectra belonging to co-eluting peaks and a comprehensive dataset is generated. This dataset must be interrogated to determine the presence and concentrations of relevant molecules. This data is stored as a complete digital record of the sample and is available for re-interrogation.

Levin *et al.*,[462] through a systematic study of the MS^E method using low-, medium-, and high-complexity samples, provide a validation of the MS^E approach for accurate quantitative analysis along with high sequence coverage of target proteins. They find that the method has a linear dynamic range spanning three orders of magnitude with a limit of quantification 61 amol/μL in low-complexity samples, and 488 amol/μL in high-complexity samples.

6.5.2 ARM — All-Reaction Monitoring

Introduced by Weisbrod *et al.*,[463] an FT-ARM experiment involves continuous, data-independent, high mass accuracy MS/MS acquisition spanning a defined m/z range. Identifications in FT-ARM are complementary

[461]Blackburn, K., Mbeunkui, F., Mitra, S. K., Mentzel, T., and Goshe, M. B. (2010) Improving protein and proteome coverage through data-independent multiplexed peptide fragmentation. *J. Proteome Res.*, **9**, 3621–3637.

[462]Levin, Y., Hradetzky, E., and Bahn, S. (2011) Quantification of proteins using data-independent analysis (MS^E) in simple and complex samples: a systematic evaluation. **11**, 3273–3287; doi:10.1002/ pm1c.201000661.

[463]Weisbrod, C. R., Eng, J. K., Hoopmann, M. R., Baker, T., and Bruce, J. E. (2012) Accurate peptide fragment mass analysis: multiplexed peptide identification and quantification. *J. Proteome Res.*, **11**, 1621–1632.

to conventional data-dependent shotgun analysis. Because of fragmenting multiple overlapping precursors, this is particularly true where the data-dependent method fails.

In the FT-ARM procedure a relatively wide m/z range of ions is accumulated and all ions within this range are subject to activation. Mass analysis of the fragment ions is performed with high mass accuracy (< 5 ppm). For each peptide against each DIA scan event of DIA-LC-MS experiment, a database of theoretical or empirical fragmentation patterns is used to calculate the dot-product score. (Dot-product $= \sum_{i=1}^{n} R_i T_i$, where R_i is the intensity of the ith m/z value of the reference spectrum, T_i is the ith m/z value of the target spectrum, and n is the number of peaks in the spectrum. Large positive dot-product scores are obtained at spectra/times when the peaks in the reference spectrum align with peaks in the target spectrum within the user-specified mass tolerance.) The program performs a dot-product calculation from each measured target spectrum acquired during chromatographic separation.

The result of this analysis is a score chromatogram for each peptide contained in the database, which is used for both identification and quantification. The false discovery rate is estimated by a search of a reverse sequence database. In the FT-ARM process, chromatographic elution profile characteristics are *not* used to cluster precursor peptide signals to their respective fragment ions; thus it is fundamentally different than other DIA approaches, including MS^E and all-ion fragmentation. FT-ARM relies upon the specificity in peptide fragmentation and high mass accuracy measurement and is, therefore, independent of precursor mass.

The FT-ARM program is written in C^{++} using Microsoft Visual Studio Developers Suite. This program is designed to accept two input data streams: theoretical or empirical reference spectra (.FASTA or .TXT peptide list format) and DIA multiplexed fragmentation target data acquired on the sample of interest.

6.5.3   AIF — All-Ion Fragmentation

Geiger *et al.*[464] have used the term "all-ion fragmentation" (AIF) for their method; however, similar acquisition mode have been used, such as in MS^E. For their AIF procedure they used Orbitrap-Exactive (high resolution 100,000 at m/z 200), stepped collision energy procedure (HCD collision cell, middle energy of 30 eV, with 20% steps above and below the middle energy) with cumulative MS readout. (Peptides were fragmented for a third of the fill time with the low, medium, and high collision energies, respectively, and

---

[464]Geiger, T., Cox, J., and Mann, M. (2010) Proteomics on an Orbitrap benchtop mass spectrometer using all-ion fragmentation. *Mol. Cell. Proteomics*, **9**, 2252–2261.

all fragments were accumulated in the cell for combined Orbitrap analysis.) Both at the precursor and product ion levels, they applied the peak recognition algorithms of MaxQuant. In an equimolar protein standard mixture the instrument could identify 45 out of 48 proteins of UPS1–Universal Protein Standard (and all of them using a small database), was able to detect peptides ranging over four orders of magnitude (using high dynamic range protein mixture), and in gel slices they could unambiguously characterize an immunoprecipated protein "barely visible" by Coomassie staining.

The same group has brought out a new peptide search engine, Andromeda,[465] which uses a probabilistic scoring model, and is claimed to be comparable to MASCOT in performance as judged by sensitivity and specificity analysis based on target decoy searches. It can handle data with very high fragment mass accuracy and able to assign and score complex patterns of post-translational modifications and can accommodate extremely large databases. It can be used in a stand-alone mode or as an integrated search engine of MaxQuant[466] platform. It can run locally on the user's computer.

6.5.4 SWATH-MS

Andrews *et al.*'s evaluation[467] of the performance characteristics of the AB Sciex TripleTOF 5600 system (in which they reported 30 K resolving power, < 3 ppm mass accuracy and 20 Hz spectral acquisition rate for the instrument) was the forerunner of the SWATH^{TM} (Sequential Window Acquisition of all Theoretical Model Spectra) procedure.[468,469,470]

SWATH-MS is a data-independent acquisition method. In a single measurement, it generates a complete recording of the fragment ion spectra of

[465] Cox, J., Neuhauser, N., Michalski, A., Scheltema, R. A., Olsen, J. V., and Mann, M. (2011) Andromeda: a peptide search engine integrated into the MaxQuant environment. *J. Proteome Res.*, **10**, 1794–1805.

[466] Cox, J. and Mann, M. (2008) MaxQuant enables high peptide identification rates, individualized p.p.b.-range mass accuracies and proteome-wide protein quantification. *Nat. Biotechnol.*, **26**, 1367–1372.

[467] Andrews, G. L., Simons, B. L., Young, J. B., Hawkridge, A. M., and Muddiman, D. C. (2011) Performance characteristics of a new hybrid quadrupole time-of-flight tandem mass spectrometer (TripleTOF 5600). *Anal. Chem.*, **83**, 5442–5446.

[468] Gillet, L. C., Navarro, P., Tate, S., Röst, H., Selevsek, N., Reiter, L., Bonner, R., and Aebersold, R. (2012) Targeted data extraction of the MS/MS spectra generated by data-independent acquisition: a new concept for consistent and accurate proteome analysis. *Mol. Cell. Proteomics*, 10.1074/mcp.O111.016717.

[469] Liu, Y., Hüttenhain, R., Surinova, S., Gillet, L. C., Mouritsen, J., Brunner, R., Navarro, P., and Aebersold, R. (2013) Quantitative measurements of N-linked glycoproteins in human plasma by SWATH-MS. *Proteomics*, **13**, 1247–1256.

[470] Röst, H. L., Rosenberger, G., Navarro, P., Gillet, L., Miladinović, S. M., Schubert, O. T., Wolski, W., Collins, B. C., Malmström, J., Malmström, L., and Aebersold, R. (2014) OpenSWATH enables automated, targeted analysis of data-independent acquisition MS data. *Nat. Biotechnol.*, **32**, 219–223.

all the analytes of the sample for which the precursor ions are within a specified m/z range and within a specified retention time window.[469] In the study of Gillet *et al.*,[468] for data acquisition they use repeated cycling through 32 consecutive 25-Da *precursor isolation* windows (1 Da overlap, 100 ms/swath, 3.2 s duty cycle). These are the *swaths* — analogous to swath acquisitions in Earth satellite scans. This scan procedure generates a 3D map (retention time – m/z – intensity) for each precursor ion window. In their work, in a single sample injection, the SWATH-MS acquisition step generates time-resolved fragment ion spectra for *all* the analytes detectable within the 400–1,200 m/z precursor range and the user-specified retention time window. For data analysis, product ion spectra acquired for each isolation window yield separate LC-MS/MS maps. Then the fragments derived from a spectral library (and defining any precursor of interest) can be extracted and analyzed against the experimentally obtained spectra to detect and quantify the targeted analytes in the sample. Thereafter, the relative intensities of the acquired product ion fragments are compared — based on mass accuracy — to that of the reference spectrum and on the co-elution of the extracted ion chromatogram of these fragments, and the confidence in peptide identification is scored. Suitable combinations of fragment ions extracted from these data sets are sufficiently specific to *identify* query peptides over a dynamic range of four orders of magnitude, even in those cases where the precursors of the queried peptides are not detectable in the survey scans. The consistency and accuracy of quantification of the queried peptides are comparable with that of selected reaction monitoring, considered standard in proteomic quantification method. Data obtained in this manner can be iteratively re-*mined* and can be subjected to open-ended quantification refinement. The method provides both broad range precursor ion fragmentation and targeted data extraction. With respect to SRM, the method obviates the need of a preliminary selection of reactions and no limit to the number of analytes in a run. In a relative current performance profiles study, they compare several figures of merit of a number of methods against SWATH-MS; they show for percentage peptides observable 89 (the percentage of peptides observable with at least five interference-free transitions in a yeast background), far higher than in the other methods.

OpenSWATH[470] is a software for *automated* targeted DIA analysis. Tens of thousands of peptides are identified in a single typical SWATH-MS data set. OpenSWATH is an *open-source* software that allows targeted analysis of DIA data in an automated high-throughput fashion. It is cross-platform software, written in C^{++}, that relies only on open data formats, allowing it to analyze DIA data from multiple instrument vendors. The five steps of the algorithm are: data conversion, retention time alignment, chromatographic extraction, peak-group scoring, and statistical analysis. The software has been benchmarked against manual analysis of $> 30,000$

chromatograms from 342 synthesized peptides and used to analyze the proteome of *Streptococcus pyogenes*.

6.5.5 Multiplexed MS/MS MSX

Egertson *et al.*,[471] stressing that DIA data are "noisy" owing to a typical five- to tenfold reduction in precursor sensitivity compared to data obtained with data-dependent acquisition or selected reaction monitoring, have suggested a multiplexing strategy, MSX, for DIA analysis that — they claim — increases precursor selectivity fivefold.

Using Q-Exactive they present a multiplexing strategy, MSX, in which five separate 4-m/z isolation windows are analyzed per spectrum. According to their description[471] these spectra are demultiplexed into the five separate 4-m/z isolation windows, resulting in data with the sampling frequency of a DIA approach using 20 20-m/z-wide windows but the selectivity of an approach using 100 4-m/z windows. Demultiplexing improves precursor selectivity by narrowing the range of potential precursors for an MS/MS spectrum from a 20-m/z window down to a 4-m/z window and generating the unmixed fragment-ion spectrum with signal from only the 4-m/z window. They implemented this demultiplexing approach in the open-source Skyline software tool,[472] which provides a useful interface for the visualization and analysis of this data.

Since multiple fills per mass analysis is involved, this multiplexing technique is best suited for instrumentation in which isolation and collisional activation of peptides is *fast* relative to mass analysis. In their procedure they randomly selected five of the 100 possible 4-m/z isolation windows in the range 500–900 m/z to be analyzed in each multiplexed scan. They modified Skyline to detect MSX spectra and demultiplex them automatically upon import for analyzing these spectra. Demultiplexing removed majority of interfering peaks (fragment ion peaks from other peptides; those originating from elution time overlapping) and this resulted in a higher dot product similarity to a DDA spectrum acquired with a 2-m/z-wide isolation window.

The lower limit of detection for 36 peptides they quantified averaged 8.66 and 4.98 fmol for MSX and MS1 respectively; all peptides showed a linear response above the limit of detection. Although the MSX method was less sensitive than the MS1 method on average, the results were notable given that the MSX data provided structural selectivity information in addition to quantification. Of the 36 peptides, seven had interference in

---

[471]Egertson, J. D., Kuehn, A., Merrihew, G. E., Bateman, N. W., MacLean, M. X., Ting, Y. S., Canterbury, J. D., Marsh, D. M., Kellman, M., Zabrouskov, V., Wu, C. C., and MacCoss, M. J. (2013) Multiplexed MS/MS for improved data-independent acquisition. *Nat. Meth.*, **10**, 744–746. doi:10.1038/NMETHOD.2528.

[472]http://skyline.maccosslab.org.

the MS1 signal, resulting in an average 3.4-fold improvement in sensitivity ranging from 1.3- to 8.3-fold by MSX quantification. Simultaneous acquisition of MS1 and MSX data combines the high sensitivity of MS1 with the structural selectivity of MSX while theoretically improving quantitative precision by combining measurements from the precursor (MS1) and fragment-ion peaks (MSX) for quantitation.

With *wide* isolation windows, increased likelihood exists of relying on fragment-ion data alone to discern between peptides with similar fragmentation patterns (such as modified forms or similar sequences); quantitation also is negatively affected by fragment ion interference even if a peptide can be unambiguously identified in such data. With narrow isolation windows other proven advantages of DIA — including increased dynamic range and more complete and reproducible MS/MS sampling across technical replicates — can contribute unhindered. Coupled with recent improvements[463,468] in the interpretation of DIA data, MSX in a proteomics experiment is thus a practical technique for global, highly selective, and reproducible relative quantitation of peptides. A summary of some of the specific parameters impacting the performance of each data acquisition technique has been given in a supplementary note[473] by Egertson *et al.*.[471]

---

[473]DDA: number of dependent scans per parent scan, MS1 resolution, MS2 resolution, maximum fill time/AGC target for trapping instruments, number of microscans, charge-state based precursor selection, precursor intensity threshold, dynamic exclusion list length, dynamic exclusion time, exclusion list settings, collision energy. SWATH: isolation window width, scan time (number of microscans for signal averaging), mass range covered, collision energy. MS^E: collision energy, scan time (number of microscans for signal averaging). MSX: isolation window width, number of isolation windows per scan, mass range to cover, MS1 resolving power, MS2 resolving power, AGC target, maximum fill time, collision energy.

INDEX

Amino acid data, 138
Amino acid structures, 138
Applications to
    Human plasma proteome, 248
    Imaging biological material, 268
    Intact apolipoproteins in
        human HDL, 228
    Mapping intact protein isoforms,
        220
    Neurexin-LRRTM2 interactions,
        245
    Purkinje cell PSDs, 243
    Rapid sequence analysis of
        conotoxins, 232
    Ribosomes, 237
    V-type ATPases, 260
    Virus capsid assembly
        intermediates, 252

Chemoproteomics, 7
Cross-linking, chemical, 200

Data mining scheme, 173
Data-dependent acquisition, 277
    MASCOT, 280
    OMSSA, 286
    PEAKS DB, 281
    Scaffold, 287
    SEQUEST, 277
    X! Tandem, 283
Data-independent acquisition, 290
    AIF — all-ion fragmentation, 292
    ARM — all-reaction
        monitoring, 291
    MS^E, 290

Multiplexed MS/MS MSX, 295
SWATH-MS, 293

Electrophoretic separations
    Gel electrophoresis, 11
    Capillary electrophoresis, 15
    Difference gel electrophoresis,
        30
    SDS PAGE, 12
ETD, 117
    ETD with CEP, 233

False discovery/positive rate, 282,
    287, 292

H-D exchange, 167
HPLC, 20

Immonium ions, 148
Ion formation
    Chain ejection model CEM, 47
    Charge residue model CRM, 42
    Ion evaporation model IEM, 42
Ion funnel, 114
Ion mobility, 105
    Cross-section, PA, PSA, 109
Ionization
    ESI–electrospray, 37
    LAESI, 66
    LILBID, 62
    MALDI, 58
    Nanospray, 51
IRMPD, 116

Liquid chromatography, 19

MALDI imaging mass
        spectrometry, 61
Manual *de novo* sequencing, 207
Mass analysis instruments
        Agilent 6560 IM QTOF, 118
        Bruker Daltonik Impact II, 120
        IMS-IMS-MS, 130
        JEOL JMS-3S000 SpiralTOF,
                122
        SCIEX TripleTOF® 6600, 123
        Thermo Fisher Scientific–Hybrid
                FT-ICR, 127
        Thermo Fisher Scientific–
                Orbitrap Fusion, 124
        Waters–Synapt G2-S, 132
Mass analyzers
        2D-Quadrupole (Q), 73
        3-D Ion Trap (IT), 76
        FT-ICR, 79
            Dynamic harmonization, 94
            High magnetic field, 90
        Linear trap (q, LT), 75
        Orbitrap, 98
        Time-of-flight (TOF), 68
        TWIMS, 105
Microfluidic chips, 53
Multidimensional separations, 27
        GELFrEE fractionation, 32
        Shotgun proteomics, 28

Nano-LC/LC-MS coupling, 27
Nanospray, 38, 51

One-third rule, 209

Peptide fragmentation
        Experiments, 165

MS/MS methods–CAD(CID),
        ECD, ETD,
        EDD, 165
Mechanism, 175
        b_x-y_z pathway, 178
        Diketopiperazine pathway, 179
        Histidine effect, 186
        Proline effect, 187
        Sequence scrambling, 189
    Mobile proton model, 175
    Nomenclature of ions, 141
Peptide mass fingerprinting, 273
Peptide sequence tag, 274
Post-translational modifications, 195
Proteins
        Molecular mass determination,
                148
        Structure, 135
Proteolysis, 35

Quantitation, 151
        AQUA, 161
        ICAT, 151
        iTRAQ, 154
        Label-free, 151
        SILAC, 153
        TMT, 158

SRM, MRM, 150

Targeted acquisition covering SRM
        and PRM, 289
Taylor cone, 39
Time-of-flight
        Reflectron, 71

Ultracentrifugation, 9

Printed in the United States
By Bookmasters